T0350060

SPATIAL BRANCHING IN RANDOM ENVIRONMENTS AND WITH INTERACTION

ADVANCED SERIES ON STATISTICAL SCIENCE & APPLIED PROBABILITY

Editor: Ole E. Barndorff-Nielsen

*To view the complete list of the published volumes in the series, please visit:
http://www.worldscientific.com/series/asssap

Advanced Series on

Statistical Science &

Applied Probability

Vol. 20

SPATIAL BRANCHING IN RANDOM ENVIRONMENTS AND WITH INTERACTION

János Engländer
University of Colorado Boulder, USA

NEW JERSEY · LONDON · SINGAPORE · BEIJING · SHANGHAI · HONG KONG · TAIPEI · CHENNAI

Published by

World Scientific Publishing Co. Pte. Ltd.
5 Toh Tuck Link, Singapore 596224
USA office: 27 Warren Street, Suite 401-402, Hackensack, NJ 07601
UK office: 57 Shelton Street, Covent Garden, London WC2H 9HE

Library of Congress Cataloging-in-Publication Data
Engländer, Janos.
  Spatial branching in random environments and with interaction / by Janos Engländer,
University of Colorado Boulder, USA.
    pages cm. -- (Advanced series on statistical science and applied probability ; vol. 20)
  Includes bibliographical references.
  ISBN 978-981-4569-83-5 (hardcover : alk. paper)
  1. Mathematical statistics. 2. Branching processes. 3. Law of large numbers. I. Title.
  QA276.E54 2014
  519.2'34--dc23
                                    2014014879

British Library Cataloguing-in-Publication Data
A catalogue record for this book is available from the British Library.

Printed in Singapore

This book is dedicated to the memory of my parents, Katalin and Tibor Engländer, Z"L

I stand at the seashore, alone, and start to think. There are the rushing waves ... mountains of molecules, each stupidly minding its own business ... trillions apart ... yet forming white surf in unison.

Richard Feynman

It is by logic that we prove, but by intuition that we discover. To know how to criticize is good, to know how to create is better.

Henri Poincaré

Preface

I felt honored and happy to receive an invitation from World Scientific to write lecture notes on the talk that I gave at the University of Illinois at Urbana-Champaign. The talk was based on certain particle models with a particular type of interaction. I was even more excited to read the following suggestion:

> Although your talk is specialized, I hope that you can write something related to your area of research...

Such a proposal gives an author the opportunity to write about his/her favorite obsession! In the case of this author, that obsession concerns spatial branching models with interactions *and* in random environments.

My conversations with biologists convinced me that even though many such models constitute serious challenges to mathematicians, they are still ridiculously simplified compared to models showing up in population biology. Now, the biologist, of course, shrugs: after all, she 'knows' the answer, by using simulations. She feels being justified by the power of modern computer clusters and the almighty Law (of the large numbers); we, mathematicians, however would still like to see *proofs*, in no small part because they give an insight into the *reasons* of the phenomena observed. Secondly, the higher the order of the asymptotics one investigates, the less convincing the simulation result.

In this volume I will present a number of such models, in the hope that it will inspire others to pursue research in this field of contemporary probability theory. (My other hope is that the reader will be kind enough to find my *Hunglish* amusing rather than annoying.)

An outline of the contents follows.

In Chapter 1, we review the preliminaries on Brownian motion and diffusion, branching processes, branching diffusion and superdiffusion, and some analytical tools. This chapter became quite lengthy, even though several results are presented without proofs. Nevertheless, the expert in probability can easily skip many well-known topics.

Chapter 2 presents a Strong Law of Large Numbers for branching diffusions and, as a main tool, the 'spine decomposition.' Chapter 3 illustrates the result through a number of examples.

Chapter 4 investigates the behavior of the center of mass for spatial branching processes and treats a spatial branching model with *interactions* between the particles.

In Chapters 5, 6 and 7, spatial branching models are considered in *random media*. This topic can be considered a generalization of the well-studied model of a Brownian particle moving among random obstacles.

Finally, Appendix A discusses path continuity for Brownian motion, while Appendix B presents some useful maximum principles for semi-linear operators.

Each chapter is accompanied by a number of exercises. The best way to digest the material is to try to solve them. Some of them, especially in the first chapter, are well known facts; others are likely to be found only here.

How to read this book (and its first chapter)?

I had three types of audience in mind:

(1) Graduate students in mathematics or statistics, with the background of, say, the typical North American student in those programs.
(2) Researchers in probability, especially those interested in spatial stochastic models.
(3) Population biologists with some background in mathematics (but not necessarily in probability).

If you are in the second category, then you will probably skip many sections when reading Chapter 1, which is really just a smorgasbord of various tools in probability and analysis that are needed for the rest of the book. However, if you are in the first or third category, then I would advise you to try to go through most of it. (And if you are a student, I recommend to read Appendix A too.) If you do not immerse yourself in the intricacies of the construction of Brownian motion, you can still enjoy the later chapters, but if you are not familiar with, say, martingales or some basic concepts for second order elliptic operators, then there is no way you can appreciate the content of this book.

As for being a 'smorgasbord': hopping from one topic to another (seemingly unrelated) one, might be a bit annoying. The author hereby apologizes for that! However, adding more connecting arguments would have resulted in inflating the already pretty lengthy introductory chapter.

What should you do if you find typos or errors? Please keep calm and send your comments to my email address below. Also, recall George Pólya's famous saying:

The traditional professor writes *a*, says *b*, means *c*; but it should be *d*.

Several discussions on these models and collaborations in various projects are gratefully acknowledged. I am thus indebted to the following colleagues: Julien Berestycki, Mine Çağlar, Zhen-Qing Chen, Chris Cosner, Bill Fagan, Simon Harris, Frank den Hollander, Sergei Kuznetsov, Andreas Kyprianou,[1] Mehmet Öz, Ross Pinsky,[2] Yanxia Ren, Nándor Sieben,[3] Renming Song, Dima Turaev and Anita Winter.

My student Liang Zhang has been great in finding typos and gaps, for which I am very grateful to him.

I am very much obliged to Ms. E. Chionh at World Scientific for her professionalism and patience in handling the manuscript.

Finally, I owe thanks to my wife, Kati, for her patience and support during the creation of this book, and to our three children for being a continuing source of happiness in our life.

Boulder, USA, 2014   *János Engländer*
janos.englander@colorado.edu

---

[1]Who even corrected my English in this preface.
[2]The author's Ph.D. advisor in the 1990s.
[3]His help with computer simulations and pictures was invaluable.

Contents

Chapter 1

Preliminaries: Diffusion, spatial branching and Poissonian obstacles

This book discusses some models involving spatial motion, branching, random environments and interactions between particles, in domains of the d-dimensional Euclidean space. These models are often easy to grasp intuitively and in fact, they dovetail very nicely with certain population models. Still, working with them requires some background in advanced probability and analysis. In this chapter, therefore, we will review the preliminaries.

I take it for granted that the reader has a measure theoretical background and is familiar with some general concepts for stochastic processes. With regard to measure theory, I am writing with the expectation that the reader has undertaken, for example, a standard graduate level measure theory course at a US university. For stochastic processes, I assume that the reader has had a graduate level probability course and has been exposed, for instance, to the concept of finite dimensional distributions of a process, Kolmogorov's Consistency Theorem, and to the fundamental notions of martingales and Markov processes in continuous time.

We start with frequently used notation.

1.1 Notation, terminology

The following notation/terminology will be used.

(1) **Topology and measures:**

- The r-*ball* in $\mathbb{R}^d$ is the (open) ball around the origin with radius $r > 0$; the boundary of this ball is the r-*sphere*. For $r = 1$, the surface area and the volume of this ball will be denoted by s_d and ω_d, respectively. An r-ball around $x \in \mathbb{R}^d$ is defined similarly, and we will denote it by $B(x, r) = B_r(x)$.

- A *domain* in $\mathbb{R}^d$ is an open and connected subset of $\mathbb{R}^d$.
- The boundary of the set $B \subset \mathbb{R}^d$ will be denoted by ∂B and the closure of B will be denoted by $\text{cl}(B)$ or $\overline{B}$, that is $\text{cl}(B) = \overline{B} := B \cup \partial B$; the interior of B will be denoted by $\dot{B}$, and $B^\epsilon := \{y \in \mathbb{R}^d : \exists x \in B \text{ s.t. } |x - y| < \epsilon\}$ will denote the ϵ-neighborhood of B. We will also use the notation $\dot{B}^\epsilon := \{y \in B : B_\epsilon + y \subset B\}$, where $B + b := \{y : y - b \in B\}$ and B_ϵ is the ϵ-ball.
- If $A, B \subset \mathbb{R}^d$ then $A \subset\subset B$ will mean that A is bounded and $\text{cl}(A) \subset B$.
- By a *bounded rational rectangle* we will mean a set $B \subset \mathbb{R}^d$ of the form $B = I_1 \times I_2 \times \cdots \times I_d$, where I_i is a bounded interval with rational endpoints for each $1 \leq i \leq d$. The family of all bounded rational rectangles will be denoted by $\mathcal{R}$.
- The symbol δ_x denotes the Dirac measure (point measure) concentrated on x.
- The symbols $\mathcal{M}_f(D)$ and $\mathcal{M}_1(D)$ will denote the space of finite measures and the space of probability measures on $D \subset \mathbb{R}^d$, respectively. For $\mu \in \mathcal{M}_f(D)$, we define $\|\mu\| := \mu(D)$. The space of locally finite measures on D will be denoted by $\mathcal{M}_{\text{loc}}(D)$, and the space of finite measures with compact support on D will be denoted by $\mathcal{M}_c(D)$. The symbols $\mathcal{M}(D)$ and $\mathcal{M}_{\text{disc}}(D)$ will denote the space of finite discrete measures on D (finitely many atoms) and the space of discrete measures on D (countably many atoms), respectvely. The Lebesgue measure of the set $B \subset \mathbb{R}^d$ will be denoted by $|B|$.
- The symbols "$\overset{w}{\Rightarrow}$" and "$\overset{v}{\Rightarrow}$" will denote convergence in the weak topology and in the vague topology, respectively.
- Given a metric space, by the 'Borels' or 'Borel sets' of that space we will mean the σ-algebra generated by the open sets.

(2) **Functions:**

- For functions $0 < f, g : (0, \infty) \to (0, \infty)$, the notation $f(x) = \mathcal{O}(g(x))$ will mean that $f(x) \leq Cg(x)$ if $x > x_0$ with some $x_0 > 0, C > 0$, while $f \approx g$ will mean that f/g tends to 1 given that the argument tends to an appropriate limit. For functions $f, g : \mathbb{N} \to (0, \infty)$, the notation $f(n) = \Theta(g(n))$ will mean that $c \leq f(n)/g(n) \leq C \; \forall n$, with some $c, C > 0$.
- If $D \subset \mathbb{R}^d$ is a Borel set, f, g are Borel-measurable functions on D and μ is a measure on some σ-algebra of Borels that

includes D, then we will denote $\langle f, \mu \rangle := \int_D f(x)\, \mu(\mathrm{d}x)$ and $\langle f, g \rangle := \int_D f(x)g(x)\, \mathrm{d}x$, where $\mathrm{d}x$ is Lebesgue measure, and so $\langle f, g\mathrm{d}x \rangle = \langle fg, \mathrm{d}x \rangle = \langle f, g \rangle$.

- The symbols $C_b^+(D)$ and $C_c^+(D)$ denote the space of non-negative bounded continuous functions on D and the space of non-negative continuous functions on D with compact support, respectively.

- As usual, for $0 < \gamma \le 1$ and for a non-empty compact set $\mathcal{K} \subset \mathbb{R}^d$, one defines the Hölder-space $C^\gamma(\mathcal{K})$, as the set of continuous bounded functions on $\mathcal{K}$ for which $\|f\|_{C^\gamma} := \|f\|_\infty + |f|_{C^\gamma}$ is finite, where

$$|f|_{C^\gamma} := \sup_{x \in \mathcal{K}, h \ne 0} \frac{|f(x+h) - f(x)|}{|h|^\gamma}.$$

Furthermore, if $D \subset \mathbb{R}^d$ is a non-empty domain, then $C^\gamma = C^\gamma(D)$ will denote the space of functions on D which, restricted to $\mathcal{K}$, are in $C^\gamma(\mathcal{K})$ for all non-empty compact set satisfying $\mathcal{K} \subset\subset D$.

- We use the notation $\mathbf{1}_B$ to denote the indicator function (characteristic function) of the set B.

(3) **Probability:**

- The sum of the *independent* random variables X and Y will be denoted by $X \oplus Y$. The symbols $\bigoplus_{i=1}^n X_i$ and $\bigoplus_{i=1}^\infty X_i$ are defined similarly.

- Stochastic processes will be denoted by the letters X, Y, Z, etc., the value of X at time t will be denoted by X_t and a 'generic path' will be denoted by $X.$; Brownian motion (see next section) is traditionally denoted by the letter B or W. The symbol $Z \oplus \widehat{Z}$ will denote the sum of the independent stochastic processes Z and $\widehat{Z}$. The symbols $\bigoplus_{i=1}^n Z_i$ and $\bigoplus_{i=1}^\infty Z_i$ are defined similarly. When the stochastic process is a branching diffusion (superdiffusion), we prefer to use the letter Z (X); the underlying motion process will be denoted by Y.

- (S,W)LLN will abbreviate the (Strong,Weak) Law of Large Numbers.

(4) **Matrices:**

- The symbol $\mathbf{I}_d$ will denote the d-dimensional unit matrix, and $\mathrm{r}(\mathbf{A})$ will denote the rank of a matrix $\mathbf{A}$.

- The transposed matrix of $\mathbf{A}$ will be denoted $\mathbf{A}^T$.

(5) **Other:**

- As usual, $\lfloor z \rfloor$ will denote the integer part of $z \in \mathbb{R}$: $z := \max\{n \in \mathbb{Z} \mid n \le z\}$.
- Labeling: We will often talk about the 'i^{th} particle' of a branching particle system. By this we will mean that we label the particles randomly, but in a way that does not depend on their spatial position.

1.2   A bit of measure theory

Let X be an abstract set. A collection $\mathcal{P}$ of subsets of X is called a π-*system* if it is closed under intersections, that is, if $A \cap B \in \mathcal{P}$, whenever $A, B \in \mathcal{P}$.

A collection $\mathcal{L}$ of subsets of X is called a λ-*system* (or Dynkin system) if

(1) $\emptyset \in \mathcal{L}$;
(2) $A^c \in \mathcal{L}$ whenever $A \in \mathcal{L}$;
(3) $\mathcal{L}$ is closed under countable disjoint unions: $\bigcup_{i \ge 1} A_i \in \mathcal{L}$, whenever $A_i \in \mathcal{L}$ for $i \ge 1$ and $A_i \cap A_j = \emptyset$ for $i \ne j$.

The following lemma is often useful in measure theoretical arguments.

Proposition 1.1 (Dynkin's π-λ-Lemma). *Let $\mathcal{P}$ be a π-system of subsets of X, and $\mathcal{L}$ a λ-system of subsets of X. Assume that $\mathcal{P} \subset \mathcal{L}$. Then $\mathcal{L}$ contains the σ-algebra generated by $\mathcal{P}$: $\sigma(\mathcal{P}) \subset \mathcal{L}$.*

For the proof, see [Billingsley (2012)] Section 1.3.

Consider a probability space $(\Omega, \mathcal{F}, P)$. We all know that a real random variable is a measurable map from Ω to the reals, and we also know how P determines the law of the random variable.

Similarly, when thinking about a (real-valued) *stochastic process* on Ω, we may want to replace the measurable map of the previous paragraph by one of the following:

(1) A collection of maps from Ω to $\mathbb{R}$, indexed by 'time' $t \in [0, \infty)$;
(2) A collection of 'paths' (that is maps from $[0, \infty)$ to $\mathbb{R}$), indexed by $\omega \in \Omega$;
(3) A map from $\Omega \times [0, \infty)$ to $\mathbb{R}$.

Although these all appear to describe the same concept, they start to differ from each other when one also requires the *measurability* of these maps. It is fairly common to adopt the first definition with the quite weak requirement that each map is measurable, that is, each map is a random variable.

Now, as far as the *law* of a stochastic process is concerned, we proceed with invoking (the simplest version of) Kolmogorov's Consistency Theorem.[1] (See [Billingsley (2012)], Section 7.36.)

To this end, consider the space $\mathbb{R}^{[0,\infty)}$ consisting of 'paths' X., equipped with its Borel sets. Here the Borel sets are the σ-algebra generated by all sets of the form

$$A := \{X. \mid X_{t_1} \in B_1, X_{t_2} \in B_2, ..., X_{t_k} \in B_k\}, \tag{1.1}$$

where the B_i are one-dimensional Borels; these A's are called 'cylindersets.' In other words, consider $\mathbb{R}^{[0,\infty)}$ as the infinite product topological space, where each term in the product is a copy of $\mathbb{R}$ equipped with the Borels and let the Borel sets of $\mathbb{R}^{[0,\infty)}$ be the product σ-algebra.

Proposition 1.2 (Kolmogorov's Consistency Theorem). *Assume that we define a family of probability measures on cylindrical sets, that is, for each fixed $k \geq 1$ and for each $t_1, t_2, ..., t_k \geq 0$ ($t_i \neq t_j$ for $i \neq j$) we assign a probability measure $\nu_{t_1, t_2, ..., t_k}$. Assume also that the definition is not 'self-contradictory,' meaning that*

(1) If π is a permutation of $\{1, 2, ..., k\}$ with $k \geq 2$, and $B_1, B_2, ..., B_k$ are one-dimensional Borels, then

$$\nu_{t_{\pi(1)}, t_{\pi(2)}, ..., t_{\pi(k)}}(X. \mid X_{t_{\pi(1)}} \in B_{\pi(1)}, X_{t_{\pi(2)}} \in B_{\pi(2)}, ..., X_{t_{\pi(k)}} \in B_{\pi(k)})$$
$$= \nu_{t_1, t_2, ..., t_k}(X. \mid X_{t_1} \in B_1, X_{t_2} \in B_2, ..., X_{t_k} \in B_k),$$

that is, the definition is invariant under permuting indices.

(2) Let $1 \leq l < k$. If $B_l = B_{l+1} = ... = B_k = \mathbb{R}$, then $\nu_{t_1, t_2, ..., t_k}(A) = \nu_{t_1, t_2, ..., t_{l-1}}(A')$, where A is as in (1.1), and

$$A' := \{X. \mid X_{t_1} \in B_1, X_{t_2} \in B_2, ..., X_{t_{l-1}} \in B_{l-1}\},$$

that is, the definition is consistent, when considering a subset of $\{t_1, t_2, ..., t_k\}$.

Then there exists a unique extension of the family of measures from the cylinder sets to all Borel sets.

Remark 1.1. (i) Clearly, conditions (1) and (2) are necessary too.

(ii) Since we can always take Ω to be $\mathbb{R}^{[0,\infty)}$ (canonical representation), the map $\Omega \to \mathbb{R}^{[0,\infty)}$ can be guaranteed to be measurable, and the process is completely described by the measures of the Borel sets of Ω.   ◇

---

[1]A.k.a. the Kolmogorov Extension Theorem.

1.3   Gronwall's inequality

The following inequality is often useful.

Lemma 1.1 (Gronwall's inequality). *Assume that $f \geq 0$ is a locally bounded Borel-measurable function on $[0, \infty)$ such that*

$$f(t) \leq a + b \int_0^t f(s) \, ds$$

for all $t \geq 0$ and some constants a, b with $b \geq 0$. Then $f(t) \leq ae^{bt}$. In particular, if $a = 0$ then $f \equiv 0$.

Proof.   Applying the inequality twice,

$$f(t) \leq a + b \left(\int_0^t \left(a + b \int_0^s f(u) \, du \right) ds \right)$$

$$= a + abt + b^2 \int_0^t (t - u) f(u) \, du \leq a + abt + b^2 t \int_0^t f(u) \, du,$$

where the equality follows by integration by parts. Applying it $n \geq 2$ times, one obtains

$$f(t) \leq a + abt + \ldots + ab^n \frac{t^n}{n!} + R_n,$$

where $R_n := \frac{b^{t+1} t^n}{n!} \int_0^t f(u) \, du$. Since f is locally bounded, $\lim_{n \to \infty} R_n = 0$, and the result follows by writing e^{bt} as a Taylor series.   □

Other names of Gronwall's inequality are 'Gronwall's lemma,' 'Grönwall's lemma' and 'Gronwall–Bellman inequality.' Often the continuity of f is assumed, but it is not needed.

1.4   Markov processes

Let D be a domain in $\mathbb{R}^d, d \geq 1$. Recall that for a time-homogeneous Markov process ξ on $(\Omega, \mathcal{F}, (\mathcal{F}_t)_{t \geq 0}, P)$ with state space D, and with transition probability function $p(t, x, dy)$, the *Chapman-Kolmogorov equation* states that

$$p(t + s, x, B) = \int_{\mathbb{R}} p(s, y, B) p(t, x, dy), \quad s, t \geq 0; B \subset D \text{ Borel}.$$

Let $(\Omega, \mathcal{F}, (\mathcal{F}_t)_{t \geq 0}, P)$ be a filtered probability space. Recall that the σ-*algebra up to the stopping time* τ (denoted by $\mathcal{F}_\tau$) is the family of sets $A \in \mathcal{F}$ which satisfy that for all $t \geq 0$, $A \cap \{\tau \leq t\} \in \mathcal{F}_t$. It is an easy exercise (left to the reader) to show that $\mathcal{F}_\tau$ is indeed a σ-algebra.

A slightly stronger notion than the Markov property is as follows.

Definition 1.1 (Strong Markov process). *Let D be a domain in $\mathbb{R}^d$. A time-homogeneous Markov process ξ on $(\Omega, \mathcal{F}, (\mathcal{F}_t)_{t \geq 0}, P)$ with state space D and transition probability function $p(t, x, \mathrm{d}y)$ is a* strong *Markov process if, for all $t \geq 0$, all $\tau \geq 0$ stopping times with respect to the canonical filtration of ξ, and all $B \subset D$, one has that $P(\xi_{\tau+t} \in B \mid \mathcal{F}_\tau) = p(t, \xi_\tau, B)$ on $\{\tau < \infty\}$.*

A strong Markov process is obviously a Markov process (take a deterministic time as a stopping time); a counterexample for the converse can be found on p. 161 in [Wentzell (1981)].

It is customary to consider time-homogeneous Markov processes as families of probability measures $\{P_x, \ x \in D\}$, where the subscript denotes the starting position of the process: $P_x(\xi_t \in \cdot) = p(t, x, \cdot)$. The corresponding expectations are then denoted by $\{E_x, \ x \in D\}$. The next important definition[2] is that of a 'Feller process.'

Definition 1.2 (Feller process). *A time-homogeneous Markov process ξ on $(\Omega, \mathcal{F}, (\mathcal{F}_t)_{t \geq 0}, P)$ with state space D is a* Feller *process if the function $x \to E_x f(\xi_t)$ is bounded and continuous for each $t \geq 0$, whenever $f : D \to \mathbb{R}$ is so.*

Another way of stating the Feller property is that the map T_t defined by $T_t(f)(x) := E_x f(\xi_t)$, leaves the space of bounded continuous functions invariant for all $t \geq 0$. Clearly, $T_t(f)$ is always bounded if f is so. Hence, yet another way of stating it is that the map $x \mapsto P_x$ is continuous if the measures $\{P_x, \ x \in \mathbb{R}\}$ are equipped with the weak topology.

Every right-continuous Feller process is a strong Markov process, but the converse is not true. (See Exercise 4.)

1.5 Martingales

1.5.1 *Basics*

Cherchez la femme[3] the French say; 'look for the martingale,' says the probabilist. (French probabilists say both.) Indeed, it is hard to overestimate the significance of martingale techniques in probability theory.

---

[2]The reader is warned that in the literature sometimes the class of bounded continuous functions is replaced in the following definition by continuous functions vanishing at Δ, where $\widehat{D} = D \cup \{\Delta\}$ is the one-point compactification (Alexandroff c.) of D.

[3]Look for the woman.

Recall that, given the filtered probability space $(\Omega, \mathcal{F}, (\mathcal{F}_t)_{t \geq 0}, P)$, a stochastic process X is called a *submartingale* if

(1) X is adapted (by which we mean that $\sigma(X_t) \subset \mathcal{F}_t$ for $t \geq 0$);
(2) $E|X_t| < \infty$ for $t \geq 0$;
(3) $E(X_t \mid \mathcal{F}_s) \geq X_s$ (*P*-a.s.) for $t > s \geq 0$.

The process X is called a *supermartingale* if $-X$ is a submartingale. Finally, if X is a submartingale and a supermartingale at the same time, then X is called a *martingale*.

It is easy to check that if one replaces the filtration by the canonical filtration generated by X (i.e. one chooses $\mathcal{F}_t := \sigma(\bigcup_0^t \sigma(X_s)))$, then the (sub)martingale property still holds. Hence, when the filtration is not specified, it is understood that the filtration is the canonical one.

Next, we recall the two most often cited results in martingale theory; they are both due to Doob.[4] The first one is his famous 'optional stopping' theorem.[5]

Theorem 1.1 (Doob's optional stopping theorem). *Given the filtered probability space* $(\Omega, \mathcal{F}, (\mathcal{F}_t)_{t \geq 0}, P)$*, let* $M = (M_t)_{t \geq 0}$ *be a martingale with right-continuous paths, and* $\tau : \Omega \to [0, \infty]$ *a stopping time. Then the process* η *defined by* $\eta_t := M_{t \wedge \tau}$ *is also a martingale with respect to the same* $(\Omega, \mathcal{F}, (\mathcal{F}_t)_{t \geq 0}, P)$*.*

Replacing the word 'martingale' by 'submartingale' in both sentences produces a true statement too.

The second one is an improvement on the Markov inequality, for submartingales.

Theorem 1.2 (Doob's inequality). *Let* M *be a submartingale with right-continuous paths and* $\lambda > 0$*. Then, for* $t > 0$*,*

$$P\left(\sup_{0 \leq s \leq t} M_s \geq \lambda \right) \leq \frac{EM_t^+}{\lambda},$$

where $x^+ := \max\{x, 0\}$*.*

---

[4]Joseph L. Doob (1910–2004), a professor at the University of Illinois, was one of the founding fathers of the modern theory of stochastic processes and probabilistic potential theory. The notion of a (sub)martingale was also introduced by him, just like most of martingale theory itself.

[5]A.k.a. 'Doob's optional sampling theorem,' although in Doob's own terminology, the latter name referred to a more general result. Another, closely related version of optional stopping concerns two stopping times $S \leq T$ and whether the defining inequality of submartingales still holds at these times. In that version though, unlike here, the martingale must be 'closable' by a last element $(M_\infty, \mathcal{F}_\infty)$.

We will also need the following slight generalization of Doob's inequality.

Lemma 1.2. *Assume that $T \in (0, \infty)$, and that the non-negative, right-continuous, filtered stochastic process $(N_t, \mathcal{F}_t, P)_{0 \leq t \leq T}$ satisfies that there exists an $a > 0$ such that*

$$E(N_t \mid \mathcal{F}_s) \geq aN_s, \ 0 \leq s < t \leq T.$$

Then, for every $\alpha \in (0, \infty)$ and $0 \leq s \leq T$,

$$P\left(\sup_{t \in [0,s]} N_t \geq \alpha \right) \leq (a\alpha)^{-1} E(N_s).$$

Proof. Looking at the proof of Doob's inequality (see Theorems 5.2.1 and 7.1.9 in [Stroock (2011)] and their proofs), one sees that, when the submartingale property is replaced by our assumption, the whole proof goes through, except that now one has to include a factor a^{-1} on the right-hand side. $\qquad \square$

A well-known inequality for conditional expectations, closely related to martingales is as follows.

Theorem 1.3 (Conditional Jensen's inequality). *Let X be a random variable on $(\Omega, \mathcal{F}, P)$ and $\mathcal{G} \subset \mathcal{F}$ be a σ-algebra. If f is a convex[6] function, then*

$$E(f(X) \mid \mathcal{G}) \geq f(E(X \mid \mathcal{G})).$$

(If the left-hand side is $+\infty$, the inequality is taken as true.)

Remark 1.2. (a) When $\mathcal{G} = \{\emptyset, \Omega\}$, one obtains the (unconditional) Jensen's inequality.

(b) The fact that a convex (concave) function of a martingale is a submartingale (supermartingale), provided it is integrable, is a simple consequence of Theorem 1.3. $\qquad \diamond$

A fundamental convergence theorem is as follows:

Theorem 1.4 (Submartingale convergence theorem). *Let M be a submartingale with right-continuous paths with respect to the filtered probability space $(\Omega, \mathcal{F}, (\mathcal{F}_t)_{t \geq 0}, P)$. Assume that $\sup_{t \geq 0} E(X_t^+) < \infty$. Then X_t has a P-almost sure limit, X_∞ as $t \to \infty$, and $E|X_\infty| < \infty$.*

---

[6]By 'convex' we mean convex from above, like $f(x) = |x|$.

Here $X^+ := \max\{0, X\}$. Letting $Y := -X$, one gets the corresponding result for supermartingales.

One often would like to know when a martingale limit exists in L^1 as well.

Theorem 1.5 (L^1-convergence theorem). *Let M be a martingale with right-continuous paths with respect to the filtered probability space $(\Omega, \mathcal{F}, (\mathcal{F}_t)_{t \geq 0}, P)$. Then the following conditions are equivalent:*

(1) $\{M_t\}_{t \geq 0}$ is a uniformly integrable family.
(2) M_t converges in L^1 as $t \to \infty$.
(3) M_t converges in L^1 as $t \to \infty$ to a random variable $M_\infty \in L^1(P)$ such that M_t is a martingale on $[0, \infty]$ with respect to $(\Omega, \mathcal{F}, (\mathcal{F}_t)_{t \in [0, \infty]}, P)$. (Here $\mathcal{F}_\infty := \sigma(\bigcup_{t \geq 0} \sigma(M_t))$, and M_∞ is the 'last element' of this martingale.)
(4) There exists a random variable $Y \in L^1(P)$ such that

$$M_t = E(Y \mid \mathcal{F}_t) \tag{1.2}$$

holds P-a.s. for all $t \geq 0$.

The last two conditions are linked by the fact that (1.2) is true for $t = \infty$ as well.

1.5.2   *Estimates for the absolute moments*

A classical result by Marcinkiewicz and Zygmund concerns independent random variables with zero mean, as follows.

Theorem 1.6 (Marcinkiewicz-Zygmund inequality; 1937). *There exist positive constants k_p, K_p for any $1 \leq p < \infty$ such that the following inequality holds for all sequences $Z_1, Z_2, \ldots$ of independent random variables in L^p, with zero mean:*

$$k_p E\left(\sum_{i=1}^n Z_i^2\right)^{p/2} \leq E\left|\sum_{i=1}^n Z_i\right|^p \leq K_p E\left(\sum_{i=1}^n Z_i^2\right)^{p/2}, \quad n \geq 1. \tag{1.3}$$

Note that $M_n := \sum_1^n Z_i$, $n \geq 1$, is a martingale. Let $[M]$ denote the *quadratic variation process*, that is, let $M_0 := 0$ and

$$[M]_n := \sum_{k=0}^{n-1} (M_{k+1} - M_k)^2 = \sum_{k=1}^n Z_i^2.$$

Then, (1.3) can be rewritten as

$$k_p E[M]_n^{p/2} \leq E|M_n|^p \leq K_p E[M]_n^{p/2}.$$

More generally, given $(\Omega, \mathcal{F}, P)$, the random variables $Z_1, Z_2, \ldots$ are called *martingale differences*, if M defined by $M_n := \sum_1^n Z_i$, $n \geq 1$, is a P-martingale.

In a more recent, famous inequality, the pth absolute moment of the martingale is replaced by the pth moment of the maximum of the $|M_k|$:

Theorem 1.7 (Burkholder-Davis-Gundy inequality; discrete time). *For $1 \leq p < \infty$, there exist positive constants c_p, C_p such that the following inequality holds for all martingales M with $M_0 = 0$, and all $n \geq 1$:*

$$c_p E[M]_n^{p/2} \leq E \max_{0 \leq k \leq n} |M_k|^p \leq C_p E[M]_n^{p/2}.$$

(Here again, $[M]_n := \sum_{k=0}^{n-1} (M_{k+1} - M_k)^2$.)

This result clearly generalizes the upper estimate in (1.3).

Even more recently, J. Biggins proved the following upper estimate for the case[7] when $1 \leq p < 2$.

Theorem 1.8 (L^p inequality of Biggins). *Let $1 \leq p < 2$. Then*

$$E|M_n|^p \leq 2^p \sum_{i=1}^{n} E|Z_i|^p, \quad n \geq 1, \tag{1.4}$$

or, equivalently,

$$\|M_n\|_p \leq 2 \|M(n,p)\|_p, \quad n \geq 1,$$

where $\|\cdot\|_p$ denotes $L^p(\Omega, P)$-norm, and $M(n,p) := (\sum_{i=1}^{n} |Z_i|^p)^{1/p}$.

(See Lemma 1 in [Biggins (1992)]; see also [Champneys et al. (1995)].)

1.6 Brownian motion

After this general review, let us proceed with discussing the building block of all stochastic analysis: *Brownian motion*.

Brownian motion is named after the Scottish botanist Robert Brown (1773–1858), because of Brown's famous 1827 experimental observations of pollen grains moving in a random, unpredictable way in water. The jittery

---

[7]It is trivially true for $p = 2$.

motion observed was assumed to be the result of a huge number of small collisions with tiny invisible particles.

One should note though, that the Dutch biologist, Jan Ingenhousz, made very similar observations in 1785, with coal dust suspended on the surface of alcohol. Moreover, as some historians pointed out, some 1900 years before Brown, the Roman poet and philosopher, Titus Lucretius Carus's six volume poetic work *'De Rerum Natura'* (On the Nature of Things) already contained a description of Brownian motion of dust particles — it is in the second volume of the work, called 'The dance of atoms.'

Following Brown, the French mathematician Louis Bachelier (1870–1946) in his 1900 PhD thesis *'Theorie de la Speculation'* (The Theory of Speculation) presented a stochastic analysis of the stock and option markets in a pioneering way involving Brownian motion.[8]

Brown's experiment was one of the motivations for Einstein's celebrated 1905 article in Volume 322 of Annalen der Physik, *'Über die von der molekularkinetischen Theorie der Wärme geforderte Bewegung von in ruhenden Flüssigkeiten suspendierten Teilchenthe,'* (On the Motion of Small Particles Suspended in a Stationary Liquid, as Required by the Molecular Kinetic Theory of Heat).

One should also mention here two other physicists' work.

The first one is Smoluchowski's 1906 paper, *'Zur kinetischen Theorie der Brownschen Molekularbewegung und der Suspensionen,'* (Towards the kinetic theory of the Brownian molecular movement and suspensions) which he wrote independently of Einstein's result.[9]

Two years later, Paul Langevin devised yet another description of Brownian motion.

The first mathematically rigorous theory of Brownian motion as a stochastic process was, however, established by MIT's famous faculty member, Norbert Wiener (1894–1964).

Although there are whole libraries written on Brownian motion (sometimes called *Wiener process*), we will just focus here on two standard approaches to the definition. In a nutshell they are the following.

(1) One defines the finite dimensional distributions and shows that they form a consistent family, which, by Kolmogorov's Consistency Theorem implies the existence of a unique probability measure on all paths.

---

[8]Bachelier's advisor was no other than Henri Poincaré, but that did not help him much in his academic career: Bachelier obtained his first permanent university position at the age of 57.

[9]Less known are his other contributions, such as his work on branching processes.

Fig. 1.1 Norbert Wiener [Wikipedia].

Then, using another theorem of A. N. Kolmogorov (the moment condition for having a continuous modification), one shows that one can uniquely transfer the previous probability measure on all paths to a probability measure on the space of *continuous* paths. (The meaning of the word 'transfer' will be explained in Appendix A.)

(2) Following P. Lévy, one constructs directly a sequence of random continuous paths on the unit interval and shows that they converge uniformly with probability one; the limiting random continuous path will be Brownian motion. Once Brownian motion is constructed on the unit time interval, it is very easy to extend it to $[0, \infty)$.

The probability distribution on continuous paths corresponding to Brownian motion is then called the *Wiener measure*.

1.6.1   *The measure theoretic approach*

Let us see now the details of the first approach.

In accordance with Proposition 1.2, consider $\widehat{\Omega} := \mathbb{R}^{[0,\infty)}$, that is, let $\widehat{\Omega}$ denote the space of all real functions on $[0, \infty)$, and let $\mathcal{B}'$ be the σ-algebra of sets generated by the cylindrical sets. (The reason for the notation $\widehat{\Omega}$ and $\mathcal{B}'$ is that Ω and $\mathcal{B}$ are reserved for certain other sets, introduced later, which will be proven much more useful.)

According to Kolmogorov's Consistency Theorem (Proposition 1.2), if we specify how to define the measure on cylindrical sets, and if that definition 'is not self-contradictory,' then the measure can uniquely be extended to $\mathcal{B}'$. We now make the particular choice that for the cylindrical sets

$$A := \{X. \mid X_{t_1} \in B_1, X_{t_2} \in B_2, ..., X_{t_k} \in B_k\},$$

where B_m, $m = 1, 2, ..., k$; $k \geq 1$ are Borels of the real line, its measure $\nu_{t_1, t_2, ..., t_k}(A)$ is the one determined by the k-dimensional Gaussian measure with zero mean, and covariance matrix given by $\mathrm{cov}(X_{t_i}, X_{t_j}) = \min(t_i, t_j)$, for $0 \leq i, j \leq k$. For this definition, both consistency requirements in Proposition 1.2 are obviously satisfied, and thus, there exists a unique extension, a probability measure ν, on $(\widehat{\Omega}, \mathcal{B}')$.

The problem however, is that the family $\mathcal{B}'$ is too small (and $\widehat{\Omega}$ is too large, for that matter) in the following sense. Recall that every 'reasonable' subset of the real line is Borel, and that in fact it requires some effort to show that there exist non-Borel sets. The situation is very different when one considers $(\widehat{\Omega}, \mathcal{B}')$! In fact, $\mathcal{B}'$ does not contain many of the sets of interest. For example, such a set is $\Omega := C[0, \infty)$, the set of continuous paths on $[0, \infty)$. This non-measurability of the set of continuous paths is clearly a source of troubles, since it implies that the innocent looking question

Q.1: What is the probability that a path is continuous?

simply does not make sense!

To explain this phenomenon, as well as the resolution to this problem, is important, but it requires a few more pages. Since this issue is not the main topic of the book, it has been relegated[10] to Appendix A. It suffices to say here that there exists a *version* (or modification) of the process which has continuous paths, and the following definition makes sense. Let $\mathcal{B}$ denote the Borel sets of Ω.

---

[10]If the reader is, say, a graduate student, then reading the appendix is recommended.

Definition 1.3 (Gaussian definition of Wiener-measure). *On the space* $(\Omega, \mathcal{B})$*, the* Wiener-measure *is the unique probability measure* μ *such that if* $0 \leq t_1 \leq ... \leq t_k$ *and*

$$A := \{X. \mid X_{t_1} \in B_1, X_{t_2} \in B_2, ..., X_{t_k} \in B_k\},$$

then $\mu(A) = \nu_{t_1,t_2,...,t_k}(B_1 \times ... \times B_k)$*, where* B_m*,* $m = 1, 2, ..., k$*;* $k \geq 1$ *are Borels of the real line.*

(See Appendix A for more elaboration.)

Another way of saying the above is that Brownian motion $B = \{B_t\}_{t \geq 0}$ is a continuous *Gaussian process* with zero mean for all times $t \geq 0$, and with covariance $\min(t, s)$ for times $t, s \geq 0$. In particular $X_0 = 0$ with probability one, and the probability density function of $B_t, t > 0$ is:

$$\frac{1}{\sqrt{2\pi t}} e^{-x^2/2t}.$$

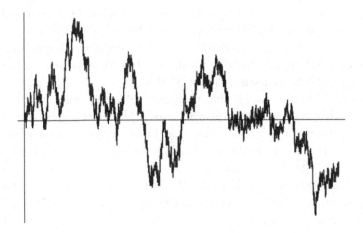

Fig. 1.2   1-dimensional Brownian trajectory.

Remark 1.3 (Wiener's method). Wiener's original approach was very different — it was the approach of a harmonic analyst. Wiener's construction gives a representation of the Brownian path on $[0, 1]$ in terms of a *Fourier series with random coefficients* as follows. Let $A_k, k = 1, 2, ...$ be independent standard normal variables on some common probability space. Then B on $[0, 1]$ given by

$$B_t = \frac{\pi}{2\sqrt{2}} \sum_1^\infty \frac{A_k}{k} \sin(\pi k t/2), \ 0 \leq t \leq 1,$$

is well defined (that is, the series converges), and it is a Brownian motion. (Wiener actually looked at his measure as a Gaussian measure on an infinite dimensional space; see Chapter 8 in [Stroock (2011)].)   ◇

1.6.2   *Lévy's approach*

The second method mentioned at the beginning is from 1948 and is due to the giant of the French probability school: Paul Lévy (1886–1971) of École Polytechnique. The main idea is as follows: We would like to construct a process B with continuous paths, such that

(1) $B_0 = 0$,
(2) $B_t - B_s$ is a mean zero normal variable with variance $t - s$, for all $0 \leq s < t$,
(3) B has independent increments.

Our motivation is coming from the fact that assumptions (1)–(3) determine the Wiener measure as the law of the process, that is, together they are in fact equivalent to Definition 1.3 – see Exercise 5.

In order to do so, define $D_n := \{k/2^n \mid 0 \leq k \leq 2^n\}$ (nth order dyadic points of the unit interval) and $D := \cup_n D_n$. We wish to approximate B with *piecewise linear* processes, such that the nth approximating process will be linear between points of D_n and the above two assumptions on the increments are satisfied *as long as the endpoints are in D_n*. (Clearly independence cannot hold on the linear pieces.)

Let $\{Z_d\}_{d \in D}$ be an independent collection of standard normal random variables on a common probability space Ω. For $n = 0$ we only have two points in D_0, and we consider the random straight line starting at the origin and ending at the point $(1, Z_1)$, that is we define $B_0 := 0$ and $B_1 = Z_1$. Now refine this random line by changing the value at the point $1/2$ to a new one by defining the new value as

$$B_{1/2} := \frac{B_1}{2} + \frac{Z_{1/2}}{2},$$

(thus mimicking what the value at $1/2$ should be if it were defined by Brownian motion at time $1/2$: it has mean zero and variance $1/2$, and the increments are independent mean zero Gaussians with variance $1/2$). Again, by using linear interpolation, we get a random polygon starting at the origin and ending at $(1, Z_1)$, and consisting of two straight pieces. Continue this in an inductive manner: once a random polygon is obtained using D_{n-1}, in the next step consider $d \in D_n \setminus D_{n-1}$ and define the random value at d by

$$B_d := \frac{B_{d_-} + B_{d_+}}{2} + \frac{Z_d}{2^{(n+1)/2}}, \qquad (1.5)$$

where d_- and d_+ are the left and right 'neighbors' of d: $d_\pm = d \pm 2^{-n}$. (Without the second term, (1.5) would simply be linear interpolation, so we can consider it a small normal 'noise.') Using induction, it is easy to check that at each step, the collection $\{B_d\}_{d \in D_n}$ is independent of the system $\{Z_d\}_{d \in D \setminus D_n}$. Furthermore, and most importantly, at each step, the construction guarantees that the new, larger family of increments we consider (with endpoints being in D_n), still consists of independent, normally distributed variables with mean zero and the 'right' variance. Once new points added, redraw the random polygon now interpolating linearly between all the values, including the new ones, getting a refinement of the previous random polygon (because straight lines are being replaced by two concatenated straight lines).

We only sketch the rest of the construction (for the details see [Mörters and Peres (2010)]). As a next step, one verifies that there is a *uniform* limit of these more and more refined random polygons for almost all $\omega \in \Omega$, and calls the limiting random continuous path between the origin and $(1, Z_1(\omega))$ a Brownian path on $[0, 1]$. (The uniform limit is essentially a consequence of the fact that the 'noise' term in (1.5) is 'small,' that is, it has a 'light' tail.) Using that D is dense in $[0, 1]$ and the continuity of the limit, it is then easy to show that all required properties concerning the increments for the limit extend from D to $[0, 1]$.

Once this is done, one can define the Brownian path on $[0, \infty)$ by induction. If we have defined it on $[0, n]$, $n \geq 1$ already, then on $[0, n+1]$ we extend the definition by

$$B_t := B_n + \widehat{B}_r^{(n)}, \ t \in (n, n+1],$$

where $\widehat{B}^{(n)}$ is a Brownian motion on the unit interval, independent of the already constructed Brownian path on $[0, n]$ and $r := t - n$.

1.6.3 *Some more properties of Brownian motion*

Lévy's construction has the great advantage over the previous one that one does not have to worry about path continuity at all. On the other hand the first method is more robust, and one understands better the general principle of defining a continuous process with given fidi's.

There are several other approaches to Brownian motion. It can be defined

- as the scaling limit[11] as $n \to \infty$ of simple random walks, where at level n, time is sped up by factor n and space is shrunk by factor $\sqrt{n}$, simultaneously, (this is called 'Donsker's Invariance Principle');
- as a time homogeneous Markov process through its transition kernel:

$$p(t, x, y) := \frac{1}{\sqrt{2\pi}} \exp\left(-\frac{(x-y)^2}{2t}\right),$$

which requires showing that the Chapman-Kolmogorov equation is satisfied by this kernel;

- as a Lévy-process[12] through its Laplace transform,

$$E_0\left(e^{i\theta W_t}\right) = \exp\left(-\frac{1}{2}t\theta^2\right), \ \theta \in \mathbb{R};$$

- as the unique solution to the so-called *martingale problem* corresponding to the operator $\frac{1}{2}\Delta$ (this will be discussed in a broader context); or
- following Wiener's original approach, which was related to Fourier analysis and Gaussian measures on infinite dimensional spaces,

just to name a few.

Some of the important properties of Brownian motion are as follows.

(1) The set of paths which are differentiable even at one point has measure zero. This means that a typical Brownian path shares the surprising property of the well-known Weierstrass function: it is nowhere differentiable although everywhere continuous. *A fortiori*, the set of paths which are of bounded variation even on one positive interval has measure zero.

(2) The set of paths which are Hölder-continuous with exponent larger than $1/2$ even on one compact interval has measure zero. On the other hand, if the exponent is less than or equal to $1/2$, then there exists a version, such that the paths are locally Hölder-continuous a.s.

(3) Brownian motion is a mean zero martingale with finite quadratic variation.

---

[11]It is quite easy to show the convergence of finite dimensional distributions; it is much more difficult to show that the corresponding laws on $C([0, \infty))$ converge weakly to a limiting law. This requires establishing the relative compactness of those laws.

[12]A Lévy-process is a stochastic process starting at zero, with stationary independent increments, and càdlàg paths.

(4) Brownian motion is a strong Markov process (and even a Feller process).

(5) Brownian motion has independent stationary increments. (The distribution of $B_t - B_s$ is normal with mean zero and covariance $t - s$.)

(6) 'Brownian scaling': the process $\tilde{B}$ defined by

$$\tilde{B}_t := aB_{t/a^2} \tag{1.6}$$

is also a Brownian motion, where $a > 0$.

(7) 'Law of Large Numbers for Brownian motion': $\lim_{s \to \infty} B_s/s = 0$ with probability one.

(8) 'Reflection principle': If $a, t > 0$ and P is Wiener measure, then

$$P\left(\sup_{0 \le s \le t} B_s \ge a\right) = 2P(B_t \ge a). \tag{1.7}$$

The really deep fluctuation result on Brownian motion is (the continuous version of) Khinchin's *Law of Iterated Logarithm*, which we mention here, although we do not need it in this book, and which says that, with probability one,

$$\limsup_{t \to +\infty} \frac{|B_t|}{\sqrt{2t \log \log t}} = 1.$$

Then of course, by Brownian scaling (1.6), we also have

$$\limsup_{h \to 0} \frac{|B_h|}{\sqrt{2h \log \log(1/h)}} = 1,$$

with probability one.

A d-*dimensional Brownian motion* is a d-dimensional stochastic process, for which all its coordinate processes are independent one-dimensional Brownian motions. That is,

$$B_t = (B_t^{(1)}, B_t^{(2)}, ..., B_t^{(d)}),$$

where $B^{(k)}$ is a one-dimensional Brownian motion, for $1 \le k \le d$, and the $B^{(k)}$'s are independent.

It is clear that if $\mathbf{x} \in \mathbb{R}^d$, then $B^{(\mathbf{x})}$ defined by

$$B_t^{(\mathbf{x})} := \mathbf{x} + B_t$$

is also a continuous Gaussian process with the same covariance structure as B and with mean value $\mathbf{x}$, starting at $\mathbf{x}$ with probability one. (Those who are more Markovian in their approach would prefer to say that we have a *family* of probability laws $\{\mu_{\mathbf{x}}; \mathbf{x} \in \mathbb{R}^d\}$ and $\mu_{\mathbf{x}}(B_0 = \mathbf{x}) = 1$.) Sometimes the $\mathbf{x} = \mathbf{0}$ case is distinguished by saying that we have a *standard* d-dimensional Brownian-motion, in which case the probability density function of $B_t, t > 0$ is:

$$f(\mathbf{x}) = \frac{1}{(2\pi t)^{d/2}} \exp\left(-|\mathbf{x}|^2/2t\right).$$

1.7   Diffusion

Starting with Brownian motion, as the fundamental building block, we now go one step further and define multidimensional diffusion processes.

In 1855, Adolf Eugen Fick (1829–1901), a German physiologist,[13] first reported his laws governing the transport of mass through diffusive means. His work was inspired by the earlier experiments of Thomas Graham, a 19th-century Scottish chemist.

The discovery that the particle density satisfies a parabolic partial differential equation[14] is due to Adriaan Fokker and Max Planck ('Fokker-Planck equation') and to Andrey Kolmogorov ('Kolmogorov forward equation'), besides Smoluchowski and Einstein.

1.7.1   *Martingale problem, L-diffusion*

We start with an assumption on the operator.

Assumption 1.1 (Diffusion operator). *L is a second order elliptic differential operator on the Euclidean domain $D \subseteq \mathbb{R}^d$ of the form*

$$L = \frac{1}{2} \sum_{i,j=1}^{d} a_{ij} \frac{\mathrm{d}^2}{\mathrm{d}x_i\,\mathrm{d}x_j} + \sum_{i=1}^{d} b_i \frac{\mathrm{d}}{\mathrm{d}x_i},$$

where the functions $a_{ij}, b_i : D \to \mathbb{R}$, $i,j = 1, ..., d$, are locally bounded and measurable, and the symmetric matrix[15] $(a_{ij}(x))_{1 \leq i,j \leq d}$ is positive definite for all $x \in D$. In addition, we assume that the functions a_{ij} are in fact continuous.

Of course, when a is differentiable, L can be written in the slightly different 'divergence form' too. For the purpose of using some PDE tools, it is useful to assume that b is smooth as well. This leads to the following, alternative assumption.

Assumption 1.2 (Divergence form). *L is a second order elliptic differential operator on $D \subseteq \mathbb{R}^d$ of the form*

$$L = \frac{1}{2} \nabla \cdot \widetilde{a} \nabla + \widetilde{b} \cdot \nabla,$$

---

[13]And Einstein's 'academic grandfather.'

[14]Considering more general, non-selfadjoint operators, one has to be a bit more careful: then the density satisfies the equation with the *formal adjoint* operator; cf. (1.8) a little later, where an equivalent formulation of this fact is given.

[15]We hope the reader forgives us for writing simply x instead of **x** in the sequel.

where the functions $\tilde{a}_{ij}, \tilde{b}_i : D \to \mathbb{R}$, $i, j = 1, ..., d$, *are in the class* $C^{1,\eta}(D)$, $\eta \in (0, 1]$ *(i.e. their first order derivatives exist and are locally Hölder-continuous), and the symmetric matrix* $(\tilde{a}_{ij}(x))_{1 \le i,j \le d}$ *is positive definite for all* $x \in D$.

In this case the non-divergence form coefficients can be expressed as $a = \tilde{a}$ and $b_i = \tilde{b}_i + \sum_{j=1}^n \frac{\mathrm{d}a_{ij}}{\mathrm{d}x_j}$.

Assumption 1.2 is more restrictive than Assumption 1.1, as it requires more smoothness. We will state in each case the assumption we will be working under. When choosing Assumption 1.2, we will simply write a and b without tildes.

Assume now that L satisfies Assumption 1.1. The operator L then corresponds to a unique *diffusion process* (or diffusion) Y on D in the following sense.[16] Take a sequence of increasing domains $D_n \uparrow D$ with $D_n \subset\subset D_{n+1}$; and let $\tau_{D_n} := \inf\{t \ge 0 \mid Y_t \notin D_n\}$ denote the first exit time from the (open) set D_n. The following result is of fundamental importance.

Proposition 1.3. *There exists a unique family of probability measures* $\{P_x, \ x \in D\}$ *on* Ω, *the space of continuous paths, describing the law of a Markov process* Y *such that*

(1) $P_x(Y_0 = x) = 1$,
(2) $f(Y_{t \wedge \tau_{D_n}}) - \int_0^{t \wedge \tau_{D_n}} (Lf)(Y_s) \, \mathrm{d}s$ *is a* P_x-*martingale, with respect to the canonical filtration, for all* $f \in C^2(D)$ *and all* $n \ge 1$.

(Our notation is in line with the Markovian approach alluded to previously.) This proposition is a generalization of the celebrated result on the 'martingale problem' by D. W. Stroock and S. R. S. Varadhan, and is due to R. Pinsky. Following his work, we say that the *generalized martingale problem* on D has a unique solution and it is the law of the corresponding diffusion process or *L-diffusion* Y on D.

Note that it is possible that the event $\lim_{n \to \infty} \tau_{D_n} < \infty$ ('explosion') has positive probability. In fact $\lim_{n \to \infty} \tau_{D_n} < \infty$ means that the process reaches Δ, a 'cemetery state' in finite time, where Δ is identified with the Euclidean boundary of D plus a point 'at infinity'. In other words, the process actually lives on $\hat{D} = D \cup \{\Delta\}$, the one-point compactification of D and once it reaches Δ it stays there forever. In fact, the word 'generalized'

---

[16]Since we define diffusions via the generalized martingale problem, there is no need to discuss stochastic differential equations, and thus the notion of Itô-integral is postponed to a subsequent section. Later, however, we will need them.

in the definition refers exactly to the fact that we allow explosion, unlike in the classical Stroock-Varadhan martingale problem.

1.7.2 *Connection to PDE's; semigroups*

The connection between diffusion processes and linear partial differential equations is well known. Let L satisfy Assumption 1.2. For a bounded continuous function f, consider the parabolic Cauchy problem:

$$\left.\begin{array}{c} \dot{u} = Lu \text{ in } (0,\infty) \times D, \\[2mm] \lim_{t\downarrow 0} u(\cdot,t) = f(\cdot) \text{ in } D. \end{array}\right\} \tag{1.8}$$

This Cauchy problem ('the generalized heat equation') is then solved[17] by $u(x,t) := T_t(f)(x) := E_x f(Y_t)$, $x \in D, t \geq 0$ and Y is the diffusion corresponding to L on D.

Furthermore, the Markov property of Y yields that is $T_{t+s} = T_t \circ T_s$ for $t, s \geq 0$, where the symbol '∘' denotes composition. One is tempted to say that $\{T_t\}_{t\geq 0}$ is a semigroup, however that is not necessarily justified, depending on the function space. Indeed, if we work with bounded continuous functions, then we need the Feller property of the underlying diffusion. If we work with bounded measurable functions, however, then calling it a semigroup is indeed correct. So, when we call $\{T_t\}_{t\geq 0}$ *the semigroup corresponding to Y (or to L) on D*, we have this latter sense in mind.

It turns out that $\{T_t\}_{t\geq 0}$ is *strongly continuous*, which means that

$$\lim_{t\to 0} T_t(f)(x) = f(x) \tag{1.9}$$

in supremum norm. Now, (1.8) gives

$$\lim_{h\downarrow 0} \frac{T_{t+h}f(x) - T_t f(x)}{h} = L(T_t f)(x), \ t > 0,$$

and formally we obtain ($t = 0$) that

$$\lim_{h\downarrow 0} \frac{T_h f - f}{h} = Lf,$$

point-wise, which can indeed be verified for a certain class of functions f, which includes the class $C_c^2(D)$ (see, for example, Section 7.3 in [Øksendal (2010)]). Hence, L is often referred to as the *infinitesimal generator* of Y.

---

[17]We do not claim that this is the unique solution. In fact, it is the minimal non-negative solution if $f \geq 0$.

Sometimes, the semigroup is given in terms of the generator, using the formula

$$T_t = e^{tL}, \ t \geq 0.$$

(Of course, this only makes sense if one defines the exponential of the operator properly, for example, by using Taylor's expansion.)

1.7.3   Further properties

One of the first results in the theory of random walks was Pólya's Theorem on recurrence/transience. Let S_n denote the position of the random walker, starting at the origin after n steps in $\mathbb{Z}^d$. The probability of the event

$$\{S_n = 0 \text{ for infinitely many } n \geq 1\}$$

is either zero or one. This follows from the well-known[18] 'Hewitt-Savage $0 - 1$ Law.' In the former case we say that the random walk is *transient*, and in the latter we say that it is *recurrent*. In fact, in the former case the walker's distance from the origin tends to infinity with probability one, and thus it may or may not ever visit back at the origin.

G. Pólya in 1921 proved that the random walk is recurrent if and only if $d \leq 2$; as S. Kakutani famously put it:

A drunk man will find his way home, but a drunk bird may get lost forever.

It turns out that an analogous result holds for the scaling limit of the d-dimensional random walk, the d-dimensional Brownian motion:

(1) If $d \leq 2$, then any ball of positive radius around the origin is hit by the process for arbitrarily large times a.s., that is, for $r > 0$ one has $P(|B_t| < r$ for arbitrarily large times$) = 1$ (recurrence).

(2) If $d > 2$, then $P(\lim_{t \to \infty} |B_t| = \infty) = 1$ (transience).

Remark 1.4 (Set vs. point recurrence). Recurrence is different from 'point recurrence.' Almost surely, a two-dimensional Brownian motion will not hit a given point for any $t > 0$. Our notion of recurrence is sometimes called 'set recurrence.'   ◇

More general diffusion processes, corresponding to operators satisfying Assumption 1.1, behave similarly. Namely, there are exactly two cases.

---

[18]See Theorem A.14 in [Liggett (2010)].

Either

$$\forall x \in D, \emptyset \neq B \subset\subset D \text{ open, } P_x(Y_t \in B \text{ for arbitrarily large } t\text{'s}) = 1, \tag{1.10}$$

or

$$\forall x \in D, \emptyset \neq B \subset\subset D \text{ open, } P_x(Y_t \notin B \text{ for all } t > T(B, \omega)) = 1. \tag{1.11}$$

Definition 1.4 (recurrence/transience). If (1.10) holds then we say that Y is *recurrent*. If (1.11) holds then we say that Y is *transient*.

A recurrent diffusion process may have an even stronger property.

Definition 1.5 (positive/null recurrence). If for all $x \in D, \emptyset \neq B \subset\subset D$ open, one has $E_x \tau_B < \infty$, where $\tau_B := \inf\{t \geq 0 \mid Y_t \in B\}$, then we say that Y is *positive recurrent* or *ergodic*. A recurrent diffusion which is not positive recurrent is called *null recurrent*.

Linear and planar Brownian motion, for instance, are null recurrent.

A useful criterion for transience in terms of the operator L will be given later in Proposition 1.9.

An important property shared by all diffusion processes is that they are strong Markov processes.

Although the family $\{P_x; x \in D\}$ even has the Feller property, one has to be a bit careful. Even though $x_n \to x$ implies $P_{x_n} \to P_x$ in the weak topology of measures, whenever $x \in D$, this property may fail for the cemetery state $x = \Delta$. This fact is related to the possibility of the so-called 'explosion inward from the boundary.' It is possible that

$$\lim_{x_n \to \Delta} P(Y_t \subset B) > 0,$$

for some $t > 0$ and $B \subset\subset D$, although, clearly, $P_\Delta(Y_t \in B) = 0$.

Finally, every diffusion process has the localization property:

Proposition 1.4 (Localization). *Let $\widehat{D}_i := D_i \cup \{\Delta\}$ for $i = 1, 2$. Let $\{P_x; x \in \widehat{D}_1\}$ solve the generalized martingale problem on $D_1 \subset \mathbb{R}^d$ for L_1 and let $\{Q_x; x \in \widehat{D}_2\}$ solve the generalized martingale problem on $D_2 \subset \mathbb{R}^d$ for L_2. Let $U \subset D_1 \cap D_2$ be a domain on which the coefficients of L_1 and L_2 coincide. Assume that these coefficients, restricted to U, satisfy Assumption 1.1. Let $\tau_U := \inf\{t \geq 0 : Y_t \notin U\}$. Then, for all $x \in U$, $P_x = Q_x$ on the σ-algebra $\mathcal{F}_{\tau_U}$.*

1.7.4   *The Ornstein-Uhlenbeck process*

The second most well-known diffusion process, after Brownian motion, is another Gaussian process, the *Ornstein-Uhlenbeck process* (O-U process, in short).

Let $\sigma, \mu > 0$ and consider

$$L := \frac{1}{2}\sigma^2 \Delta - \mu x \cdot \nabla \text{ on } \mathbb{R}^d.$$

The corresponding diffusion process is called a d-dimensional Ornstein-Uhlenbeck process (sometimes called 'mean-reverting process'), and it is a positive recurrent process in any dimension. Similarly to the Brownian case, the ith coordinate process is a one-dimensional Ornstein-Uhlenbeck process, corresponding to the operator

$$L := \frac{1}{2}\sigma^2 \frac{d}{dx_i^2} - \mu x_i \frac{d}{dx_i} \text{on } \mathbb{R}^d.$$

In fact, for this Gaussian process one has mean $E_x(Y_t) = xe^{-t}$, and covariance

$$\text{cov}(Y_s, Y_t) = E_x[Y_t - E_x(Y_t)][Y_s - E_x(Y_s)] = \frac{\sigma^2}{2\mu}(e^{\mu(s-t)} - e^{-\mu(t+s)}), \ s < t.$$

In particular, no matter what x is, the time t mean and variance tend rapidly to zero and $\frac{\sigma^2}{2\mu}$, respectively, as $t \to \infty$. In fact one can show the stronger statement that no matter what x is, $\lim_{t\to\infty} P_x(Y_t \in B) = \int_B \pi(x)\,dx$, for B Borel, where π is the normal density with mean zero and variance $\frac{\sigma^2}{2\mu}$:

$$\pi(x) = \left(\frac{\mu}{\pi\sigma^2}\right)^{d/2} \exp\left\{-\frac{\mu}{\sigma^2}\,x^2\right\}.$$

It turns out that π is not only the limiting density, but also the *invariant density* (or 'stationary density') for the process. What we mean by this is that $\int_{\mathbb{R}^d} P_x(Y_t \in B)\,\pi(x)dx = \int_B \pi(x)\,dx$ for all $t \geq 0$. In words: if R is a random variable on $\mathbb{R}^d$ with density π, and if we start the process at the random location R, then the density of the location of the process is π at *any* time.

Similarly to Brownian motion, the one-dimensional Ornstein-Uhlenbeck process may also be obtained as a scaling limit of discrete processes. Instead of simple random walks, however, one uses the so-called *Ehrenfest Urn Model*. The model was originally proposed as a model for dissipation of heat, with this formulation: two boxes contain altogether n particles and at each step a randomly chosen particle moves to the opposite box. The following formulation is equivalent: consider an urn containing black and white balls, n balls altogether. At each step a ball is chosen at random and replaced by a ball of the opposite color.

Let now $N_m^{(n)}$ be the number of black balls in the urn after m steps for $m \geq 1$. For $n \geq 1$, consider the process $X^{(n)}$ defined by

$$X_t^{(n)} := \frac{N_{\lfloor nt \rfloor}^{(n)} - \frac{n}{2}}{\sqrt{n}}, \ t \geq 0,$$

which, for n even, can be considered a rescaled, non-symmetric[19] random walk, living on $[-n/2, n/2]$. One can show that as $n \to \infty$, the processes $X^{(n)}$, $n = 1, 2, \ldots$ converge in law to a one-dimensional Ornstein-Uhlenbeck process, corresponding to the operator $\frac{1}{2}\frac{d^2}{dx^2} - x \cdot \frac{d}{dx}$ on $\mathbb{R}$.

Definition 1.6 ('Outward' O-U process). Let $\sigma, \mu > 0$ and consider

$$L := \frac{1}{2}\sigma^2 \Delta + \mu x \cdot \nabla \text{ on } \mathbb{R}^d.$$

The corresponding diffusion process is often referred to as the *'outward' Ornstein-Uhlenbeck process.*

Although we have just switched the sign of the drift, this process exhibits a long time behavior which could not differ from the classical ('inward') Ornstein-Uhlenbeck process's behavior more. Namely, while the classical O-U process is positive recurrent, the 'outward' O-U process is a transient process. As the linearly growing outward drift suggests, it has a large radial speed.

1.7.5   *Transition measures and h-transform*

We have encountered the notion of the transition measure for Markov processes, and in particular, for diffusions. In the case of a diffusion process corresponding to the operator L on D, the transition measure is thus associated with an elliptic operator.

Sometimes it is necessary to extend the notion of transition measure to operators with a potential part, that is, to operators of the form $L + \beta$. If $\beta \leq 0$, then this has a clear intuitive meaning, as $L + \beta$ corresponds to an L-diffusion with spatially dependent *killing* at rate $|\beta|$. Otherwise, we do not associate $L + \beta$ with a single diffusion process, yet we define the concept of transition measure for such operators.

---

[19]It is clear that if the walkers's position has a large absolute value, then she prefers to step to the neighboring site which has a smaller absolute value.

One technical reason which makes this unavoidable is that we will be working with a transformation (h-transform) which leaves the family of elliptic operators with potential terms invariant, but for which the sub-family of diffusion operators ($\beta \equiv 0$) is *not* invariant, unless we impose a severe restriction and use harmonic functions only.

With the above motivation, we now present an important notion.

Definition 1.7 (Transition measure). *Let L satisfy Assumption 1.2 on $D \subset \mathbb{R}^d$, and let $\beta \in C^\eta(D)$. Let Y under $\{P_x\ x \in D\}$ be the diffusion process corresponding to L on D (in the sense of the generalized martingale problem). Define*

$$p(t, x, B) := E_x \left[\exp \left(\int_0^t \beta(Y_s)\,\mathrm{d}s \right) \mathbf{1}_{\{Y_t \in B\}} \right],$$

for $B \subset D$ measurable. If $p(t, x, B) < \infty$ for all $B \subset\subset D$, then we call the σ-finite measure $p(t, x, \cdot)$ the transition measure for $L + \beta$ on D at t, starting from x. (Otherwise, the transition measure is not defined at t starting from x.)

Remark 1.5. From the physicist's perspective, we are re-weighting the paths of the process, using a 'Feynman-Kac term.' Indeed, *probabilistic potential theory* was inspired, to a large extent, by physics.   ◇

When considering all $t \geq 0$ and $x \in D$, we are talking about *transition kernel.* Note that for $\beta \equiv 0$, one has $p(t, x, D) = P_x(Y_t \in D)$, which is the probability that the process has not left D by t, and this is not necessarily one.

Clearly, the transition measure corresponding to $L + \beta$ satisfies

$$\int_D p(t, x, \mathrm{d}y)g(y) = E_x \left[\exp \left(\int_0^t \beta(Y_s)\,\mathrm{d}s \right) g(Y_t) \right], \quad t \geq 0, x \in D,$$

for any compactly supported measurable g, with the convention that $g(\Delta) := 0$ and by defining $\beta(\Delta)$ in an arbitrary way.

Definition 1.8 (Doob's h-transform). If $0 < h \in C^{2,\eta}(D)$ with $\eta \in (0, 1]$, then changing the operator $L + \beta$ to

$$(L + \beta)^h(\cdot) := \frac{1}{h}(L + \beta)(h \cdot)$$

is called an h-transform. Writing out the new operator in detail, one obtains

$$(L + \beta)^h = L + a\frac{\nabla h}{h} \cdot \nabla + \beta + \frac{Lh}{h}.$$

Note that if L satisfies Assumption 1.2 on D, then L^h satisfies it as well.

A straightforward computation reveals that if $p(t, x, y)$ (resp. $p^h(t, x, y)$) is the transition density corresponding to $L + \beta$ (resp. $(L + \beta)^h$), then

$$p^h(t, x, y) = \frac{h(y)}{h(x)} \cdot p(t, x, y), \ t \geq 0, x, y \in D.$$

The probabilistic impact of the h-transform will be clear in Theorem 1.12 later.

Another way to see the probabilistic significance is via conditioned processes. Here we just discuss the simplest example, for illustration.[20] Let $d \geq 2$, and $\Theta \subset\subset \mathbb{R}^d$ be a smooth non-empty bounded subdomain.[21] Let Y be a diffusion process on $\mathbb{R}^d$ with transition density $p(t, x, y)$, corresponding to the second order elliptic operator L satisfying Assumption 1.2 on $\mathbb{R}^d$, and denote the probabilities by $\{P_x\}$. Restricting Y to the exterior domain Θ^c, we have a diffusion process on this new domain. (Recall that upon exiting Θ^c, the process is put into a cemetery state forever.) With a slight abuse of notation, we will still denote it by Y, and keep the notation $p(t, x, y)$, L, and $\{P_x\}$ too. Note that, considering L on the exterior domain, its coefficients are smooth up to $\partial\Theta$.

For $x \in \Theta^c$, define $h(x) := P_x(\sigma_\Theta < \infty)$, where σ_Θ is the entrance time of Θ, that is $\sigma_\Theta := \inf\{t \geq 0 \mid Y_t \in \Theta\}$. Of course, if Y is recurrent on the original domain $\mathbb{R}^d$, then $h \equiv 1$. In any case, one can show that h solves

$$\left. \begin{array}{c} Lh = 0 \text{ in } \Theta^c, \\[2mm] \lim_{x \to \partial\Theta} h(x) = 1, \\[2mm] 0 \leq h \leq 1. \end{array} \right\} \tag{1.12}$$

In fact, h is the *minimal* solution to this problem. This is because $h = \lim_{n \to \infty} u_n$, where u_n is the unique solution to

$$\left. \begin{array}{c} Lu = 0 \text{ in } \Theta^c, \\[2mm] \lim_{x \to \partial\Theta} u(x) = 1, \\[2mm] \lim_{x \to \partial B_n(0)} u(x) = 0, \\[2mm] 0 \leq u \leq 1. \end{array} \right\} \tag{1.13}$$

(Here we assume that n is so large that the n-ball $B_n(0)$ contains $\overline{\Theta}$.) The existence of the limit follows from the fact that u_n is monotone increasing

---

[20]The probabilistic audience will hopefully appreciate this explanation, besides the analytic description of h-transforms. However, we will not need this tool, only the Girsanov transform.

[21]Since $d \geq 2$, it is connected.

in n, which, in turn, follows from the elliptic maximum principle. That h is minimal, follows from the fact that if v is another solution to (1.12) then $u_n \leq v$ holds on $B_n(0) \setminus \Theta$ for each (large enough) n, again, because of the elliptic maximum principle.

Since h is harmonic (that is, $(L + \beta)h = 0$) in the exterior domain Θ^c, we know that L^h has no potential (zeroth order) part, and thus, it corresponds to a diffusion process on the domain Θ^c. Let $p^h(t, x, y)$ denote the transition probability for this latter diffusion. Then

$$p^h(t, x, y) = \frac{h(y)}{h(x)} \cdot p(t, x, y), \ t \geq 0, x, y \in \Theta^c.$$

The probabilistic content of the h-transform is now compounded in the following fundamental fact of Doob's h-transform theory:

$$p(t, x, \mathrm{d}y) = P_x(Y_t \in \mathrm{d}y \mid \sigma_\Theta < \infty),$$

that is, the harmonic h-transform is tantamount to conditioning the diffusion to hit the set Θ (at which instant it is killed). It is a remarkable fact that the conditioned diffusion is a diffusion process as well.

We note that if the boundary condition $h = 1$ is replaced by a more general one on $\partial\Theta$, then the transformation with the corresponding h is no longer merely conditioning on hitting Θ, but rather, it is *conditioning in an appropriate manner, which depends on the boundary condition.* (See Chapter 7 in [Pinsky (1995)] for more elaboration.)

1.8 Itô-integral and SDE's

Another approach to diffusions is to consider them as the unique solutions of 'stochastic' differential equations (SDE's), when those equations have nice coefficients. In fact, those SDE's will be interpreted as integral equations, involving 'stochastic integrals.' To this end, one has to attempt to define an integral of a function (deterministic or random) against Brownian motion.

The naive approach, namely a path-wise Lebesgue-Stieltjes integral, obviously does not work. The reason is the roughness of Brownian paths. Since the paths are almost surely nowhere differentiable, this immediately implies that, on a given time interval, the probability of having bounded variation is zero! Although this fact would still allow one to integrate against $\mathrm{d}B_s$ if the *integrand* were sufficiently smooth (if one defines the integral via integration by parts), typically one needs to define integrals like $\int_0^1 B_s \, \mathrm{d}B_s$, for which both the integrand and the integrator lack the bounded variation property.

The resolution of this problem was due[22] to the Japanese mathematician, K. Itô. The main idea is as follows. Let B be a standard Brownian motion on (Ω, P), where P is Wiener-measure and let the corresponding expectation be E. Take, for simplicity, a deterministic nonnegative continuous function f on $[0, T]$. Approximate this function by the step functions $f_n \uparrow f$. Since step functions are piecewise smooth (constant), it is no problem at all to define $\int_0^T f_n(s) \, dB_s$ path-wise as a Lebesgue-Stieltjes (even Riemann-Stieltjes) integral. Now, the key observation is that one has to give up path-wise convergence, but one can *replace it by the $L^2(\Omega, P)$-convergence* of the random variables $\int_0^T f_n(s) \, dB_s$. That is, it turns out that the limit *in mean square*

$$\int_0^T f(s) \, dB_s := \lim_{n \to \infty} \int_0^T f_n(s) \, dB_s$$

can serve as the definition of the stochastic integral. In other words, there exists a P-square-integrable random variable M on Ω, such that

$$\lim_{n \to \infty} E \left(M - \int_0^T f_n(s) \, dB_s \right)^2 = 0,$$

and $\int_0^T f(s) \, dB_s := M$.

To carry out this program rigorously, let g be a stochastic process,[23] adapted to the canonical filtration of B, which we denote by $\{\mathcal{F}_t^B\}_{t \geq 0}$. We write $g \in \mathcal{L}^2([a, b])$ if $E \int_a^b g^2(s) ds < \infty$; if $g \in \mathcal{L}^2([0, b])$ for all $b > 0$, then we write $g \in \mathcal{L}^2$.

A simple (or elementary) function $g \in \mathcal{L}^2([a, b])$ is such that $g(s) = g(t_k)$ for $s \in (t_k, t_{k+1}]$, for some division of the interval $a = t_0 < t_1 < ... < t_n = b$. (Here $g(a) = g_{t_0}$.) In this case the stochastic integral is defined as

$$\int_a^b g(s) \, dB_s := \sum_{k=0}^{n-1} g(t_k)(B_{t_{k+1}} - B_{t_k}).$$

For a generic $g \in \mathcal{L}^2([a, b])$, there exists an approximating (in mean square) sequence of simple processes, $g_n \in \mathcal{L}^2([a, b])$, that is, a sequence for which

$$\lim_{n \to \infty} E \int_a^b [g(s) - g_n(s)]^2 \, ds = 0.$$

---

[22]Unknown to Itô, and to the world until 2000, W. Döblin (the novelist Alfred Döblin's son) achieved similar results, including the famous result that today is called *Itô's formula*. Döblin's tragic story during WWII is well known today, and so is his work that had been hidden away in a sealed envelope in the safe of the French Academy of Sciences for sixty years, before it was finally opened in 2000.

[23]For now, we do not use capital letter G, and also suppress the dependence on ω in the notation as we would like to stress that g is the integrand.

One then defines the stochastic integral (Itô-integral) by

$$\int_a^b g(s)\,\mathrm{d}B_s := l.i.m._{n\to\infty} \int_a^b g_n(s)\,\mathrm{d}B_s,$$

where $l.i.m.$ means $L^2(\Omega, P)$-limit. Thus the Itô-integral between a and b is a P-square integrable random variable on Ω, only determined up to null-sets (and not ω-wise).

It can be shown that $l.i.m.$ always exists and does not depend on the choice of the approximating sequence. This is essentially a consequence of the following important property.

Proposition 1.5 (Itô-isometry). *If $g \in \mathcal{L}^2([a, b])$, then*

$$\left\| \int_a^b g(s)\,\mathrm{d}B_s \right\| = \|g\|_{[a,b]}$$

holds, where the norm on the left-hand side is the $L^2(\Omega, P)$-norm, and the norm on the right-hand side is the usual L^2-norm on $[a, b]$.

(The isometry property is first used to define the stochastic integral besides simple functions, and then one proves that the isometry is 'inherited' to all square-integrable integrands.)

The Itô integral enjoys some pleasant properties. Firstly, the stochastic integral is a *linear operator,* more precisely, it is linear with respect to the integrand g and additive with respect to the interval of integration.

Secondly, we can consider it as a stochastic process M, where $M_t := \int_{t_0}^t g(s)\,\mathrm{d}s$ on $[t_0, t_1]$, with some $0 \le t_0 < t_1$ fixed. It is easy to show that this notion is *consistent,* that is, for $[c, d] \subset [a, b]$,

$$\int_c^d g(s)\,\mathrm{d}s = \int_a^b g(s)\mathbf{1}_{[c,d]}(s)\,\mathrm{d}s.$$

Working a bit harder one can show that M on $[t_0, t_1]$ is a P-square integrable martingale on Ω, adapted to $(\mathcal{F}_t^B)_{t_0 \le t \le t_1}$, which possesses a continuous version.

Remark 1.6 (Importance of left endpoint). It might not seem too important that we used the 'left endpoint' of the interval in the definition of the integral for elementary functions, but it is in fact of great significance. Using, for example, the middle point $1/2(t_k + t_{k+1})$ instead, would lead to a different integral, called the *Stratonovich integral.* It has certain advantages over the Itô integral, because working with it mimics the rules of classical calculus, but it also has certain disadvantages. The reason one usually goes with the Itô definition is because the process $M_t =: \int_0^t g(s)\,\mathrm{d}s$ is adapted to the Brownian filtration and it is a martingale; this is not the case for the Stratonovich integral.   ◇

We conclude this brief review on the one-dimensional Itô-integral by noting that it can be extended to integrands which are merely measurable, $\mathcal{F}_t^B$-adapted and have square integrable paths.

An important example is the following representation result for O-U processes, introduced in Subsection 1.7.4.

Example 1.1 (O-U process with Itô-integral). *Fix $\sigma, \mu > 0$. Given a Brownian motion B in $\mathbb{R}$, the process Y defined by*

$$Y_t = Y_0 e^{-\mu t} + \int_0^t \sigma e^{\mu(s-t)}\ \mathrm{d}B_s, \qquad (1.14)$$

is an (inward) Ornstein-Uhlenbeck process with parameters σ, μ.

The first term implies that the process converges in expectation very rapidly to zero. This deterministic decay of Y_t is being perturbed by the second term, introducing some variance due to the diffusive motion. The mean and variance can be read off from this form, the latter with using the Itô-isometry.

One can then also define *multidimensional* Itô-integrals with respect to the d-dimensional Brownian motion $(B^1, ..., B^d)$, as follows. Let $\mathbb{R}^{d \times d}$ denote the space of $d \times d$ matrices with real entries equipped with the Euclidean norm $\|\sigma\| := (\sum_{1 \le i \le n; 1 \le j \le n} \sigma_{i,j}^2)^{1/2}$ for $\sigma \in \mathbb{R}^{d \times d}$. Let the filtered probability space $(\Omega, \mathcal{F}, (\mathcal{F}_t)_{t \ge 0}, P)$ be given, and let $B = (B^1, ..., B^d)$ be a given (adapted) Brownian motion on this space. If $g = (g_{i,j})_{1 \le i \le n; 1 \le j \le n}$ is 'matrix-valued,' that is $g(t, \omega) \in \mathbb{R}^{d \times d}$ for $t \ge 0, \omega \in \Omega$, then the integral $I = \int_S^T g(t) \cdot \mathrm{d}B_t$ (given $0 \le S \le T$) is a d-dimensional random vector, whose i-th component ($1 \le i \le d$) equals

$$I_i := \sum_{j=1}^d \int_S^T g_{i,j}(t)\,\mathrm{d}B_t^j,$$

provided the right-hand side is defined.

(We could also define the integral, a bit more generally, for non-square matrix-valued integrands, in a similar fashion, but we do not need that in the sequel.)

The representation for the O-U process still goes through in higher dimensions.

Example 1.2 (O-U process as a multidimensional Itô-integral).
Fix $\sigma, \mu > 0$. Given a Brownian motion B in $\mathbb{R}^d$, the process Y defined by

$$Y_t = Y_0 e^{-\mu t} + \int_0^t e^{\mu(s-t)} \sigma I_d \cdot \mathrm{d}B_s, \qquad (1.15)$$

is a d-dimensional O-U process with parameters σ and μ.

Once we defined the stochastic integral, it is natural[24] to consider integral equations of the form

$$X_t = X_0 + \int_0^t b(X_s) \, \mathrm{d}s + \int_0^t \sigma(X_s) \cdot \mathrm{d}B_s, \qquad (1.16)$$

or more generally, of the form

$$X_t = X_0 + \int_0^t b(\omega, s) \, \mathrm{d}s + \int_0^t \sigma(\omega, s) \cdot \mathrm{d}B_s,$$

where $b : \mathbb{R}^d \to \mathbb{R}^d$ and $\sigma : \mathbb{R}^d \to \mathbb{R}^{d \times d}$ are 'nice' functions ($b : \Omega \times [0, \infty) \to \mathbb{R}^d$ and $\sigma : \Omega \times [0, \infty) \to \mathbb{R}^{d \times d}$ are 'nice' processes).

Although it is customary to abbreviate these equations in the 'differential form'

$$\mathrm{d}X_t = b \, \mathrm{d}t + \sigma \cdot \mathrm{d}B_t,$$

and call them *stochastic differential equations*, we should keep in mind that, strictly speaking, one is dealing with integral equations. The matrix σ is called the *diffusion matrix* (diffusion coefficient for $d = 1$) and b is called the *drift*.

For (1.16), the following existence and uniqueness result is standard. (See e.g. [Pinsky (1995)].)

Theorem 1.9 (Existence/uniqueness for SDEs). *Let the filtered probability space $(\Omega, \mathcal{F}, (\mathcal{F}_t)_{t \geq 0}, P)$ be given, and let B be a given (adapted) Brownian motion on this space.*

Existence: Assume that $b : \mathbb{R}^d \to \mathbb{R}^d$ and $\sigma : \mathbb{R}^d \to \mathbb{R}^{d \times d}$ are given Lipschitz functions, that is, there exists a $K > 0$ such that

$$\|b(x) - b(y)\| + \|\sigma(x) - \sigma(y)\| < K \|x - y\|, \ x, y \in \mathbb{R}^d,$$

where $\| \cdot \|$ denotes the Euclidean norm of vectors and matrices. Then, for each $x \in \mathbb{R}^d$ there exists a solution X^x to (1.16) with P-a.s. continuous paths. Furthermore, the solution is adapted to the canonical Brownian filtration, that is, for $t \geq 0$, $\sigma(X_t^x) \subset \sigma\{B_s, 0 \leq s \leq t\}$.

---

[24]One of the natural approaches that leads to these equations is to consider deterministic differential equations, perturbed by 'white noise.' The latter is the derivative of Brownian motion, although only in a weak sense.

Uniqueness: The solution is unique in both of the following senses:

(a) It is unique up to indistinguishability on $(\Omega, \mathcal{F}, (\mathcal{F}_t, P)_{t \geq 0})$, that is, any other $\mathcal{F}_t$-adapted solution to (1.16) with P-a.s. continuous paths is undistinguishable from X^x. (With $x \in \mathbb{R}^d$ given.)

(b) The law of the process on $C([0, \infty))$ is unique. That is, let P_x denote the law of X^x on $C([0, \infty))$: $P_x := P \circ (X^x)^{-1}$, where the $\mathcal{F}_t$-adapted Brownian motion B lives on $(\Omega, \mathcal{F}, (\mathcal{F}_t)_{t \geq 0}, P)$. If we can replace $(\Omega, \mathcal{F}, (\mathcal{F}_t)_{t \geq 0}, P)$, B, the solution X^x and P_x above by some $(\Omega^*, \mathcal{F}^*, (\mathcal{F}_t^*)_{t \geq 0}, P^*)$, B^*, and $X^{x,*}$, respectively, but b, σ (and x) are unchanged, then $P_x = Q_x := P^* \circ (X^{x,*})^{-1}$.

Remark 1.7. The measurability of X_t^x with respect to $\sigma\{B_s, 0 \leq s \leq t\}$ is important. This is the distinguishing mark of a *strong solution*. Intuitively: the realization of the Brownian motion up to t 'completely determines' the realization of the solution ('output'), in accordance with the 'principle of causality' for dynamical systems. The Brownian motion is often called the *driving Brownian motion*. ◇

Remark 1.8 (SDEs and martingale problems). Recall the notion of 'martingale problems' and Proposition 1.3. A fundamentally important connection is that the probability laws $\{P_x; x \in \mathbb{R}^d\}$ defined via SDE will actually solve the martingale problem on $\mathbb{R}^d$ with

$$L = \frac{1}{2} \sum_{i,j=1}^{d} a_{ij} \frac{\mathrm{d}^2}{\mathrm{d}x_i \, \mathrm{d}x_j} + \sum_{i=1}^{d} b_i \frac{\mathrm{d}}{\mathrm{d}x_i},$$

where $a = \sigma \sigma^T$. In fact, the martingale problem is an equivalent characterization of the laws $\{P_x; x \in \mathbb{R}^d\}$. This remains true if $\mathbb{R}^d$ is replaced by one of its subdomains. The 'martingale problem approach' of D. W. Stroock and S. R. S. Varadhan, developed in the 1970s, has proven to be much more fruitful than the SDE point of view, when one seeks to establish results for broader classes of coefficients. Furthermore, the law P_x is defined 'directly,' and not via some auxiliary process (Brownian motion) and probability space. More on this subject can be found in the fundamental monograph [Stroock and Varadhan (2006)], as well as in Chapter 1 of [Pinsky (1995)].

Note that a given symmetric positive definite matrix a may have more than one representation $a = \sigma \sigma^T$. (In general, this can even happen with non-square matrices — the case we skipped.) In this case, however, the solutions of the corresponding SDE's all share the same law. On the other hand, a uniquely determines a corresponding symmetric square-matrix σ. Indeed, since a is positive definite, one can always choose σ to be the unique square root of a. In that sense L corresponds to a unique SDE. $\diamond$

1.8.1 *The Bessel process and a large deviation result for Brownian motion*

In light of the scaling property (see (1.6)), the 'typical' displacement for Brownian motion during time t is on the order $\sqrt{t}$. Our last result in this section is a *large deviation* result, telling us that it is exponentially unlikely for a Brownian particle to reach distance const·t in time t, when t is large. This result will be important when we consider random environments.

Lemma 1.3 (Linear distances are unlikely). *Let B be a Brownian motion in $\mathbb{R}^d$ (starting at the origin), with corresponding probability P, and let $k > 0$. Then, as $t \to \infty$,*

$$P\left(\sup_{0 \leq s \leq t} |B_s| \geq kt\right) = \exp\left\{-\frac{k^2 t}{2}(1 + o(1))\right\}. \qquad (1.17)$$

Furthermore, for $d = 1$, even the following, stronger statement is true: let $m_t := \min_{0 \leq s \leq t} B_s$ and $M_t := \max_{0 \leq s \leq t} B_s$. (The process $M_t - m_t$ is the range process of B.) Then, as $t \to \infty$,

$$P(M_t - m_t \geq kt) = \exp\left\{-\frac{k^2 t}{2}(1 + o(1))\right\}. \qquad (1.18)$$

Remark 1.9. If $\gamma > 0$ and we replace time t by time γt, then by Brownian scaling (1.6), the right-hand sides of (1.17) and (1.18) become $\exp\left\{-\frac{k^2 t}{2\gamma}(1 + o(1))\right\}$. $\diamond$

Proof. First, let $d = 1$. The relation (1.17) is a consequence of the reflection principle (1.7) (see formula (7.3.3) in [Karlin and Taylor (1975)]). Thus, to verify (1.18), it is sufficient to estimate it from above. To this end, define

$$\theta_c := \inf\{s \geq 0 \mid M_t - m_t = c\},$$

and use that, according to p. 199 in [Chaumont and Yor (2012)] and the references therein, the Laplace transform of θ_c satisfies

$$E \exp\left(-\frac{\lambda^2}{2}\theta_c\right) = \frac{2}{1 + \cosh(\lambda c)}, \quad \lambda > 0.$$

Hence, by the exponential Markov inequality,

$$P(\theta_c < t) \le \exp\left(\frac{\lambda^2}{2}t\right) E \exp\left(-\frac{\lambda^2}{2}\theta_c\right) = \exp\left(\frac{\lambda^2}{2}t\right) \frac{2}{1 + \cosh(\lambda c)}.$$

Taking $c = kt$, one gets

$$P\left(M_t - m_t \ge kt\right) = P(\theta_{kt} < t) \le \exp\left(\frac{\lambda^2}{2}t\right) \frac{2}{1 + \cosh(\lambda kt)}$$

$$\sim 4 \exp\left\{\frac{\lambda^2}{2}t - \lambda kt\right\},$$

as $t \to \infty$.

Optimizing, we see that the estimate is the sharpest when $\lambda = k$, in which case, we obtain the desired upper estimate.

For $d \ge 2$, let us consider, more generally, the Brownian motion $B^{(\mathbf{x})}$, $\mathbf{x} \in \mathbb{R}^d$ and let $r := |\mathbf{x}|$. We are going to use the well-known fact (see Example 4.2.2 in [Øksendal (2010)]) that if $R = |B|$ and $r > 0$, then the process R, called the d-*dimensional Bessel process*, is the strong solution of the one-dimensional stochastic differential equation on $(0, \infty)$:

$$R_t = r + \int_0^t \frac{d-1}{2R_s}\, ds + W_t, \tag{1.19}$$

where W is standard Brownian motion. (Note that the existence of a solution does not follow from Theorem 1.9, since the assumption about the Lipschitz property is violated.)

Using the strong Markov property of Brownian motion, applied at the first hitting time of the ρ-sphere, τ_ρ, it is clear, that in order to verify the lemma, it is sufficient to prove it when B is replaced by $B^{(\mathbf{x})}$, and $r = |\mathbf{x}| = \rho > 0$. (Simply because $\tau_\rho \ge 0$.)

Next, define the sequence of stopping times $0 = \tau_0 < \sigma_0 < \tau_1 < \sigma_1, \dots$ with respect to the filtration generated by R (and thus, also with respect to the one generated by W) as follows:

$$\tau_0 := 0; \; \sigma_0 := \inf\{s > 0 \mid R_s = \rho/2\},$$

and for $i \ge 1$,

$$\tau_{i+1} := \inf\{s > \sigma_i \mid R_s = \rho\}; \; \sigma_{i+1} := \inf\{s > \tau_{i+1} \mid R_s = \rho/2\}.$$

Note that for $i \ge 0$ and $s \in [\tau_i, \sigma_i]$,

$$R_s = \rho + \int_{\tau_i}^s \frac{d-1}{2R_z}\, dz + W_s - W_{\tau_i} \le \rho + \rho^{-1}(d-1)\Delta_i^s + W_s - W_{\tau_i}, \tag{1.20}$$

where $\Delta_i^s := s - \tau_i$.

Since $R_s \leq \rho$ for $\sigma_i \leq s \leq \tau_{i+1}$ and $i \geq 0$, it is also clear that for $t > \rho/k$, the relation $\sup_{0 \leq s \leq t} R_s \geq kt$ is tantamount to

$$\sup_{i \geq 0} \sup_{\tau_i \wedge t < s \leq \sigma_i \wedge t} R_s \geq kt.$$

Putting this together with (1.20), it follows that if $\{P_r; r > 0\}$ are the probabilities for R (or for W), then $P_\rho \left(\sup_{0 \leq s \leq t} R_s \geq kt \right)$ can be upper estimated by

$$P_\rho \left(\exists i \geq 0, \, \exists \tau_i \wedge t < s \leq \sigma_i \wedge t : \, W_s - W_{\tau_i} \geq kt - \rho^{-1}(d-1)\Delta_i^s - \rho \right).$$

This can be further upper estimated by

$$P_\rho \left(M_t - m_t \geq [k - \rho^{-1}(d-1) - \delta]t \right),$$

for any $\delta > 0$, as long as $t \geq \rho/\delta$. To complete the proof, fix $\rho, \delta > 0$, let $t \to \infty$, and use the already proven relation (1.18); finally let $\delta \to 0$ and $\rho \to \infty$. $\qquad \square$

1.9 Martingale change of measure

A fundamental tool in the theory of stochastic processes is called 'change of measure' and it is intimately related to nonnegative martingales. Two examples of changes of measures will be especially useful for us, the Girsanov change of measure and the Poisson change of measure.

1.9.1 *Changes of measures, density process, uniform integrability*

Theorem 1.10 (General change of measure). *Let X be a stochastic process on the filtered probability space $(\Omega, \mathcal{F}, \{\mathcal{F}\}_{t \geq 0}, P)$ and let M be a nonnegative P-martingale with unit mean, adapted to the filtration. Define the new probability measure Q on the same filtered probability space by*

$$\left. \frac{dQ}{dP} \right|_{\mathcal{F}_t} = M_t, \, t \geq 0.$$

Then $(\Omega, \mathcal{F}, \{\mathcal{F}\}_{t \geq 0}, Q)$ defines a stochastic process Y.

Conversely, if for the stochastic process Y on $(\Omega, \mathcal{F}, \{\mathcal{F}\}_{t \geq 0}, Q)$, we have $Q << P$ on $(\Omega, \mathcal{F}_t)$ for all $t \geq 0$ and

$$M_t := \left. \frac{dQ}{dP} \right|_{\mathcal{F}_t}, \, t \geq 0,$$

then M is a nonnegative $\mathcal{F}_t$-adapted P-martingale with unit mean.

Remark 1.10 (Density process). From the general theory of continuous martingales, it is known that the martingale M has a càdlàg version, which is unique in the sense of indistinguishability, that is, if M^* is another càdlàg version, then $P(M_t = M_t^*, \forall t \geq 0) = 1$. (Because càdlàg versions are indistinguishable.) Then, of course $Q(M_t = M_t^*, \forall t \geq 0) = 1$ too. The unique càdlàg version is called the *density process*. ◇

Proof. Assume first that M is a unit mean nonnegative P-martingale, adapted to the filtration. Since M has unit mean, according to the Kolmogorov consistency theorem, $(\Omega, \mathcal{F}, \{\mathcal{F}\}_{t \geq 0}, Q)$ defines a stochastic process if and only if for $A \in \mathcal{F}_s$, its measure $Q(A)$ is the same whether we define Q on $\mathcal{F}_s$ or on $\mathcal{F}_t$, for $0 \leq s < t$. If E corresponds to P, then by the martingale property, $E(M_t \mid \mathcal{F}_s) = M_s$, and so, indeed, $E(M_s; A) = E(M_t; A)$.

Conversely, if M is the density process, then $E(M_t \mid \mathcal{F}_s) = M_s$, exactly because $E(M_s; A) = E(M_t; A)$ holds for $A \in \mathcal{F}_s$, since Q generates consistent measures on different $\mathcal{F}_t$ σ-algebras. Since Q is a probability measure, M_t must have unit P-mean. □

Let $\mathcal{F}_\infty := \sigma \left(\bigcup_{t \geq 0} \mathcal{F}_t \right) \subset \mathcal{F}$ be the σ-algebra generated by $\{\mathcal{F}_t; \ t \geq 0\}$. It is important to point out, that we do not know whether $Q << P$ holds on $(\Omega, \mathcal{F}_\infty)$, even though we have absolute continuity up to all finite times. The following theorem gives a criterion for absolute continuity 'up to time infinity.'

Theorem 1.11 (Uniform integrability). *Let M be the (càdlàg) density process for the measures P and Q as above. Then the following are equivalent.*

(1) $Q << P$ on $(\Omega, \mathcal{F}_\infty)$;
(2) $Q(\sup_{t \geq 0} M_t < \infty) = 1$;
(3) M is P-uniformly integrable.

Proof. We give a cyclical proof.
Assume (1). Then the a.s. finite limit of M under P is also a.s. finite under Q, giving (2). (Since càdlàg functions are bounded on compacts.)
Assume (2). Then

$$E(M_s \mathbf{1}_{\{M_s > n\}}) = Q(M_s > n) \leq Q \left(\sup_{t \geq 0} M_t > n \right) \to 0,$$

as $n \to \infty$, giving (3).
Assume (3). By uniform integrability, the P-a.s. finite limit M_∞ is in $L^1(\Omega, \mathcal{F}, P)$, and for $A \in \mathcal{F}_t$, one has $Q(A) = E(\mathbf{1}_A M_t) = E(\mathbf{1}_A M_\infty)$.

Since it's easy to see that the class $\{A \in \mathcal{F} \mid Q(A) = E(\mathbf{1}_A M_\infty)\}$ is a σ-algebra, $M_\infty = \mathrm{d}Q/\mathrm{d}P$ on $(\Omega, \mathcal{F}_\infty)$, giving (1). $\qquad\square$

1.9.2 Two particular changes of measures: Girsanov and Poisson

An example for the change of measure discussed in the previous subsection is the following particular version of 'Girsanov's Theorem.'[25]

Theorem 1.12 (Girsanov transform). *Let L be a second order elliptic operator on $D \subset \mathbb{R}^d$, satisfying Assumption 1.2, and let $\beta \in C^\eta(D)$ be bounded from above. Assume that the diffusion process Y on D under the laws $\{\mathbb{P}_x; x \in D\}$ corresponds to L, it is adapted to some filtration $\{\mathcal{G}_t : t \geq 0\}$, and it is conservative, that is, $P_x(Y_t \in D) = 1$ for $x \in D$ and $t \geq 0$. Let $0 < h \in C^{2,\eta}(D)$ satisfy $(L + \beta)h = 0$ on D. Under the change of measure*

$$\left.\frac{\mathrm{d}\mathbb{P}_x^h}{\mathrm{d}\mathbb{P}_x}\right|_{\mathcal{G}_t} = \frac{h(Y_t)}{h(x)} e^{\int_0^t \beta(Y_s)\mathrm{d}s} = \frac{h(Y_t)}{h(x)} e^{-\int_0^t (Lh/h)(Y_s)\mathrm{d}s}, \quad t \geq 0, \qquad (1.21)$$

the process $(Y, \mathbb{P}_x^h)$ is an L_0^h-diffusion on D, where

$$L_0^h := L^h - \frac{Lh}{h} = (L + \beta)^h = L + a\frac{\nabla h}{h} \cdot \nabla.$$

Needless to say, the right-hand side of (1.21) is a $\mathcal{G}_t$-adapted martingale with unit mean. It is called the *Girsanov density (process)*. As mentioned already, if $p(t, x, \mathrm{d}y)$ is the transition measure for $L + \beta$ (not for L!), then

$$p^h(t, x, \mathrm{d}y) := \frac{h(y)}{h(x)} p(t, x, \mathrm{d}y)$$

is the transition measure for the diffusion $(Y, \mathbb{P}_x^h)$.

Note that, unless β vanishes everywhere, h is not harmonic with respect to L, and that is the reason that, in order to obtain a new diffusion operator, we have to incorporate the exponential integral in the transformation.

The assumption that $\sup_D \beta < \infty$ is not essential, because one can show the finiteness of the expectations involved, using the fact that h is harmonic with respect to $L + \beta$.

---

[25] Also called the Cameron-Martin-Girsanov Theorem. In fact, the first result of this type was proved by R. H. Cameron and W. T. Martin in the 1940s, while I. V. Girsanov generalized the result in 1960.

Proof.   First note that if N_t is a functional of the path $Y(\omega)$, defined as

$$N_t(Y(\omega)) := \frac{h(Y_t(\omega))}{h(Y_0(\omega))} \exp\left(\int_0^t \beta(Y_z(\omega))\mathrm{d}z \right)$$

and $\theta_t, t > 0$ is the *shift operator* on paths, defined by $(\theta_t Y)(z,\omega) := Y(t + z,\omega)$, then $N = \{N_t\}_{t \geq 0}$ is a so-called *multiplicative functional*,[26] meaning that

$$N_{t+s} = N_t \cdot (N_s \circ \theta_t). \tag{1.22}$$

Indeed,

$$\frac{h(Y_{t+s}(\omega))}{h(Y_0(\omega))} \exp\left(\int_0^{t+s} \beta(Y_z(\omega))\mathrm{d}z \right)$$

$$= \frac{h(Y_t(\omega))}{h(Y_0(\omega))} \exp\left(\int_0^t \beta(Y_z(\omega))\mathrm{d}z \right) \cdot \frac{h(Y_{t+s}(\omega))}{h(Y_t(\omega))} \exp\left(\int_t^{t+s} \beta(Y_z(\omega))\mathrm{d}z \right).$$

It is then easy to show that the stochastic process corresponding to $\mathbb{P}_x^h$ is a Markov process; to do so, one has to establish the Chapman-Kolmogorov equation. The computation is left to the reader – see Ex. 9 at the end of this chapter.

By the Markov property, it is enough to determine the transition kernel for $\mathbb{P}_x^h$.

Now let $B \subset\subset D$ and pick $g(y) := \frac{\phi(y)}{\phi(x)}\mathbf{1}_B(y)$. Let $q(t,x,\mathrm{d}y)$ be the transition kernel corresponding to $\mathbb{P}^h$. We have

$$q(t,x,B) = \mathbb{P}_x^h(Y_t \in B) = \mathbb{E}_x\left[\frac{h(Y_t)}{h(x)} \exp\left(\int_0^t \beta(Y_s)\,\mathrm{d}s \right) \mathbf{1}_B(Y_t) \right]$$

$$= \int_D p(t,x,\mathrm{d}y)g(y) = \int_B p^h(t,x,\mathrm{d}y) = p^h(t,x,B), \ t \geq 0, x \in D.$$

That is, $q(t,x,B) = p^h(t,x,B)$ for all $B \subset\subset D, t \geq 0, x \in D$, and hence $q(t,x,\mathrm{d}y) = p^h(t,x,\mathrm{d}y)$.   □

We close this section with another particular change of measure, which will be proven handy when analyzing branching diffusions.

Recall that a measurable and locally integrable function $g : [0,\infty) \to [0,\infty)$ defines a *Poisson point process*, which is a random collection of points on $[0,\infty)$, satisfying that if $X(B)$ is the number of points in a Borel set $B \subset [0,\infty)$, then

---

[26]It is traditionally called a 'functional', although it should be called an 'operator,' as to each path of Y it assigns another path.

(1) $X(B_1), X(B_2), ..., X(B_k)$ are independent if $B_i \cap B_j = \emptyset$ for $1 \le i \ne j \le k$, for $k \ge 1$,

(2) $X([a, b])$ is Poisson distributed with parameter $\int_a^b g(z) \, dz$.

The process N defined by $N_t := X([0, t])$ is then of independent increments and is called the *Poisson process* with rate function g. By construction, it has right-continuous paths with left limits.

The following 'rate doubling' result for Poisson processes[27] will be useful.

Theorem 1.13 (Rate doubling for Poisson process). *Given a continuous function $g \ge 0$ on $[0, \infty)$, consider the Poisson process $(N, \mathbb{L}^g)$, with rate function g, and assume that N is adapted to the filtration $\{\mathcal{G}_t\}_{t \ge 0}$. Then, $M_t := 2^{N_t} \exp\left\{ -\int_0^t g(s) \, ds \right\}$ is an $\mathbb{L}^g$-martingale with respect to the same filtration, with unit mean, and under the change of measure*

$$\left. \frac{d\mathbb{L}^{2g}}{d\mathbb{L}^g} \right|_{\mathcal{G}_t} = M_t$$

the process $(N, \mathbb{L}^{2g})$ is a Poisson process with rate function $2g$.

Proof. Let us first show that M is a martingale with unit mean. Let E denote the expectation corresponding to $\mathbb{L}^g$. It is sufficient to show that $EM_t = 1$ for $t \ge 0$, because then, for $0 \le s < t$,

$$E(M_t \mid \mathcal{F}_s) = M_s E \left(2^{N_t - N_s} \exp\left(-\int_s^t g(z) \, dz \right) \middle| \mathcal{F}_s \right),$$

and

$$E \left(2^{N_t - N_s} \exp\left(-\int_s^t g(z) \, dz \right) \middle| \mathcal{F}_s \right) = 1.$$

Indeed, by the Markov property, defining $g^*(z) := g(z + s)$; $N_z^* := N_{s+z} - N_s$, we can rewrite this as

$$E \left(2^{N_{t-s}^*} \exp\left(-\int_0^{t-s} g^*(z) \, dz \right) \right) = 1,$$

which is tantamount to $EM_{t-s} = 1$ for the process N^* with rate g^*.

To show that $EM_t = 1$ for $t \ge 0$, consider a strictly dyadic branching process[28] Z (setting $Z_0 = 1$) with time-inhomogeneous exponential rate g,

---

[27]In the exercises at the end of this chapter, the reader is asked to generalize this result for the case when the rate changes to k times the original one.

[28]If the reader is not familiar with the notion of a branching process, then (s)he should take a look at Section 1.13.

where the particles' clocks are not independent but synchronized, that is, the branching times are defined by the underlying Poisson point process N and at every time point all particles alive simultaneously split into two. Then $Z_t = 2^{N_t}$, and so $E2^{N_t} = EZ_t = \exp\left\{\int_0^t g(s)\,\mathrm{d}s\right\}$, where the second equation follows from the fact that if $u(t) := EZ_t$, then $u'(t) = g(t)u(t)$. This, in turn, follows from the fact that the probability of not branching in the interval $(t, t + \Delta t]$ is

$$\exp\left(-\int_t^{t+\Delta t} g(s)\,\mathrm{d}s\right) = 1 - \int_t^{t+\Delta t} g(s)\,\mathrm{d}s + o(\Delta t) = 1 - \Delta t \cdot g(t) + o(\Delta t),$$

as $\Delta t \to 0$, and thus

$$E(Z_{t+\Delta t}) = EZ_t \cdot (1 - g(t)\Delta t) + 2EZ_t \cdot g(t)\Delta t + o(\Delta t),$$

that is,

$$E(Z_{t+\Delta t}) = EZ_t + EZ_t \cdot g(t)\Delta t + o(\Delta t).$$

Finally, we show that M is the density between $\mathbb{L}^{2g}$ and $\mathbb{L}^g$. For this, it is enough to show that the process under $\mathbb{L}^{2g}$ has independent increments and $N_t - N_s$ has Poisson distribution with parameter $\int_s^t 2g(z)\mathrm{d}z$. This, in turn, follows from the fact that conditioned on $\mathcal{F}_s$, the conditional moment generating function is $u \mapsto \exp(\int_s^t 2g(z)\mathrm{d}z \cdot (e^u - 1))$. Indeed, for $u \in \mathbb{R}$,

$$\mathbb{L}^{2g}\left[e^{u(N_t-N_s)} \mid \mathcal{F}_s\right] = \mathbb{L}^g\left[e^{u(N_t-N_s)}\frac{M_t}{M_s} \mid \mathcal{F}_s\right]$$

$$= \mathbb{L}^g\left[e^{u(N_t-N_s)}2^{N_t-N_s} \cdot e^{-\int_s^t g(z)\mathrm{d}z} \mid \mathcal{F}_s\right]$$

$$= e^{-\int_s^t g(z)\mathrm{d}z} \cdot \mathbb{L}^g\left[e^{(N_t-N_s)(u+\log 2)} \mid \mathcal{F}_s\right]$$

$$= e^{-\int_s^t g(z)\mathrm{d}z} \cdot \exp\left(\int_s^t g(z)\mathrm{d}z \cdot (e^{u+\log 2} - 1)\right)$$

$$= \exp\left(\int_s^t 2g(z)\mathrm{d}z \cdot (e^u - 1)\right). \qquad \square$$

Remark 1.11 (Compound Poisson process). Let $g > 0$ be constant and let the $\{\xi_i, i \geq 1\}$ be independent, identically distributed random variables, which are independent of N as well. If $F(\mathrm{d}x)$ denotes the common probability distribution (it is convenient to assume that F has no atom at zero) and

$$K_t := \sum_{i=1}^{N_t} \xi_i, \ t \geq 0,$$

then K is called a *compound Poisson process* with jump distribution F. (One gets back the Poisson process when $\xi_i = 1$ a.s.) By construction, this process starts at zero, has right-continuous paths with left limits, and its increments are independent and stationary. Such processes[29] are called *Lévy processes* after P. Lévy. The reader interested in the general theory of these processes is referred to the recent monograph [Kyprianou (2014)]. ⋄

1.10 The generalized principal eigenvalue for a second order elliptic operator

It turns out that in the theory of branching diffusions and superdiffusions, and in problems concerning Poissonian obstacles,[30] a spectral theoretical quantity plays a central role. To be really precise though, instead of spectral theory, we should refer to the *criticality theory* of second order elliptic operators, developed by Y. Pinchover and R. Pinsky.[31] We now give the definition of this important notion.

Let $D \subseteq \mathbb{R}^d$ be a non-empty domain and, as usual, write $C^{i,\eta}(D)$ to denote the space of i times $(i = 1, 2)$ continuously differentiable functions with all their ith order derivatives belonging to $C^\eta(D)$. Similarly, write $C^{i,\eta}(\overline{D})$ to denote the space of i times $(i = 1, 2)$ continuously differentiable functions with all their ith order derivatives belonging to $C^\eta(\overline{D})$. (Recall that $C^\eta(D)$ and $C^\eta(\overline{D})$ are the Hölder spaces with $\eta \in (0, 1]$.) Let the operator

$$L = \frac{1}{2}\nabla \cdot a\nabla + b \cdot \nabla \quad \text{on } D, \tag{1.23}$$

satisfy Assumption 1.2. Furthermore, let $\beta \in C^\eta(D)$.

A function u on D is called *harmonic* with respect to $L + \beta - \lambda$ if $(L + \beta - \lambda)u = 0$ holds on D. Now define

$$\lambda_c := \lambda_c(L+\beta, D) := \inf\{\lambda \in \mathbb{R} : \exists u > 0 \text{ satisfying } (L+\beta-\lambda)u = 0 \text{ in } D\},$$

in other words,

$$\lambda_c := \inf\{\lambda \in \mathbb{R} : \exists \text{ positive harmonic function for } L + \beta - \lambda \text{ on } D\},$$

and call λ_c the *generalized principal eigenvalue* for $L + \beta$ on D. (The subscript c refers to the word 'critical.')

---

[29]Notice, that another example for such a process is Brownian motion.
[30]See Section 1.12.
[31]Pinchover used a purely analytical approach; the theory was reformulated using probabilistic notions and tools by Pinsky.

Remark 1.12 (The Berestycki-Rossi approach). In the recent paper [Berestycki and Rossi (2015)], the authors introduce three different notions[32] of the generalized principal eigenvalue for second order elliptic operators in unbounded domains, and discuss the relations between these principal eigenvalues, their simplicity and several other properties. The validity of the maximum principle and the existence of positive eigenfunctions for the Dirichlet problem are also investigated. In this book, however, we are following the 'criticality theory' approach.   ◇

The following properties of this quantity can be found (with detailed proofs) in Chapter 4 in [Pinsky (1995)].

(1) **Symmetric case:** When $b = a\nabla Q$ with some $Q \in C^{2,\alpha}(D)$, the operator L can be written in the form $L = \frac{1}{2}h^{-1}\nabla \cdot ah\nabla$, where $h := \exp(2Q)$, and, with the domain $C_c^\infty(D)$, it is symmetric on the Hilbert space $L_2(D, hdx)$. In this case, λ_c coincides with the supremum of the spectrum of the self-adjoint operator (having real spectrum), obtained via the Friedrichs extension from $L + \beta$,[33] hence the word 'generalized.' (See Proposition 4.10.1 in [Pinsky (1995)].)

When λ_c is actually an eigenvalue, it is simple (that is, the corresponding eigenspace is one-dimensional), and the corresponding eigenfunctions are positive. (This is related to the fact that the transition measure is *positivity improving* — see p. 195 in [Pinsky (1995)].) Thus λ_c plays a similar role as the 'Perron-Frobenius eigenvalue' in the theory of positive matrices.

(2) $\lambda_c \in (-\infty, \infty]$.

(3) If $\beta \equiv 0$, then $\lambda_c \leq 0$. (Because $h \equiv 1$ is a positive harmonic function.)

(4) If $\beta \equiv B$, then $\lambda_c(L + \beta) = \lambda_c(L) + \beta$. (This is evident from the definition of λ_c.)

(5) There is monotonicity in β, that is, if $\hat\beta \leq \beta$, then

$$\lambda_c(L + \hat\beta, D) \leq \lambda_c(L + \beta, D).$$

Because of the previous two properties, it then follows that if $\beta \leq K$, then $\lambda_c \leq K$, for $K \in \mathbb{R}$.

(6) There is monotonicity in the domain, that is, if $\widehat{D} \subset D$, then

$$\lambda_c(L + \beta, \widehat{D}) \leq \lambda_c(L + \beta, D).$$

---

[32] One of them is essentially the same as our definition, although the assumptions on the operator are different from ours. See formula (1) in [Berestycki and Rossi (2015)].

[33] Of course, we are cheating a little, as $L + \beta$ is not a positive operator. So the Friedrichs extension actually applies to $-L - \beta$ instead of $L + \beta$ — this inconvenience is the price of our probabilistic treatment of operators.

(Because the restriction of a positive harmonic function to a smaller domain is still positive harmonic.)

(7) Let $D_1 \subset D_2 \subset ...$ be subdomains in $D \subset \mathbb{R}^d$ and $\cup_{n \geq 1} D_n =: D$. Then

$$\lambda_c(L + \beta, D) = \lim_{n \to \infty} \lambda_c(L + \beta, D_n).$$

Remark 1.13 (Spectrum versus positive harmonic functions). At first sight, it might be confusing that λ_c is defined both as a supremum (of the spectrum) and an infimum in the symmetric case. The explanation is as follows. When $\lambda \geq \lambda_c$, there are positive harmonic functions for $L + \beta - \lambda$, but for $\lambda > \lambda_c$, they are not in L^2, so they are not actually eigenfunctions. An L^2-eigenfunction for $L + \beta - \lambda$ with $\lambda < \lambda_c$, on the other hand, cannot be everywhere positive. In other words, λ_c separates the spectrum from those values which correspond to positive (but not L^2) harmonic functions. ⋄

1.10.1 *Smooth bounded domains*

Recall that a non-empty domain D of $\mathbb{R}^d$ has a $C^{2,\alpha}$-*boundary* if for each point $x_0 \in \partial D$, there is a ball B around x_0 and a one-to-one mapping ψ of B onto $A \subset \mathbb{R}^d$, such that $\psi(B \cap D) \subset \{x \in \mathbb{R}^d : x_n > 0\}, \psi(B \cap \partial D) \subset \{x \in \mathbb{R}^d : x_n = 0\}$ and $\psi \in C^{2,\alpha}(B), \psi^{-1} \in C^{2,\alpha}(A)$. So ∂D can be thought of as locally being the graph of a Hölder-continuous function.

Recall also that $L + \beta$ is called *uniformly elliptic* on $\overline{D}$ if there exists a number $c > 0$, such that $\sum_{i,j=1}^d a_{ij}(x)v_iv_j \geq c\|v\|^2$ holds for all $x \in \overline{D}$ and all $v = (v_1, v_2, ..., v_d) \in \mathbb{R}^d$, where $\|v\| = (\sum_{i=1}^d v_i^2)^{\frac{1}{2}}$.

Regarding the definition of λ_c, it is important to discuss the particular case when

(1) $D \subset\subset \mathbb{R}^d$ is a bounded domain with a $C^{2,\alpha}$-boundary;
(2) L, defined on $\overline{D}$, is uniformly elliptic;
(3) $L + \beta$ has uniformly Hölder-continuous coefficients on D.

Assume now (1)–(3) above. The following is a sketch of the relevant parts of Chapter 3 in [Pinsky (1995)], where full proofs can be found.

Let $\gamma := \sup_D \beta$. In the Dirichlet problem

$$\left.\begin{array}{c} (L + \beta - \gamma)u = f \text{ in } D, \\ \\ u = 0 \text{ on } \partial D, \end{array}\right\} \tag{1.24}$$

the potential term $\beta - \gamma$ is non-positive, and thus, it is well known that the problem has a unique solution $u \in C^{2,\alpha}(\overline{D})$, whenever $f \in C^\alpha(\overline{D})$.

Consider now

$$\mathcal{B}_\alpha := \{u \in C^\alpha(\overline{D}) \mid u = 0 \text{ on } \partial D\}.$$

Then $\mathcal{B}_\alpha$ is a Banach space with the norm

$$\|u\|_{0,\alpha;D} := \sup_{x \in D} |u(x)| + \sup_{x,y \in D; x \neq y} \frac{|f(x) - f(y)|}{|x - y|^\alpha}.$$

By the previous comment on the unique solution of the Dirichlet problem, the inverse operator $(L + \beta - \gamma)^{-1} : \mathcal{B}_\alpha \to C^{2,\alpha}(\overline{D}) \cap \mathcal{B}_\alpha$ is well defined. In fact it is a compact operator,[34] and if $\mathcal{D}_\alpha \subset C^{2,\alpha}(\overline{D}) \cap \mathcal{B}_\alpha$ denotes its range, then $\mathcal{D}_\alpha$ is dense in $\mathcal{B}_\alpha$.

Recall that the *resolvent set* consists of those $\lambda \in \mathbb{C}$, for which the operator $(L + \beta - \lambda)^{-1}$ exists, is a bounded linear operator, and is defined on a dense subspace. Let $\sigma(L + \beta)$ denote the spectrum (which, by definition, is the complement of the resolvent set) of the operator $L + \beta$ on $\mathcal{D}_\alpha$; it is always a closed set, and for a symmetric operator, it is a subset of $\mathbb{R}$. It can be shown that $\sigma(L + \beta)$ consists only of eigenvalues, and that $\lambda \in \sigma(L + \beta)$ if and only if $-1/(\gamma - \lambda)$ belongs to the spectrum of $(L + \beta - \gamma)^{-1}$ on $\mathcal{B}_\alpha$. Thus, from the spectral properties of the latter operator, one can deduce the structure of $\sigma(L + \beta)$. This is important, because it turns out that λ_c can be described using the spectrum of $L + \beta$:

$$\lambda_c = \sup\{\Re(z) \mid z \in \sigma(L + \beta)\},$$

where $\Re(z) := x$ for $z = x + iy$.

Furthermore, we encounter the 'Perron-Frobenius-type' behavior once again: $\lambda_c \in \sigma(L + \beta)$ and the corresponding eigenspace is one-dimensional, and consists of functions which are positive on D, and all other eigenfunctions of the operator change sign in D (see Theorem 3.5.5 in [Pinsky (1995)]).

As mentioned before, if L happens to be symmetric on D then λ_c can also be described as the supremum of the L^2-spectrum.

1.10.2   *Probabilistic representation of λ_c*

Below is a *probabilistic* representation of λ_c in terms of the diffusion process corresponding to L on D and in terms of stopping times of compactly embedded sets of D. (For the proof, see Theorem 4.4.4 in [Pinsky (1995)].)

---

[34] A bounded linear operator A on a Banach space is called compact if the image of a bounded set is pre-compact (i.e., its closure is compact). Their spectral theory was first developed by F. Riesz, as a generalization of the corresponding theory for square matrices.

Proposition 1.6 (Probabilistic representation of λ_c). *Let $\{D_n\}_{n\geq 1}$ be a sequence of subdomains such that $D_n \subset\subset D$ and $D_n \uparrow D$. Let Y be the diffusion corresponding to L on D (which is assumed to satisfy Assumption 1.2) with expectations $\{E_x; x \in D\}$, and*

$$\tau_n := \inf_{t\geq 0}\{t : Y_t \notin D_n\}, \ n \geq 1.$$

Then

$$\lambda_c(L + \beta, D) = \sup_n \lim_{t\to\infty} \frac{1}{t} \log \sup_{x\in D_n} E_x(e^{\int_0^t \beta(Y_s)\,\mathrm{d}s}; t \leq \tau_n)$$

$$= \lim_{n\to\infty} \lim_{t\to\infty} \frac{1}{t} \log \sup_{x\in D_n} E_x(e^{\int_0^t \beta(Y_s)\,\mathrm{d}s}; t \leq \tau_n). \quad (1.25)$$

When $\beta \equiv 0$, the previous proposition tells us that λ_c describes the 'asymptotic rate' at which the diffusion leaves compacts. For recurrent diffusions it is zero, and for transient ones, it measures 'the extent of transience' of the process. In the transient case both $\lambda_c = 0$ and $\lambda_c < 0$ are possible. For example $\lambda_c = 0$ for a d-dimensional Brownian motion, for all $d \geq 1$, although for $d \geq 3$ the process is transient. In the latter case, the process leaves compacts 'quite slowly.' On the other hand, for an 'outward' O-U process (see Definition 1.6) one has $\lambda_c > 0$, and the process leaves compacts 'fast.'

1.11   Some more criticality theory

Just like in the previous section, assume that D is a non-empty domain in $\mathbb{R}^d$, that L on D satisfies Assumption 1.2. and that $\beta \in C^\eta(D)$. Assume also that $p(t, x, \mathrm{d}y)$, the transition measure for $L + \beta$ on D, exists (i.e., assume that it is σ-finite for all $t \geq 0$ and $x \in D$).

Definition 1.9 (Green's measure). *The measure*

$$G(x, \mathrm{d}y) := \int_0^\infty p(t, x, \mathrm{d}y)\,\mathrm{d}t$$

is called the Green's measure *if $G(x, B) < \infty$ for all $x \in D$ and $B \subset\subset D$. Otherwise, one says that the Green's measure does not exist. In fact it can be shown that in this case $G(x, B) = \infty$ for all $x \in D$ and $B \subset\subset D$.*

It is a standard fact that the Green's measure actually possesses a density, called the *Green's function*: $G(x, \mathrm{d}y) = G(x, y)\mathrm{d}y$.

Before making the following crucial definition, we state an important fact.

Proposition 1.7. *Assume that the operator $L + \beta$ on D possesses a Green's measure. Then there exists a positive harmonic function, that is, a function $h > 0$ such that $(L + \beta)h = 0$.*

The following definition is fundamental in criticality theory.

Definition 1.10 (Criticality). The operator $L + \beta$ on D is called

(a) *subcritical* if the operator possesses a Green's measure,
(b) *supercritical* if the associated space of positive harmonic functions is empty,
(c) *critical* if the associated space of positive harmonic functions is non-empty but the operator does not possess a Green's measure.

(The notions 'critical, sub- and supercritical' were initially suggested by B. Simon in studying perturbations of the Laplacian.)

Note: When $\lambda_c < \infty$, it is sometimes more convenient to define the above properties for $\beta - \lambda_c$ instead of β, as we will choose to do later.

In the critical case (c), the space of positive harmonic functions is in fact one-dimensional.[35] This unique (up to positive constant multiples) function is called the *ground state*. Moreover, the space of positive harmonic functions of the adjoint operator is also one-dimensional.

Assume now that the operator $L + \beta - \lambda_c$ on D is critical. Choose representatives for the ground states of this operator and of its adjoint to be ϕ and $\widetilde{\phi}$, respectively, and make the following definition.

Definition 1.11 (Product-criticality). The operator $L + \beta - \lambda_c$ is called *product critical* (or 'product L^1 critical'), if $\langle \psi, \widetilde{\psi} \rangle < \infty$, and in this case we pick ϕ and $\widetilde{\phi}$ with the normalization $\langle \phi, \widetilde{\phi} \rangle = 1$.

A crucially important fact is that the above notions are *invariant under h-transforms*. That is, if the operator $L + \beta$ on D is critical (supercritical, subcritical, product-critical), then so is any other operator obtained by h-transform from $L + \beta$.

A measurable function $f > 0$ on D is called an *invariant function* if

$$\int_D f(y)p(t, \cdot, \mathrm{d}y) = f(\cdot), \ t > 0.$$

---

[35]This result is due to S. Agmon. Also, cf. Proposition 8.1 in [Berestycki and Rossi (2015)].

The following result is Theorem 4.8.6. in [Pinsky (1995)].

Proposition 1.8. *In the critical case, the space of nonnegative invariant functions is a one-dimensional cone: it contains the multiples of the ground state only.*

In the particular case when $\beta \equiv 0$, the corresponding diffusion Y on D is recurrent (resp. transient) if and only if the operator is critical (resp. subcritical). (Clearly the function $u \equiv 1$ is a positive harmonic function for L, and so supercriticality is ruled out.) Furthermore, the diffusion Y is positive recurrent if and only if L is product-critical.

Remark 1.14 (Harmonic h-transform). Assume that $(L+\beta)h = 0$ for some $h > 0$ on D. (This is only possible if $L + \beta$ is subcritical or critical.) Then the new zeroth order (potential) term is $\beta^h = (L+\beta)h/h = 0$, that is, the h-transform 'knocks the potential out.' Hence, there is a diffusion process on D, say Y, corresponding to the new operator $(L + \beta)^h$. We conclude that Y is recurrent (resp. transient, positive recurrent) on D if and only if the operator $L + \beta$ on D is critical (resp. subcritical, product-critical). ◇

It is useful to have the following analytical criterion (see Theorem 4.3.9 in [Pinsky (1995)]) for transience.

Proposition 1.9. *Let Y be an L-diffusion on $D \subset \mathbb{R}^d$. There exists a function $0 < w \in C^{2,\alpha}(D)$ on D such that $Lw \leq 0$ and Lw is not identically zero if and only if Y is transient on D.*

(Using an argument, similar to that of Remark 1.14, one can easily show *sufficiency*: An h-transform with $h := w$ results in a potential term $(Lw)/w$, which is non-positive and not identically zero. Such an operator is known to be subcritical. By invariance, L is subcritical too, and so Y is transient.)

An important result on *perturbations* is given in the following proposition (Theorem 4.6.3 in [Pinsky (1995)]):

Proposition 1.10. *If L corresponds to a recurrent diffusion on the domain $D \subset \mathbb{R}^d$ and $0 \leq \beta \in C^\eta(D)$ with $\beta \not\equiv 0$, then $\lambda_c(L + \beta, D) > 0$.*

(Although in Theorem 4.6.3 in [Pinsky (1995)] β has to be compactly supported too, the result immediately follows by monotonicity.)

Remark 1.15. The conclusion in Proposition 1.10 fails to hold when L is transient on D. (See Section 4.6 in [Pinsky (1995)].)

Finally we note that when $\lambda_c > 0$, the operator $L + \beta$ on D is always supercritical.

1.12   Poissonian obstacles

Although we are now switching topic, we note that Poissonian obstacles *are* related to generalized principal eigenvalues — see Remark 1.18 later.

In the definition of the Poisson process we have already recalled how a measurable, and locally integrable function $g : [0, \infty) \to [0, \infty)$ defines a Poisson point process, which is a random collection of points on $[0, \infty)$. The construction in $\mathbb{R}^d$ is similar. Let ν be a locally finite measure on $\mathbb{R}^d$. A d-dimensional Poisson point process (PPP) with intensity measure ν is a random collection of points ω on $\mathbb{R}^d$, satisfying that if $X(B)$ is the number of points in a Borel set $B \subset \mathbb{R}^d$, then

(1) $X(B_1), X(B_2), ..., X(B_k)$ are independent if $B_i \cap B_j = \emptyset$ for $1 \le i \ne j \le k$, for $k \ge 1$,

(2) $X(B)$ is Poisson distributed with parameter $\nu(B)$.

That (1) and (2) uniquely defines a point process, is well known. (See e.g. [Ethier and Kurtz (1986)] or [Kingman (1993)].)

In particular, using a slight abuse of notation, if $\nu(B) := \int_B \nu(x) \mathrm{d}x$ for some nonnegative, measurable and locally integrable function ν on $\mathbb{R}^d$, then the PPP can be determined by the density function ν as well.

Remark 1.16. The reason Poisson point processes arise naturally is the Poisson approximation of the Binomial distribution. The reader can easily see this for the very simple case of constant density $\lambda > 0$ on the unit interval, using the following heuristic argument. Divide each interval into n subintervals, put into each one of them a point with probability λ/n, independently of each other (say, in the middle). As $n \to \infty$, the distribution of the number of points in a given interval is then going to tend to the Poisson distribution, the parameter of which equals λ times the length of the interval. The independence of the number of points on disjoint intervals is also clear, and so is the extension to the real line, by performing this procedure independently on every unit interval with integer endpoints. The argument is similar for $\mathbb{R}^d$.

For a general measure, the only difference is that one has to use the more general version of the Poisson Approximation Theorem, where the sums of different probabilities converge.   ◇

Definition 1.12. The random set

$$K := \bigcup_{x_i \in \text{supp}(\omega)} \overline{B}(x_i, a)$$

is called a *trap configuration* (or hard obstacle) attached to ω. (And $\overline{B}(x, a)$ is the closed ball with radius a centered at $x \in \mathbb{R}^d$.) A set $B \subset \mathbb{R}^d$ is a *clearing* if it is obstacle-free, that is, if $B \subset K^c$.

One usually works with ball-shaped clearings. The probability that a given Borel set B is a clearing is of course $\exp(-\nu(B^a))$ (where B^a is the a-neighborhood of B).

Remark 1.17. We will identify ω with K, that is an ω-wise statement will mean that it is true for all trap configurations (with a fixed). ◇

Denote by $B(0, 1)$ the d-dimensional unit (open) ball, and by $-\lambda_d$ the principal eigenvalue of $\frac{1}{2}\Delta$ on it ($\lambda_d = -\lambda_c(\frac{1}{2}\Delta; B(0, 1))$), and let ω_d be the volume of $B(0, 1)$:

$$\omega_d = \frac{\pi^{d/2}}{\Gamma(\frac{d}{2} + 1)},$$

where Γ is Euler's gamma function.

The following proposition is important for calculating survival probabilities among obstacles. It tells us how far a particle has to travel in order to be able to find a clearing of a certain size; it is Lemma 4.5.2 (appearing in the proof of Theorem 4.5.1) in [Sznitman (1998)]:

Proposition 1.11 (Size of clearings within given distance). *Consider a PPP on $\mathbb{R}^d$ with constant density $\nu > 0$, and with probability $\mathbb{P}_\nu$. Abbreviate*

$$R_0 = R_0(d, \nu) := \sqrt[d]{d/(\nu\omega_d)},$$

and

$$\rho(l) := R_0(\log l)^{1/d} - (\log \log l)^2, \quad l > 1. \tag{1.26}$$

Define the event $\mathcal{C} = \mathcal{C}(d, \nu)$ as

$$\mathcal{C} := \exists \, l_0(\omega) > 0 \text{ such that } \forall l > l_0(\omega) \, \exists \text{ clearing } B(x_0, \rho(l)) \text{ with } |x_0| \le l.$$

Then,

$$\mathbb{P}_\nu(\mathcal{C}) = 1. \tag{1.27}$$

1.12.1   *Wiener-sausage and obstacles*

In the classic paper [Donsker and Varadhan (1975)], the authors described the asymptotic behavior of the volume of the so-called 'Wiener-sausage.' If W denotes Brownian motion (Wiener-process) in d-dimension, with expectation E, then for $t, a > 0$,

$$W_t^a := \bigcup_{0 \leq s \leq t} \overline{B}(W_s, a) \tag{1.28}$$

is called the *Wiener-sausage* up to time t. As usual, $|W_t^a|$ denotes the d-dimensional volume of W_t^a. By the classical result of Donsker and Varadhan, its Laplace-transform obeys the following asymptotics:

$$\lim_{t \to \infty} t^{-d/(d+2)} \log E_0 \exp(-\nu |W_t^a|) = -c(d, \nu), \quad \nu > 0, \tag{1.29}$$

for any $a > 0$, where, for $\nu > 0$ and $d = 1, 2, ...$, one defines

$$c(d, \nu) := \nu^{2/(d+2)} \left(\frac{d+2}{2} \right) \left(\frac{2\lambda_d}{d} \right)^{d/(d+2)}. \tag{1.30}$$

Note that the limit does not depend on the radius a.

The lower estimate for (1.29) had been known by M. Kac and J.M. Luttinger earlier, and in fact the upper estimate turned out to be much harder. This latter one was obtained in [Donsker and Varadhan (1975)] by using techniques from the theory of large deviations.

Let us now make a short detour, before returning to Wiener sausages. Let ω be a PPP with probability $\mathbb{P}$ on $\mathbb{R}^d$, and let K be as in Definition 1.12. Define the 'trapping time' $T_K := \inf \{s \geq 0, \ W_s \in K\}$. The problem of describing the distribution of T_K is called a 'trapping problem.'

The motivation for studying 'trapping problems' comes from various models in chemistry and physics. In those models particles move according to a random motion process in a space containing randomly located traps (obstacles), which may or may not be mobile. Typically the particles and traps are spheres or points and in the simplest models the point-like particles are annihilated when hitting the immobile and sphere-shaped traps. In the language of reaction kinetics: when molecule A (particle) and molecule B (trap) react, A is annihilated while B remains intact. The basic object of interest is the probability that a single particle avoids traps up to time t. This is sometimes done by averaging over all trap configurations, and sometimes for fixed 'typical' configurations. For several particles, we assume independence and obtain the probability that no reaction has occurred between the two types of molecules up to time t. The reader can find more material on the subject in e.g. [den Hollander and Weiss (1994); Merkl and Wüthrich (2002); Sethuraman (2003); van den Berg, Bolthausen and den Hollander (2005)].

Now, returning to the result of Donsker and Varadhan discussed above, we note the following elementary but important connection between the Wiener-sausage and trapping problems. Let $\mathbb{P}_\nu$ have intensity measure $\nu \, dl$ (dl is the Lebesgue-measure), $\nu > 0$, and denote the expectation by $\mathbb{E}_\nu$.

Proposition 1.12 (Wiener-sausage via obstacles).

$$E_0 \exp(-\nu|W_t^a|) = (\mathbb{E}_\nu \otimes P_0)(T_K > t), \text{ for } t > 0. \tag{1.31}$$

By (1.31), the law of $|W_t^a|$ can be expressed in terms of the 'annealed' or 'averaged' probabilities that the Wiener-process avoids the Poissonian traps of size a up to time t. Using this interpretation of the problem, Sznitman [Sznitman (1998)] presented an alternative proof for (1.29). His method, called the 'enlargement of obstacles' turned out to be extremely useful and resulted in a whole array of results concerning similar questions (see the fundamental monograph [Sznitman (1998)], and references therein).

Remark 1.18 (Poissonian obstacles and GPE). By the probabilistic representation of the generalized principal eigenvalue (formula (1.25)), we know that for a fixed K, the probability that Brownian motion stays in a specific clearing for large times, is related to the generalized principal eigenvalue of $\Delta/2$ on that clearing. So, the problem of avoiding obstacles for large times is linked to estimating generalized principal eigenvalues of certain *random domains*. (See Section 4.4 in [Sznitman (1998)] for more on this connection.)   ◇

1.12.2   *'Annealed' and 'quenched'; 'soft' and 'hard'*

As mentioned above, the term 'annealed' means 'averaged,' that is, we are using the product measure $\mathbb{P}_\nu \otimes P_0$ to evaluate the probability of events.[36] This term, just like its counterpart 'quenched,' are coming from metallurgy. The latter means 'frozen,' that is we fix (freeze) the environment and want to say something about the behavior of the system in that fixed environment — the goal is to obtain statements which are valid for $\mathbb{P}$-almost every environment.

Annealed and quenched results are quite different. Indeed, suppose that we want to construct an event which is part of the 'trap-avoiding event'

---

[36]A statistical physicist would say that we are 'averaging over disorder.'

$\{T_K > t\}$. Our goal is to find an event like that, which is simple enough, so that we can calculate its probability, and at the same time, we try to make this probability as large as possible. (This approach will lead to a lower estimate, and the event we found is called a 'strategy.')

Now, in the quenched case, we take the environment 'frozen' (fixed), and whatever (lucky or unlucky) environment is given, we try to find a way to make the Brownian motion avoid the traps. On the other hand, in the annealed case, we are to describe an event which is a result of the *joint behavior* of the Brownian motion *and* the Poissonian environment. For example, in the annealed setting, traps can be avoided up to a given time by having an appropriate 'clearing' around the origin, and at the same time confining the path of the particle to that region.

In fact the quenched asymptotics for Brownian survival among Poissonian traps is different from the corresponding annealed one, given in (1.29).

The obstacles we have worked with so far are called 'hard obstacles,' meaning that the Brownian particle is immediately annihilated (or trapped) upon contact. Another type of obstacle one considers often is called 'soft obstacle,' because it does not annihilate (trap) the particle instantly.

To define soft obstacles mathematically, take a nonnegative, measurable, compactly supported function ('shape function') V on $\mathbb{R}^d$. Next, define $U(x) = U(x, \omega) := \sum_{y \in \omega} V(x - y)$, that is *sum up* the shape functions about all the Poissonian points. The object of interest now is $E_0 \exp\left[-\int_0^t U(W_s)\, ds\right]$, because this is the survival probability up to time t if the particle is killed according to the exponential rate U. The particle has to pass through 'soft obstacles' which, although do not necessarily kill the particle immediately, make survival increasingly difficult if the particle spends a long time in the support of U.

The fundamental quenched result is that for $\mathbb{P}_\nu$ almost every ω,

$$E_0 \exp\left[-\int_0^t U(W_s)\, ds\right] = \exp\left[-k(d, \nu)\frac{t}{(\log t)^{2/d}}(1 + o(1))\right], \quad (1.32)$$

as $t \to \infty$, where

$$k(d, \nu) := \lambda_d \left(\frac{d}{\nu \omega_d}\right)^{-2/d}. \quad (1.33)$$

Remark 1.19 (Robustness). It is important to stress the (perhaps surprising) fact that these asymptotic results are robust (that is, they do not depend on the shape function). This is even true in the 'extreme' case: When '$V = \infty$' on $B(0, a)$ and vanishes outside of it, one obtains the quenched 'hard obstacle' situation which is the quenched analog of (1.29), and for which (1.32) still holds, if one conditions on the event that the origin is located in an unbounded trap-free region.

Similarly, the annealed asymptotics (1.29) stays correct for the expectation with soft obstacles, that is, for $(\mathbb{E}_\nu \otimes E_0) \exp\left[-\int_0^t U(W_s)\,\mathrm{d}s\right]$.

See again [Sznitman (1998)] for more elaboration on these results. ◇

1.13 Branching

In this section we give a review on some basic properties of branching processes. Suppose that we start with a single ancestor, and any individual has $X = 0, 1, 2, \ldots$ offspring with corresponding probabilities $p_0, p_1, p_2, \ldots$ and suppose that branching occurs at every time unit. Let h be the *gener-*

Fig. 1.3 Sir Francis Galton [Wikipedia].

ating function[37] of the offspring distribution,

$$h(z) := Ez^X = p_0 + p_1 z + p_2 z^2 + \dots.$$

Being a power series around zero, with $h(1) = 1$, it is clear that h is finite and analytic on the unit interval. If R is the radius of convergence, then $R \geq 1$; in the sequel, for the sake of simplicity, we assume $R = \infty$. Then,

$$h'(1) = p_1 + 2p_2 + 3p_3 + \dots =: m,$$

and m is the expected (or mean) offspring number. Higher moments are obtained by differentiating the function h more times. For example, for the variance σ^2 of the offspring distribution, it yields

$$\sigma^2 = h''(1) + m - m^2 = h''(1) + h'(1) - [h'(1)]^2.$$

Now, suppose that all the offspring of the original single individual also give birth to a random number of offspring, according to the law of X, their offspring do the same as well, and continue this in an inductive manner, assuming that all these mechanisms are independent of each other. The model may describe the evolution of certain bacteria, or some nuclear reaction, for example.

Let Z_n denotes the size of the nth generation for $n \geq 0$. (We set $Z_0 = 1$, as we start with a single particle.) Using induction, it is an easy exercise (left to the reader) to verify another very handy property of the generating function: the generating function of Z_n satisfies

$$Ez^{Z_n} = h(h(\dots(z)\dots)), \ n \geq 1, \tag{1.34}$$

where on the right-hand side one has precisely the nth iterate of the function h.

The process $Z = (Z_n)_{n \geq 0}$ is called a *Galton-Watson process*. Its study was initiated when Darwin's cousin, Sir Francis Galton (1822–1911), the English Victorian scientist, became interested in the statistical properties of family names. A concern amongst some Victorians was that aristocratic family names were becoming extinct. Galton posed the question regarding the probability of such an event in 1873 in the Educational Times – the answer to Galton's question was provided by Reverend Henry William Watson. Next year, they published their results on what is today known as Galton-Watson (or Bienaymé-Galton-Watson[38]) process. The iterative formula (1.34) had already been known to them.

---

[37]In combinatorics, this would be the generating function of the sequence of the p_k's. Substituting e^t for z yields the moment generating function.

[38]Independently of Galton and Watson, the French statistician I. J. Bienaymé obtained similar results.

We will always assume that the process is non-degenerate, that is that $p_1 < 1$. Then, there are three, qualitatively different, cases for such a process:

(1) $m < 1$ and Z dies out in finite time, a.s. (*subcritical* case)
(2) $m = 1$ and Z dies out in finite time, a.s., though its mean does not change (*critical* case)
(3) $m > 1$ and Z survives with positive probability (*supercritical* case)

For example, one is talking about *strictly dyadic* branching when $p_2 = 1$, and in this case $h(z) = z^2$ and $m = 2$. In the supercritical case, the expected offspring number m_n at time n, satisfies $m_n = m^n$ and if we further assume that $p_0 > 0$, then one can show that $d = P(\text{extinction})$ is the only root of $h(z) = z$ in $(0, 1)$. (The function h can be shown to be concave upward.)

1.13.1 *The critical case; Kolmogorov's result*

The critical case has the somewhat peculiar property that, while the expected size of any generation is one, nevertheless the process becomes extinct almost surely. This can be explained intuitively by the fact that, although 'most probably' the process is extinct by time n, nevertheless, there is still a 'very small' probability of having a 'very large' nth generation.

A classic result due to Kolmogorov (Formula 10.8 in [Harris(2002)]) that we will need later, gives the asymptotic decay rate of survival for critical branching.

Theorem 1.14 (Survival for critical branching). *For critical unit time branching with generating function h, as $n \to \infty$,*

$$P(\text{survival up to } n) \sim \frac{2}{nh''(1)}.$$

(Recall that $h''(1) = EZ_1^2 - EZ_1 > 0$.)

1.13.2 *The supercritical case; Kesten-Stigum Theorem*

In the supercritical case ($m > 1$), a direct computation shows that if $m < \infty$ and $W_n := Z_n/m^n$, $n \geq 0$, then W is a martingale with respect to the canonical filtration, and thus $0 \leq W := \lim_{n \to \infty} W_n$ is well defined a.s. It is natural to ask whether we 'do not loose mass in the limit,' that is, whether $E(W) = 1$ holds.

A celebrated result of H. Kesten and B. P. Stigum, obtained in 1966, answers this question.

Theorem 1.15 (Kesten and Stigum's Scaling Limit). *Assume that $1 < m < \infty$, and recall that d is the probability of extinction. Then the following are equivalent.*

(1) $P(W = 0) = d$.
(2) $E(W) = 1$.
(3) $W = \lim_{n \to \infty} W_n$ *in* $L^1(P)$.
(4) $E(X \log^+ X) = \sum_{k=1}^{\infty} p_k k \log k < \infty$.

(For the continuous analog of the Kesten-Stigum result, see the next subsection.)

The first condition means that $W > 0$ almost surely on the survival set. The last condition says that the tail of the offspring distribution is 'not too heavy.' This very mild moment-like condition is called the '$X \log X$-condition.'

The Kesten-Stigum result was later reproved in [Lyons, Pemantle and Peres (1995)] in a 'conceptual' way, using a size biasing change of measure, and extended to supercritical branching random walks by J. Biggins. (See [Biggins (1992); Stam (1966); Watanabe (1967)].)

1.13.3   *Exponential branching clock*

Instead of unit time branching, one often considers random branching times with *exponential* distribution. It is left to the reader to check that the memoryless property of the exponential distribution guarantees the Markov property for the branching process Z. In this case the above classification of subcritical, critical and supercritical branching is the same as before. Moreover, if the exponential rate is $\beta > 0$, then the probability of not branching up to time t is obviously $e^{-\beta t}$.

If Z_t is the population size at t, then the continuous analog of the Kesten-Stigum Theorem holds (see Chapter IV, Theorem 2.7 in [Asmussen and Hering (1983)]):

Theorem 1.16 (Scaling limit in the supercritical case). *In the supercritical case, almost surely on the survival set,*

$$\exists \lim_{t \to \infty} e^{-\beta t} Z_t =: N > 0,$$

whenever the offspring number X *satisfies* $E(X \log^+ X) < \infty$. *(Where* $\beta > 0$ *is the exponential branching rate.)*

For the dyadic (precisely two offspring) case, we will later need the distribution of Z_t. The following lemma is well known (see e.g. [Karlin and Taylor (1975)], equation (8.11.5) and the discussion afterwards):

Lemma 1.4 (Yule's process). *Let* Z *under* **P** *be a pure birth process (also called* Yule's *process) with parameter* $\beta > 0$.

(i) If $Z_0 = 1$ *then*

$$\mathbf{P}(Z_t = k) = e^{-\beta t}(1 - e^{-\beta t})^{k-1}, \qquad k \in \mathbb{N},\ t \geq 0.$$

(ii) When $Z_0 = m$ *and* $m \geq 2$, Z_t *is obtained as the independent sum of* m *pure birth processes at* t, *each starting with a single individual. Hence, the time* t *distribution of the pure birth process is always negative binomial.*

In a more general setting, the branching rate changes in time, so that, with some measurable $g : [0, \infty) \to [0, \infty)$, the probability of branching on the time interval $[s, t]$ is $1 - e^{-\int_s^t g(z)\,dz}$, which is approximately $g(t)(t - s)$, if $t - s$ is small. The function g is called the *rate function* and $g(t)$ is called the *instantaneous rate* at t.

1.14 Branching diffusion

Let us now try to combine diffusive motion and branching for a system of particles. Let the operator L satisfy Assumption 1.2 on the non-empty Euclidean domain D. Consider $Y = \{Y_t;\ t \geq 0\}$, the diffusion process with probabilities $\{\mathbb{P}_x,\ x \in D\}$ and expectations $\{\mathbb{E}_x,\ x \in D\}$ corresponding to L on D. At this point, we do not assume that Y is conservative, that is, for $\tau_D := \inf\{t \geq 0 \mid Y_t \notin D\}$, the exit time from D, $\tau_D < \infty$ may hold with positive probability. Intuitively, this means that Y may get killed at the Euclidean boundary of D or 'run out to infinity' in finite time.

1.14.1 *When the branching rate is bounded from above*

Let us first assume[39] that

$$0 \leq \beta \in C^{\eta}(D), \ \sup_D \beta < \infty, \ \beta \not\equiv 0. \tag{1.35}$$

The (strictly dyadic) $(L, \beta; D)$-*branching diffusion* is the Markov process Z with motion component Y and with spatially dependent rate β, replacing particles by precisely two offspring when branching and starting from a single individual. Informally, starting with an initial particle at $x \in D$, it performs a diffusion corresponding to L (with killing at ∂D) and the probability that it does not branch until $t > 0$ given its path $\{Y_s; 0 \leq s \leq t\}$ is $\exp(- \int_0^t \beta(Y_s) \, ds)$. When it does branch, it dies and produces two offspring, each of which follow the same rule, independently of each other and of the parent particle's past, etc. The convention is that at the instant of branching we already have two offspring particles at the same location (right continuity), namely, at the location of the death of their parent. The formal construction of Z is well known (see Section 9.4 in [Ethier and Kurtz (1986)]).

This stochastic process can be considered living on

(1) the space of 'point configurations,' that is, sets which consist of finitely many (not necessarily different) points in D; or
(2) $\mathcal{M}(D)$, the space of finite discrete measures on D.

A discrete point configuration $\{Z_t^1, ..., Z_t^{N_t}\}$ at time $t \geq 0$, is associated with the discrete measure $\sum_1^{N_t} \delta_{Z_t^i} \in \mathcal{M}(D)$, where $N_t = |Z_t|$ is the number of points (with multiplicity) in D at time t. In other words, Z can be viewed as a set-valued as well as a measure-valued process. (But in the set-valued view we allow the repetition of the same point.)

Notation 1.1. Even though we are going to view the process as a measure-valued one, we will write, somewhat sloppily, P_x (instead of the more correct P_{δ_x}) for the probability when Z starts with a single particle at $x \in D$. Since, for $t \geq 0$ given, we consider Z_t as a random discrete measure, we adopt the notation $Z_t(B)$ for $B \subset D$ Borel to denote the mass in B, and use $\langle f, Z_t \rangle$ to denote integral of f against the (random) measure Z_t. The total population size is $\|Z_t\| = Z_t(D)$, but we will sometimes write $|Z_t|$.

---

[39]The smoothness of β is not important at this point. It just makes the use of PDE tools easier.

1.14.2   The branching Markov property

The *branching Markov property* (BMP) for Z is similar to the Markov property for a single particle, but at time $t > 0$ we have a number of particles and the branching trees emanating from them will all contribute to the further evolution of the system. The following result is well known (see e.g. [Asmussen and Hering (1983)]), and it is a consequence of the Markov property of the underlying motion and the memoryless property of the exponential branching clock.

Lemma 1.5 (BMP). *Fix $t \geq 0$ and $B \subset D$ Borel. Conditionally on Z_t, if $\{Z_s^{Z_t^i}, s \geq 0\}, i = 1, ..., N_t$ are independent copies of $Z = \{Z_s, s \geq 0\}$ starting at $Z_t^i, i = 1, ..., N_t$, respectively, then the distribution of $\bigoplus_{i=1}^{N_t} Z_s^{Z_t^i}(B)$ is the same as that of $Z_{t+s}(B)$ under $P_x(\cdot \mid Z_t)$, for $s \geq 0$.*

Let $0 \leq \beta \leq \widehat{\beta}$. An easy coupling argument (left to the reader[40]) shows that if Z and $\widehat{Z}$ are branching diffusions on the same domain D, with the same motion component Y and with branching rates β and $\widehat{\beta}$, respectively, then for any given $t > 0$ and $B \subset D$, the random variable $Z_t(B)$ is stochastically smaller than $\widehat{Z}_t(B)$. That is, if the corresponding probabilities are denoted by $\{P_x, x \in D\}$ and $\{\widehat{P}_x, x \in D\}$, then

$$P_x(Z_t(B) > a) \leq \widehat{P}_x(\widehat{Z}_t(B) > a),$$

for all $x \in D$ and $a \geq 0$.

In particular, compare β with $\widehat{\beta} \equiv \sup_D \beta$. First suppose that the underlying motion Y is conservative. It is then clear that the total population process $t \mapsto |\widehat{Z}_t|$ is just a non-spatial branching process with temporarily constant branching (Yule's process). Thus, since Yule's process is almost surely finite for all times, the same is true for Z. By Theorem 1.15, $e^{-\widehat{\beta}t}\widehat{Z}_t$ tends to a finite nontrivial nonnegative random variable, say N, $\widehat{P}_x$-almost surely as $t \to \infty$. In particular, for any $a \geq 0$,

$$\limsup_{t \to \infty} P_x(e^{-\widehat{\beta}t}|\widehat{Z}_t| \geq a) \leq P_x(N \geq a).$$

Hence, using that N is almost surely finite, along with the comparison between β and $\widehat{\beta}$, it follows that for any function $f : [0, \infty) \to \mathbb{R}$ satisfying $\lim_{t \to \infty} f(t) = \infty$, one has $\lim_{t \to \infty} P_x(|Z_t| \geq f(t)e^{\beta t}) = 0$, that is,

$$|Z_t| \text{ grows at most exponentially in time.} \tag{1.36}$$

---

[40]Hint: Consider Z and attach independent, new trees to it at rate $\widehat{\beta} - \beta$, where the new trees are copies of $\widehat{Z}$, launched at different space-time points.

Then, *a fortiori*, (1.36) is true for a non-conservative motion on D, as killing decreases the population size.

Finally, one may replace the single initial individual with a finite configuration of individuals (or, finite discrete measure) in the following way. Consider independent branching diffusions, emanating from single (not necessarily differently located) individuals $x_1, ..., x_n \in D$ and at every time $t > 0$ take the union of the point configurations (sum of discrete measures) belonging to each of those branching diffusions. If $\mu = \sum_{i=1}^n \delta_{x_i}$, then the corresponding probability will be denoted by P_μ, except for $n = 1$, when we use P_{x_1}.

Then, the integrated form of BMP is as follows.

Proposition 1.13. *Let $\mu \in \mathcal{M}(D)$. For $t, s \geq 0$,*

$$P_\mu(Z_{t+s} \in \cdot) = E_\mu P_{Z_t}(Z_s \in \cdot).$$

A particular case we will study more closely is when $L = \Delta/2$, in which case the branching diffusion is *branching Brownian motion* (BBM).

The next result is sometimes called the *many-to-one formula*. As the name suggests, it enables one to carry out computations concerning functionals of a *single* particle instead of working with the whole system. At least, this is the case when one is only interested in the *expectation* of the process.

Lemma 1.6 (Many-to-One Formula). *Let Y, the diffusion process on $D \subseteq \mathbb{R}^d$, with expectations $\{\mathbb{E}_x\}_{x \in D}$, correspond to the operator L, where L satisfies Assumption 1.2. Let β be as in (1.35). If Z is the $(L, \beta; D)$-branching diffusion, and $f \geq 0$ is a bounded measurable function on D, then*

$$E_x \langle f, Z_t \rangle = \mathbb{E}_x \left(f(Y_t) \exp \left[\int_0^t \beta(Y_s) \mathrm{d}s \right] \mathbf{1}_{\{Y_t \in D\}} \right). \tag{1.37}$$

Remark 1.20 (Slight reformulation). The previous lemma states that if $\{T_t\}_{t \geq 0}$ denotes the semigroup corresponding to the generator $L + \beta$ on D, then $E_x \langle f, Z_t \rangle = (T_t f)(x)$, which, by the Feynman-Kac formula, means that $u(x, t) := E_x \langle f, Z_t \rangle$ is the minimal solution of the parabolic problem:

$$\left. \begin{array}{c} \dot{u} = (L + \beta)\, u \text{ on } D \times (0, \infty), \\[1ex] \lim_{t \downarrow 0} u(\cdot, t) = f(\cdot), \text{ in } D, \\[1ex] u \geq 0. \end{array} \right\} \tag{1.38}$$

Indeed, the Feynman-Kac formula is essentially the fact that the right-hand side of (1.37) is the minimal solution[41] to (1.38) — see Section 2.4 in [Pinsky (1995)].

Sometimes (1.38) is called the Cauchy problem for the *generalized heat equation*. ◇

We now present the proof of Lemma 1.6.

Proof. Let $u(x,t) := E_x\langle f, Z_t \rangle, x \in D, t \geq 0$; then $u(x,0) = f(x)$. By right continuity, the first time of fission, S, is a stopping time with respect to the canonical filtration, and it is exponentially distributed with path dependent rate $\beta(Y)$. Condition on S. Intuitively, by the 'self-similarity' built into the construction, the expected population size is the sum of two other expected population sizes, where those two populations are descending from the two particles created at S. (Rigorously, the strong BMP is used — see Remark 1.22 a little later.) This observation yields the integral equation

$$u(x,t) = 2\mathbb{E}_x \int_0^t u(Y_s, t-s)\beta(Y_s) \exp\left(-\int_0^s \beta(Y_z)\,dz\right)\,ds.$$

By straightforward computation, the function

$$u(x,t) = \mathbb{E}_x \left(f(Y_t) \exp\left[\int_0^t \beta(Y_s)ds\right] \mathbf{1}_{\{Y_t \in D\}}\right)$$

solves this integral equation, and so we just have to show that the solution, with the initial condition f, is unique.

Let v and w be two continuous solutions and $g(t) := \sup_{x \in D} |(v - w)(x,t)|$. Then, since β is bounded from above, $|(v-w)(x,t)| \leq C \int_0^t g(s)\,ds$ for all $x \in D$ and $t \geq 0$, and thus $g(t) \leq C \int_0^t g(s)\,ds$ $(C = 2\sup_D \beta)$. Gronwall's inequality (Lemma 1.1) now implies that $g \equiv 0$, and we are done. □

1.14.3 *Requiring only that $\lambda_c < \infty$*

Recall the notion of the generalized principal eigenvalue (GPE) from Section 1.10. We now relax the condition $\sup_D \beta < \infty$ and replace it by the following, milder assumption.

---

[41]More precisely, in order to obtain the minimal solution, one approximates the domain with smooth compactly embedded subdomains and uses the Feynman-Kac formula for the unique solution with zero Dirichlet condition.

Assumption 1.3 (Assumption GPE). *In addition to the assumptions on L, β and D given in Section 1.10, let us assume that*

$$\lambda_c(L + \beta, D) < \infty. \tag{1.39}$$

We already know that (1.39) is always satisfied when β is bounded from above. In fact, (1.39) is substantially milder than $\sup_D \beta < \infty$. For example, if D is a smooth bounded domain and $L = \Delta/2$, then (1.39) holds as long as β is locally bounded and

$$\beta(x) \leq \frac{1}{8}(\text{dist}(x, \partial D))^{-2},$$

for x near ∂D [Marcus, Mizel and Pinchover (1998)].

On the other hand, it is not hard to find cases when (1.39) breaks down. For example, when L on $\mathbb{R}^d$ has constant coefficients, even a 'slight unboundedness' makes (1.39) impossible, as the following lemma shows.

Lemma 1.7. *Assume that L on $\mathbb{R}^d$ has constant coefficients and that there exists an $\epsilon > 0$ and a sequence $\{x_n\}$ in $\mathbb{R}^d$ such that*

$$\lim_{n \to \infty} \inf_{x \in B(x_n, \epsilon)} \beta(x) = \infty.$$

Then (1.39) does not hold. (Here $B(y, r)$ denotes the open ball of radius r around y.)

Proof.   By the assumption, for every $K > 0$ there exists an $n = n(K) \in \mathbb{N}$ such that $\beta \geq K$ on $B_\epsilon(x_n)$. Let $\lambda^{(\epsilon)}$ denote the principal eigenvalue of L on a ball of radius ϵ. (Since L has constant coefficients, $\lambda^{(\epsilon)}$ is well defined.) Since

$$\lambda_c = \lambda_c(L + \beta, \mathbb{R}^d) \geq \lambda_c(L + \beta, B_\epsilon(x_n)) \geq \lambda^{(\epsilon)} + K,$$

and $K > 0$ was arbitrary, it follows that $\lambda_c = \infty$.   □

For more on the (in)finiteness of the GPE, see section 4.4.5 in [Pinsky (1995)].

We now show that under Assumption GPE the construction of the $(L, \beta; D)$-branching diffusion is still possible. We will see that this is essentially a consequence of the existence of a positive harmonic function, that is, a function $0 < h \in C^2(D)$ with $(L + \beta - \lambda_c)h = 0$ on D. Indeed, using this fact and following an idea of S. C. Harris and A. E. Kyprianou, one constructs the branching process as a limit of certain branching processes, by successively adding branches such that the nth process is an

$(L, \beta^{(n)}; D)$-branching diffusion, where $\beta^{(n)} := \min(\beta, n)$, $n = 1, 2, \ldots$. That is, once the nth branching process $Z^{(n)}$, an $(L, \beta^{(n)}; D)$-branching diffusion, has been defined, we add branches by adding branching points with spatial intensity $\beta^{(n+1)} - \beta^{(n)} \geq 0$, independently on each branch, and by launching $(L, \beta^{(n+1)}; D)$-branching diffusions from those, independently from each other, and from the already constructed tree. Using the fact that the sum of independent Poisson processes is again a Poisson process with the intensities summed up, the resulting process is an $(L, \beta^{(n+1)}; D)$-branching diffusion, as required.

Now, although, by monotonicity, the limiting process Z is clearly well defined, one still has to check possible *explosions*, that is the possibility of having infinitely many particles at a finite time. (For the case when β is bounded from above, this is not an issue, because, as we have seen, one can use comparison with the Yule's process.) We now show how the existence of h will guarantee that the limiting process Z is an almost surely locally finite process, that is, that no 'local explosion' occurs.

To this end, we will need the following lemma.

Lemma 1.8. *Assume that β is bounded from above and that there exists a real number λ and a positive function h satisfying that $(L + \beta - \lambda)h \leq 0$ on D. Then*

$$E_x \langle h, Z_t \rangle \leq e^{\lambda t} h(x),$$

for $x \in D, t \geq 0$.

Proof.   Our argument will rely on the following slight generalization[42] of the many-to-one formula. If $u(x, t) := E_x \langle h, Z_t \rangle$, then u is the minimal nonnegative solution to (1.38), with h in place of f. In other words,

$$u(x, t) = \mathbb{E}_x \left(h(Y_t) e^{\int_0^t \beta(Y_s)\, ds} \mathbf{1}_{\{Y_t \in D\}} \right).$$

(The proof is similar to that of the many-to-one formula. See pp. 154-155 in [Asmussen and Hering (1983)].)

Now the statement of the lemma follows from the minimality of u as follows. Since the function v defined by $v(x, t) := e^{\lambda t} h(x)$ is a non-negative 'super-solution' (that is $(L + \beta - \lambda)v - \partial_t v \leq 0$) to (1.38), with h in place of f, the well-known parabolic maximum principle (see e.g. Chapter 2 in [Friedman (2008)]) yields $u(x, t) \leq v(x, t)$ for all $x \in D$ and $t \geq 0$. (Indeed, by the parabolic maximum principle, the minimal non-negative solution is

---

[42]Here h is not necessarily bounded.

obtained by approximating the domain with smooth compactly embedded subdomains and considering the unique solution with zero Dirichlet condition on each of those subdomains; those solutions will tend to the minimal solution on D, in a monotone non-decreasing way. By the same principle, each of those solutions are bounded from above by v, restricted to the subdomain.) $\qquad\square$

Returning to the problem of no local explosion, denote $N_t^{(n)} := |Z_t^{(n)}|$. As before, $N_t := |Z_t|$, but note that $N_t = \infty$ is not ruled out if β is unbounded from above, although it is clear by construction, that $\mathrm{supp}(Z_t)$ consists of countably many points.

Note that $(L + \beta^{(n)} - \lambda_c)h \le 0$ for $n \ge 1$. Using monotone convergence, it follows that for any fixed $t \ge 0, x \in D$ and $B \subset\subset D$, if h is chosen so that $\min_B h \ge 1$, then

$$E_x Z_t(B) \le E_x \sum_{i \le N_t} h(Z_t^i) = \lim_{n \to \infty} E_x^{(n)} \sum_{i \le N_t^{(n)}} h(Z_t^{n,i}) \le e^{\lambda t} h(x), \quad (1.40)$$

where $Z_t^{(n)} = \sum_{i \le N_t} \delta_{Z_t^{n,i}}$ with corresponding expectations $\{E_x^{(n)}; x \in D\}$, and the last inequality follows from Lemma 1.8. Fix $T > 0$ and consider the time integral (occupation measure) $\int_0^T Z_t(B)\,dt$. By Fubini,

$$E_x \int_0^T Z_t(B)\,dt = \int_0^T E_x Z_t(B)\,dt \le \frac{e^{\lambda T} - 1}{\lambda} \cdot h(x).$$

Thus, *the occupation measure has finite expectation* for any $B \subset\subset D$. In particular, $P_x \left(\int_0^T Z_t(B)\,dt < \infty \right) = 1$, and so if $I_B := \{t > 0 : Z_t(B) = \infty\}$, then $P_x (|I_B| = 0) = 1$. By monotonicity this is even true simultaneously for every $B \subset\subset D$.

If we want to show that actually, $P_x(Z_t(B) < \infty,\ \forall t > 0) = 1$, then we can argue as follows. Clearly, it is enough to show that

$$P_x(Z_t(B) < \infty, 0 < t \le 1) = 1. \quad (1.41)$$

Fix a $\widehat{B}$ Borel set with $B \subset\subset \widehat{B} \subset\subset D$. Now (1.41) follows from the following fact (which, in turn follows from basic properties of the underlying L-diffusion, and by ignoring the branching after t).

$$\forall 0 < t \le 1 : \inf_{y \in B} P_y \left(Z_{2-t}(\widehat{B}) \ge 1 \right) > 0. \quad (1.42)$$

Indeed, first, by (a version of the) Borel-Cantelli lemma, (1.42) implies that

$$\forall 0 < t \le 1 : P_x \left(Z_2(\widehat{B}) = \infty \mid Z_t(B) = \infty \right) = 1. \quad (1.43)$$

Now suppose that (1.41) is not true, that is, $P_x(A) > 0$, where

$$A := \{\omega \in \Omega : \exists 0 < t \leq 1 \text{ s.t. } Z_t(\omega, B) = \infty\}.$$

Then, for a.e. $\omega \in A$, say, on A', the set $K(\omega) := \{0 < t \leq 1 : Z_t(\omega, B) = \infty\}$ is a Lebesgue measurable subset of $(0, 1]$. For each $\omega \in A'$ we define the random time $T(\omega, \omega')$ by picking a random point of $K(\omega)$, independently of everything else, using some arbitrary distribution $Q = Q^{(\omega)}$ on $K(\omega)$; for $\omega \notin A'$, set $T^{(\omega)} \equiv \infty$.

That is, on A', we have picked a random time $T \in (0, 1]$, when B contains infinitely many particles.[43] Then, by conditioning on $T = t$, (1.43) leads to $P_x\left(Z_2(\widehat{B}) = \infty \mid A\right) = 1$; a contradiction. (To be precise, we are using the fact that, after conditioning, the spatial motion of the particles after t, is still an L-diffusion, that is, conditioning on having infinitely many particles in B at t, does not change the statistics for their future motion. This follows from the Markov property of diffusions.)

Remark 1.21 (Harmonic supermartingale). Notice that the argument yielding (1.40) also shows that the process W^h defined by $W_t^h := e^{-\lambda_c t}\langle h, Z_t \rangle = e^{-\lambda_c t} \sum_{i \leq N_t} h(Z_t^i)$, $t \geq 0$, is a supermartingale. In particular, since it is nonnegative and right-continuous, it has an almost sure limit as $t \to \infty$. ◇

1.14.4 *The branching Markov property; general case*

We now check that the branching Markov property remains valid when we only assume the finiteness of λ_c. When reading the following statement, keep in mind that $|Z_t| = N_t = \infty$ is possible.

Claim 1.1 (BMP; general case). *Fix $t \geq 0$ and $B \subset\subset D$. Conditionally on Z_t, if $\{Z_s^{Z_t^i}, s \geq 0\}, 1 \leq i \leq N_t$ are independent copies of $Z = \{Z_s, s \geq 0\}$ starting at $Z_t^i, 1 \leq i \leq N_t$, respectively, then the distribution of $\bigoplus_{i=1}^{N_t} Z_s^{Z_t^i}(B)$ is the same as that of $Z_{t+s}(B)$ under $P_x(\cdot \mid Z_t)$, for $s \geq 0$.*

Proof. Recall the recursive construction: once the nth branching process has been defined, we add branches by adding branching points with spatial

---

[43]Technically, restricted to A', T is a finite random variable on $\{(\omega, \omega') : \omega \in A', \omega' \in K(\omega)\}$ such that $T(\omega, \omega') = \omega'$. Its distribution is a mixture of the $Q^{(\omega)}$ distributions, according to P conditioned on A'.

intensity $\beta^{(n+1)} - \beta^{(n)} \geq 0$, and by launching independent $(L, \beta^{(n+1)}; D)$-branching diffusions from those. Thus Z_t is defined by adding more and more points (particles) for getting each $Z_t^{(n)}$, and then taking the union of all those points; Z_{t+s} is defined in a similar manner.

Now a given point in Z_{t+s} was added at some step, say n. But then, it has an ancestor at time t. Even though this ancestor might have been added at a step earlier than n, we can consider her just as well as the one who generated the given point time s later by emanating a branching tree with rate $\beta^{(n)}$. This should be clear by recalling from the construction that by successively adding branches, the nth process is precisely an $(L, \beta^{(n)}; D)$-branching diffusion.

In the rest of the proof let us consider Z_t and $Z_t^{(n)}$ as set of points. Let the random set $M_{Z_t^{(n)}}(s)$ $(M_{Z_t}(s))$ denote the descendants, time s later, of the points in $Z_t^{(n)}$ (Z_t), keeping in mind that each line of descent is a result of branching at rate $\beta^{(n)}$. Then the argument in the previous paragraph shows that, conditionally on Z_t,

$$Z_{t+s} \subset \bigcup_{n \geq 1} M_{Z_t^{(n)}}(s).$$

On the other hand,

$$\bigcup_{n \geq 1} M_{Z_t^{(n)}}(s) \subset Z_{t+s}$$

is obvious, because $M_{Z_t^{(n)}}(s)$ will be included in Z_{t+s} during the nth step of the construction of Z. Since, again, by construction, $Z_t = \bigcup_{n \geq 1} Z_t^{(n)}$, one has $Z_{t+s} = M_{Z_t}(s)$.   □

Remark 1.22 (Strong Branching Markov Property). The branching Markov property (Property BMP) can actually be strengthened to *strong* branching Markov property, that is, t in Property BMP can be replaced by any nonnegative stopping time τ with respect to the canonical filtration of the process. See [Asmussen and Hering (1983)], Chapter V, Sections 1–2.   ◇

1.14.5   *Further properties*

Clearly, $t \mapsto Z_t(B)$ is measurable whenever $B \subset\subset D$, since $Z_t(B) = \sup_n Z_t^{(n)}(B)$. Furthermore, the paths of Z are almost surely right-continuous in the vague topology of measures, that is, the following holds with probability one: if $f \in C_c(D)$ and $t \geq 0$, then $\langle f, Z_{t+\epsilon} \rangle \to \langle f, Z_t \rangle$ as $\epsilon \downarrow 0$.

This follows from the fact that $P_x(Z_t(B) < \infty, \ \forall t > 0) = 1$ for all $x \in D$, and from the fact that by (1.9), $\mathbb{E}_{x'} f(Y_s) \to f(x')$ as $s \downarrow 0$ for all $x' \in D$. (Keeping in mind that the nth tree in the construction has a.s. right continuous paths.)

Remark 1.23 (General Many-to-One Formula). The useful many-to-one formula remains valid in this more general setting as well. Indeed, it is valid for $E_x \langle f, Z_t^{(n)} \rangle$ for all $n \geq 1$ (where $Z_t^{(n)}$ is as in the previous proof and $f \geq 0$), that is (1.37) is valid when β is replaced by $\beta^{(n)}$ and Z_t by $Z_t^{(n)}$. Now apply the Monotone Convergence Theorem on both sides of the equation.   ◇

1.14.6   *Local extinction*

The following notion is very important. Intuitively, *local extinction* means that the particle configuration leaves any compactly embedded domain in some finite (random) time, never charging it again. We stress that such a random time cannot be defined as a stopping time.

Recall that $B \subset\subset D$ means that B is bounded and $\overline{B} \subset D$.

Definition 1.13 (Local extinction). *Fix $\mu \in \mathcal{M}(D)$. We say that Z exhibits local extinction under P_μ if for every Borel set $B \subset\subset D$, there exists a random time τ_B such that*

$$P_\mu(\tau_B < \infty, \text{ and } Z_t(B) = 0 \text{ for all } t \geq \tau_B) = 1.$$

1.14.7   *Four useful results on branching diffusions*

In this section we prove a number of useful facts about the dyadic branching Brownian motion and about general branching diffusions.

Let Z be a dyadic (always precisely two offspring) branching Brownian motion in $\mathbb{R}^d$, with constant branching rate $\beta > 0$, and let $|Z_t|$ denote the total number of particles at time t. Assume that Z starts at the origin with a single particle, and let **P** denote the corresponding probability.

The first result says that 'overproduction' is super-exponentially unlikely.

Proposition 1.14 (Overproduction). *Let $\delta > 0$. Then*

$$\lim_{t \to \infty} \frac{1}{t} \log \mathbf{P}\left(|Z_t| > e^{(\beta + \delta)t}\right) = -\infty. \tag{1.44}$$

Proof. Since $|Z_t|$ under $\mathbf{P}$ is a pure birth process (Yule's process) with $|Z_0| = 1$, we have, by Lemma 1.4, that

$$\mathbf{P}(|Z_t| = k) = e^{-\beta t}(1 - e^{-\beta t})^{k-1}, \qquad k \in \mathbb{N}, \ t \geq 0. \tag{1.45}$$

Hence

$$\mathbf{P}(|Z_t| > l) = (1 - e^{-\beta t})^l, \qquad l \in \mathbb{N}, \ t \geq 0, \tag{1.46}$$

giving (1.44). □

Next, for $B \subset \mathbb{R}^d$ open or closed, let η_B and $\hat{\eta}_B$ denote the first exit times from B for one Brownian motion W, resp. for the BBM Z, that is, the $(\frac{1}{2}\Delta, \beta, \mathbb{R}^d)$-branching diffusion, with constant $\beta > 0$:

$$\eta_B = \inf\{t \geq 0 \colon W_t \in B^c\},$$
$$\hat{\eta}_B = \inf\{t \geq 0 \colon Z_t(B^c) \geq 1\}. \tag{1.47}$$

The following result makes a comparison between these two quantities.

Proposition 1.15. *Let P_x denote the law of Brownian motion starting at x, and $\mathbf{P}_x$ the law of Z, starting at δ_x. For any $B \subset \mathbb{R}^d$ open or closed and any $x \in B$,*

$$\mathbf{P}_x\left(\hat{\eta}_B > t \mid |Z_t| \leq k\right) \geq [P_x(\eta_B > t)]^k, \qquad k \in \mathbb{N}, t \geq 0. \tag{1.48}$$

Proof. By an obvious monotonicity argument, it is enough to show that

$$\mathbf{P}_x\left(\hat{\eta}_B > t \mid |Z_t| = k\right) \geq [P_x(\eta_B > t)]^k, \qquad k \in \mathbb{N}, t \geq 0. \tag{1.49}$$

We will prove this inequality by induction on k. The statement is obviously true for $k = 1$. Assume that the statement is true for $1, 2, \ldots, k - 1$. Let σ_1 be the first branching time:

$$\sigma_1 = \inf\{t \geq 0 \colon |Z_t| \geq 2\}. \tag{1.50}$$

By the strong BMP, it suffices to prove the assertion conditioned on the event $\{\sigma_1 = s\}$ with $0 \leq s \leq t$ fixed. To that end, let $p_{x,s} = P_x(\eta_B > s)$ and

$$\tilde{p}(s, x, dy) = P_x\left(W_s \in dy \mid \eta_B > s\right), \tag{1.51}$$

where W is standard Brownian motion. By the strong BMP, after time s the BBM evolves like two independent BBM's Z^1, Z^2 starting from Z_s. For $i = 1, \ldots, k-1$ and $y \in \mathbb{R}^d$, let

$$q_{i,k}(s, t-s) = \mathbf{P}_y\Big(|Z^1(t-s)| = i, |Z^2(t-s)| = k-i$$

$$\big|\, |Z_t| = k, \sigma_1 = s\Big) \qquad (1.52)$$

(which does not depend on y). Write $\hat{\eta}_B^1, \hat{\eta}_B^2$ to denote the analogues of $\hat{\eta}_B$ for Z^1, Z^2. Then

$$\mathbf{P}_x\Big(\hat{\eta}_B > t \mid |Z_t| = k, \sigma_1 = s\Big)$$

$$= \int_B P_x(\eta_B > s, W_s \in dy)$$

$$\times \mathbf{P}_x\Big(\hat{\eta}_B^1 > t-s, \hat{\eta}_B^2 > t-s \mid |Z_t| = k, \sigma_1 = s\Big)$$

$$= p_{x,s} \int_B \tilde{p}(s, x, dy) \sum_{i=1}^{k-1} q_{i,k}(s, t-s)$$

$$\times \mathbf{P}_y\Big(\hat{\eta}_B^1 > t-s \mid |Z^1(t-s)| = i\Big)$$

$$\times \mathbf{P}_y\Big(\hat{\eta}_B^2 > t-s \mid |Z^2(t-s)| = k-i\Big)$$

$$\geq p_{x,s} \int_B \tilde{p}(s, x, dy) \sum_{i=1}^{k-1} q_{i,k}(s, t-s)$$

$$\times [P_y(\eta_B > t-s)]^i [P_y(\eta_B > t-s)]^{k-i}$$

$$= p_{x,s} \int_B \tilde{p}(s, x, dy)[P_y(\eta_B > t-s)]^k$$

$$\geq p_{x,s} \left[\int_B P_y(\eta_B > t-s)\tilde{p}(s, x, dy)\right]^k, \qquad (1.53)$$

where we use the induction hypothesis and Jensen's inequality. Replacing $p_{x,s}$ by $(p_{x,s})^k$, we obtain

$$\mathbf{P}_x\Big(\hat{\eta}_B > t \mid |Z_t| = k, \sigma_1 = s\Big)$$

$$\geq \left[p_{x,s} \int_B \tilde{p}(s, x, dy) P_y(\eta_B > t-s)\right]^k. \qquad (1.54)$$

By the Markov property of Brownian motion, the right-hand side precisely equals $[P_x(\eta_B > t)]^k$, giving (1.48). $\qquad \square$

Let $R_t := \cup_{s \in [0,t]} \mathrm{supp}(Z_s)$ be the range of Z up to time t. Let

$$M^+(t) := \sup R_t \qquad \text{for } d = 1,$$
$$M^-(t) := \inf R_t \qquad \text{for } d = 1,$$
$$M(t) := \inf\{r > 0 : R_t \subseteq B_r(0)\} \text{ for } d \geq 1,$$

be the right-most and left-most point of R_t (resp. the radius of the minimal ball containing R_t). The following result identifies the typical behavior of these quantities as $t \to \infty$.

Proposition 1.16. *(i) For $d = 1$, $M^+(t)/t$ and $-M^-(t)/t$ converge to $\sqrt{2\beta}$ in $\mathbf{P}_0$-probability as $t \to \infty$.*
(ii) For $d \geq 1$, $M(t)/t$ converges to $\sqrt{2\beta}$ in $\mathbf{P}_0$-probability as $t \to \infty$.

We note that almost sure speed results exist too (see e.g. [Kyprianou (2005)] for a proof with martingale techniques), but for our purposes, convergence in probability suffices.

Proof. For (i), the reader is referred to the articles [McKean (1975, 1976)]; see also [Freidlin (1985)], Section 5.5 and equation (6.3.12).

Turning to (ii), first note that the projection of Z onto the first coordinate axis is a one-dimensional BBM with branching rate β. Hence, the lower estimate for (ii) follows from (i) and the inequality

$$\mathbf{P}_0\Big(M(t)/t > \sqrt{2\beta} - \varepsilon\Big)$$
$$\geq \mathbf{P}_0^*\Big(M^+(t)/t > \sqrt{2\beta} - \varepsilon\Big) \qquad \forall \varepsilon > 0, t > 0, \qquad (1.55)$$

where $\mathbf{P}_0^*$ denotes the law of the one-dimensional projection of Z. To prove the upper estimate for (ii), pick any $\varepsilon > 0$, abbreviate $B = B_{(\sqrt{2\beta}+\varepsilon)t}(0)$, and pick any $\delta > 0$ such that

$$\frac{1}{2}(\sqrt{2\beta} + \varepsilon)^2 > \beta + \delta. \qquad (1.56)$$

Estimate (recall (1.47))

$$\mathbf{P}_0\Big(M(t)/t > \sqrt{2\beta} + \varepsilon\Big)$$
$$\leq \mathbf{P}_0\Big(|Z_t| > \lfloor e^{(\beta+\delta)t} \rfloor\Big)$$
$$+ \mathbf{P}_0\Big(\hat{\eta}_B \leq t \mid |Z_t| \leq \lfloor e^{(\beta+\delta)t} \rfloor\Big). \qquad (1.57)$$

By Proposition 1.14, the first term on the right-hand side of (1.57) tends to zero super-exponentially fast. To handle the second term, we use Proposition 1.15 to estimate

$$\mathbf{P}_0\Big(\hat{\eta}_B > t \mid |Z_t| \leq \lfloor e^{(\beta+\delta)t} \rfloor\Big) \geq [P_0(\eta_B > t)]^{\lfloor e^{(\beta+\delta)t} \rfloor}. \qquad (1.58)$$

Using the large deviation result concerning linear distances (Lemma 1.3), we find that for t large enough:

$$\mathbf{P}_0\Big(\hat{\eta}_B > t \mid |Z_t| \le \lfloor e^{(\beta+\delta)t} \rfloor\Big)$$

$$\ge \left[1 - \exp\left(-\frac{[(\sqrt{2\beta}+\varepsilon)t]^2}{2t}[1+o(1)] \right) \right]^{\lfloor e^{(\beta+\delta)t} \rfloor} . \quad (1.59)$$

By (1.56), the right-hand side of (1.59) tends to 1 exponentially fast as $t \to \infty$, so that (1.57) yields

$$\lim_{t\to\infty} \mathbf{P}_0\Big(M(t)/t > \sqrt{2\beta} + \varepsilon\Big) = 0 \qquad \forall \varepsilon > 0, \quad (1.60)$$

which completes the proof. $\qquad\square$

In Proposition 1.16, (i) is stronger than (ii) for $d=1$, since it says that the BBM reaches both ends of the interval $[-\sqrt{2\beta}t, \sqrt{2\beta}t]$.

Our final result concerning branching diffusions will be a consequence of this abstract lemma:

Lemma 1.9. *Given the probability triple $(\Omega, \mathcal{F}, P)$, let $A_1, A_2, ..., A_N \in \mathcal{F}$ be events that are positively correlated in the following sense. If $k \le N$ and $\{j_1, j_2, ..., j_k\} \subseteq \{1, 2, ..., N\}$ then $\mathrm{cov}\left(\mathbf{1}_{A_{j_1} \cap A_{j_2} \cap A_{j_3} ... \cap A_{j_{k-1}}}, \mathbf{1}_{A_{j_k}}\right) \ge 0$. Then*

$$P\left(\bigcap_{i=1}^N A_i\right) \ge \prod_{i=1}^N P(A_i).$$

Proof. We use induction on N. Let $N = 2$. Then $P(A_1 \cap A_2) \ge P(A_1)P(A_2)$ is tantamount to $\mathrm{cov}(\mathbf{1}_{A_1}, \mathbf{1}_{A_2}) \ge 0$.

If $N+1$ events are positively correlated in the above sense then any subset of them is positively correlated as well. Given that the statement is true for $N \ge 2$, one has

$$P\left(\bigcap_{i=1}^{N+1} A_i\right) \ge P\left(\bigcap_{i=1}^N A_i\right) P(A_{N+1}) \ge P(A_{N+1}) \prod_{i=1}^N P(A_i) = \prod_{i=1}^{N+1} P(A_i),$$

and so the statement is true for $N+1$. $\qquad\square$

Corollary 1.1. *Consider Z, the $(L, \beta; \mathbb{R}^d)$-branching diffusion where L satisfies Assumption 1.2, and corresponds to the diffusion process Y on $\mathbb{R}^d$, and the branching rate $\beta = \beta(\cdot) \ge 0$ is not identically zero. For $t > 0$ let N_t denote the number of particles at t, and G_t an open set containing* *the origin. Denote the probabilities for* Y *by* $\{Q_x,\ x \in \mathbb{R}^d\}$, *and for* Z *by* $\{\mathbf{P}_x,\ x \in \mathbb{R}^d\}$.

Finally, let the function $g : \mathbb{R}_+ \to \mathbb{N}_+$ *be so large that* $\lim_{t\to\infty} P(N_t \le g(t)) = 1$. *Then, as* $t \to \infty$, *the lower estimate*

$$\mathbf{P}_0\left[\operatorname{supp}(Z_s) \in G_t,\ 0 \le s \le t\right] \ge [Q_0(Y_s \in G_t,\ 0 \le s \le t)]^{g(t)} - o(1)$$

holds.

Proof. As usual, let us label the particles in a way that does not depend on their motion. We get N_t (correlated) trajectories of Y: $Y^{(i)}$, $1 \le i \le N_t$. Denote $A_i := (Y_s^{(i)} \in B_t,\ 0 \le s \le t)$. When $N_t < g(t)$, consider some additional (positively correlated) 'imaginary' particles — for example by taking $g(t) - N_t$ extra copies of the first particle. We have

$$P_0[\operatorname{supp}(Z_s) \in B_t,\ 0 \le s \le t] = P_0\left(\bigcap_{i=1}^{N_t} A_i\right) \ge P_0\left(\bigcap_{i=1}^{g(t)} A_i \cap \{N_t \le g(t)\}\right)$$

$$\ge P_0\left(\bigcap_{i=1}^{g(t)} A_i\right) - P_0\left(N_t > g(t)\right).$$

It is easy to check that $A_1,\ A_2, ..., A_{g(t)}$ are positively correlated, hence, by Lemma 1.9, one can continue the lower estimate with

$$\ge \prod_{i=1}^{g(t)} P(A_i) - o(1) = [\mathbf{Q}_0(Y_s \in B_t,\ 0 \le s \le t)]^{g(t)} - o(1),$$

completing the proof. $\qquad\qquad\qquad\qquad\qquad\qquad\qquad\qquad\qquad\qquad\square$

1.14.8 *Some more classes of elliptic operators/branching diffusions*

Let Z be an $(L, \beta; D)$-branching diffusion. Assuming product-criticality for $L + \beta$, we now define the classes $\mathcal{P}_p(D)$ and $\mathcal{P}_p^*(D)$. We will want to talk about spatial spread on a generic domain D, and so we fix an arbitrary family of domains $\{D_t,\ t \ge 0\}$ with $D_t \subset\subset D$, $D_t \uparrow D$. (For $D = \mathbb{R}^d$, D_t can be the t-ball, but we can take any other family with $D_t \subset\subset D$, $D_t \uparrow D$ too.)

Recall Definition 1.11, and consider the following one.

Definition 1.14. Assuming that $L + \beta$ is product-critical on D, for $p \ge 1$, we write $L + \beta \in \mathcal{P}_p(D)$ if

(i) $\lambda_c = \lambda_c(L + \beta, D) > 0$,

(ii) $\langle \phi^p, \widetilde{\phi} \rangle < \infty$, in which case we say that $L + \beta - \lambda_c$ is *product p-critical*.

Let $q(x, y, t)$ be transition density of $L + \beta$ and

$$Q(x, y, t) := q(x, y, t) - e^{\lambda_c t} \widetilde{\phi}(y)\phi(x).$$

We write $L + \beta \in \mathcal{P}_p^*(D)$ when the following additional conditions hold for each given $x \in D$ and $\emptyset \neq B \subset\subset D$.

(iii) There exists a function $a : [0, \infty) \to [0, \infty)$ such that for all $\delta > 0$,

$$P_x \left(\exists n_0, \forall n > n_0 \; : \; \mathrm{supp}(Z_{n\delta}) \subset D_{a_{n\delta}} \right) = 1.$$

(iv) There exists a function $\zeta : [0, \infty) \to [0, \infty)$ such that, as $t \uparrow \infty$,

    (1) $\zeta(t) \uparrow \infty$,

    (2) $\zeta(a_t) = \mathcal{O}(t)$,

    (3)

$$\alpha_t := \sup_{z \in D_t, y \in B} \frac{|Q(z, y, \zeta(t))|}{\widetilde{\phi}(y)\phi(z)} = o(e^{\lambda_c t}).$$

Let $p(x, y, t)$ denote the transition density of the diffusion corresponding to the operator $(L + \beta - \lambda_c)^\phi$. Then $p(x, y, t) = e^{-\lambda_c t}\phi(y)\phi^{-1}(x)q(x, y, t)$, and thus, (iv) is equivalent to

(iv*) With the same ζ as in (iv),

$$\lim_{t \to \infty} \sup_{z \in D_t, y \in B} \left| \frac{p(z, y, \zeta(t))}{\phi\widetilde{\phi}(y)} - 1 \right| = 0.$$

Although a depends on x and ζ, α depend on x and B through (2) and (3), we will suppress this dependency, because in the proofs we will not need uniformity in x and B. As a matter of fact, ζ and α often *do not depend* on x or B, as will be demonstrated in the examples of the third chapter, where explicit cases of these quantities are discussed.

1.14.9 *Ergodicity*

Recall that criticality is invariant under h-transforms. Moreover, an easy computation shows that ϕ and $\widetilde{\phi}$ transform into 1 and $\phi\widetilde{\phi}$ respectively when turning from $(L + \beta - \lambda_c)$ to the h-transformed ($h = \phi$) operator $(L + \beta - \lambda_c)^\phi = L + a\phi^{-1}\nabla\phi \cdot \nabla$. Therefore product-criticality is invariant under h-transforms too (this is not the case with product p-criticality when $p > 1$).

Recall next, that for operators with no zeroth order term, product-criticality is equivalent to the positive recurrence (ergodicity) of the corresponding diffusion process. In particular then, by h-transform invariance, $(L + \beta - \lambda_c)^\phi$ corresponds to an ergodic diffusion process Y, provided $(L + \beta - \lambda_c)$ is product-critical, and the invariant density is $\Phi := \phi\tilde{\phi}$. (Recall that by our choice, $\langle \Phi, 1 \rangle = 1$.) See [Pinsky (1995)], Section 4.9 for more on the topic.

The following statement[44] will be important in the next chapter.

Lemma 1.10 (Ergodicity). *Let $x \in D$. With the setting above, assuming product-criticality for $L + \beta - \lambda_c$, one has $\lim_{t \to \infty} \mathbb{E}_x^\phi(f(Y_t)) = \langle f, \Phi \rangle$ for every $f \in L^1(\Phi(x)\,\mathrm{d}x)$.*

Proof.   Let the transition density for Y be $p(x, y, t)$. By Theorem 1.3 (ii) in [Pinchover (2013)], $\lim_{t \to \infty} p(x, y, t) = \Phi(y)$. The following crucial inequality can be found in the same paper, right after Lemma 2.5 and in (3.29-30) in [Pinchover (1992)]: There exists a function c on D such that

$$p(x, y, t) \leq c(x)\Phi(y) \quad \text{for all } x, y \in D \text{ and } t > 1.$$

Hence, by dominated convergence, as $t \to \infty$,

$$\mathbb{E}_x^\phi(f(Y_t)) = \int_D f(y)p(x, y, t)\,\mathrm{d}y \to \int_D f(y)\Phi(y)\,\mathrm{d}y. \qquad \square$$

1.15   Super-Brownian motion and superdiffusions

Just like Brownian motion *super-Brownian motion* also serves as a building block in stochastic analysis. And just like Brownian motion is a particular case of the more general concept of diffusion processes, super-Brownian motion is a particular *superdiffusion*. Superdiffusions are measure-valued Markov processes, but here, unlike for branching diffusions, the values taken by the process for $t > 0$ are no longer discrete measures. Intuitively, such a process describes the evolution of a random cloud, or random mass distributed in space, moving and creating more mass at some regions while annihilating mass at some others.

We now give two definitions for superdiffusions:

---

[44]It is well known for bounded functions. The point is that we consider $L^1(\Phi(x)\,\mathrm{d}x)$-functions here.

(1) as measure-valued Markov processes via their Laplace functionals,
(2) as scaling limits of branching diffusions.

We start with the first approach.

1.15.1   *Superprocess via its Laplace functional*

Let $\emptyset \neq D \subseteq \mathbb{R}^d$ be a domain, and let L on D satisfy Assumption 1.2. In addition, let $\alpha, \beta \in C^\eta(D)$, and assume that α is positive, and β is bounded from above.[45]

As usual, write $\mathcal{M}_f(D)$ and $\mathcal{M}_c(D)$ for the class of finite measures (with the weak topology) resp. the class of finite measures with compact support on D; the spaces $C_b^+(D)$ and $C_c^+(D)$ are the spaces of non-negative bounded continuous resp. non-negative continuous functions $D \to \mathbb{R}$, having compact support.

To see that the following definition makes sense, see [Engländer and Pinsky (1999)].

Definition 1.15 (($L, \beta, \alpha; D$)-**superdiffusion**). With D, L, β and α as above, $(X, \mathbf{P}_\mu, \ \mu \in \mathcal{M}_f(D))$ will denote the $(L, \beta, \alpha; D)$-*superdiffusion*, where μ denotes the starting measure X_0. What we mean by this is that X is the unique $\mathcal{M}_f(D)$-valued continuous (time-homogeneous) Markov process which satisfies, for any $g \in C_b^+(D)$, that

$$\mathbf{E}_\mu \exp \langle -g, X_t \rangle = \exp \langle -u(\cdot, t), \mu \rangle, \tag{1.61}$$

where u is the minimal nonnegative solution to

$$\left.\begin{array}{l} u_t = Lu + \beta u - \alpha u^2 \quad on \ D \times (0, \infty), \\[2mm] \lim_{t \downarrow 0} u(\cdot, t) = g(\cdot). \end{array}\right\} \tag{1.62}$$

Remark 1.24. The fact that one can pick a version of the process with continuous paths is, of course, a highly non-trivial issue, similarly to the continuity of Brownian paths. Here continuity is meant in the weak topology of measures.   ◇

The equation (1.61) is called the *log-Laplace equation*, while (1.62) is called the *cumulant equation*. Often (1.62) is written in the form of an integral equation.

---

[45]The boundedness of β from above can in fact be significantly relaxed in the construction of superdiffusions. See [Engländer and Pinsky (1999)].

Definition 1.16. One usually refers to β as *mass creation* and α as the *intensity parameter* (or variance).

The Markov property is in fact equivalent to the property that the 'time shift' defined by (1.62) defines a semigroup, which in turn follows from the minimality of the solution (see [Engländer and Pinsky (1999)]). The branching property is captured by the equations

$$\mathbf{E}_{\mu+\nu} \exp \langle -g, X_t \rangle = \mathbf{E}_\mu \exp \langle -g, X_t \rangle \cdot \mathbf{E}_\nu \exp \langle -g, X_t \rangle, \ \mu, \nu \in \mathcal{M}_f(D),$$

and

$$\log \mathbf{E}_\mu \exp \langle -g, X_t \rangle = \langle \log \mathbf{E}_{\delta_x} \exp \langle -g, X_t \rangle, \mu(\mathrm{d}x) \rangle, \ \mu \in \mathcal{M}_f(D),$$

which are consequences of (1.61) and (1.62).

The 'Many-to-One formula' (which we have encountered for branching diffusions) now takes the form

$$\mathbf{E}_{\delta_x} \langle f, X_t \rangle = E_x \left(f(Y_t) \exp \left[\int_0^t \beta(Y_s) \mathrm{d}s \right] 1_{\{Y_t \in D\}} \right), \ x \in D, \quad (1.63)$$

where Y is the underlying diffusion process on D, under the probability P_x, corresponding to L, and $f \geq 0$ is a bounded measurable function on D.

Remark 1.25. At this point the reader should have noticed the remarkable fact that the expectations of $\langle f, X_t \rangle$ and $\langle f, Z_t \rangle$ agree, when started from the same point measure. (Here Z is the $(L, \beta; D)$-branching diffusion.) $\diamond$

1.15.2 *The particle picture for the superprocess*

Previously we defined the $(L, \beta, \alpha; D)$-superprocess X analytically, through its Laplace-functional. In fact, X also arises as the short lifetime and high density diffusion limit of a *branching particle system*, which can be described as follows: in the nth approximation step each particle has mass $1/n$ and lives a random time which is exponential with mean $1/n$. While a particle is alive, its motion is described by a diffusion process corresponding to the operator L (on $\widehat{D} = D \cup \{\Delta\}$). At the end of its life, the particle located at $x \in D$ dies and is replaced by a random number of particles situated at the parent particle's final position. The law of the number of descendants is spatially varying such that the mean number of descendants is $1 + \frac{\beta(x)}{n}$, while the variance is assumed to be $2\alpha(x)$. All these mechanisms are independent of each other.

More precisely, for each positive integer n, consider N_n particles, each of mass $\frac{1}{n}$, starting at points $x_i^{(n)} \in D, i = 1, 2, \ldots, N_n$, and performing independent branching diffusion according to the motion process Y, with branching rate $cn, c > 0$, and branching distribution $\{p_k^{(n)}(x)\}_{k=0}^{\infty}$, where the expectation at level n is 'close to critical' in the sense that

$$e_n(x) := \sum_{k=0}^{\infty} k p_k^{(n)}(x) = 1 + \frac{\gamma(x)}{n},$$

and

$$v_n^2(x) := \sum_{k=0}^{\infty} (k-1)^2 p_k^{(n)}(x) = m(x) + o(1)$$

as $n \to \infty$, uniformly in x; $m, \gamma \in C^{\eta}(D), \eta \in (0,1]$ and $m(x) > 0$. Let

$$\mu_n := \frac{1}{n} \sum_{i=1}^{N_n} \delta_{x_i^{(n)}}.$$

Let $N_n(t)$ denote the number of particles alive at time $t \geq 0$ and denote their positions by $\{Z_t^{(i,n)}\}_{i=1}^{N_n(t)}$. Denote by $\mathcal{M}_f(D)$ $(\mathcal{M}_f(\widehat{D}))$ the space of finite measures on D $(\widehat{D})$. Define an $\mathcal{M}_f(\widehat{D})$-valued 'weighted branching diffusion' $Z^{(n)}$ by

$$Z_t^{(n)} := \frac{1}{n} \sum_{i=1}^{N_n(t)} \delta_{Z_t^{(i,n)}}, \ t \geq 0.$$

Denote by $P_{\mu_n}^{(n)}$ the probability measure on the Skorokhod-space[46] $D([0,\infty), \mathcal{M}_f(\widehat{D}))$, induced by $Z^{(n)}$. Assume that m and γ are bounded from above. Then the following hold.

Proposition 1.17.

(i) Let $\mu_n \overset{w}{\Rightarrow} \mu \in \mathcal{M}_f(D)$ as $n \to \infty$. Then there exists a law P_{μ}^* such that

$$P_{\mu_n}^{(n)} \overset{w}{\Rightarrow} P_{\mu}^* \text{ on } D\left([0,\infty), \mathcal{M}_f(\widehat{D})\right). \tag{1.64}$$

(ii) The P_{μ}^*-outer measure of $C([0,\infty), \mathcal{M}_f(\widehat{D}))$ is one and thus P_{μ}^* can be transferred to $C([0,\infty), \mathcal{M}_f(\widehat{D}))$.

(iii) Define P_{μ} on $C([0,\infty), \mathcal{M}_f(D))$ by $P_{\mu}(\cdot) = P_{\mu}^*(\cdot \cap D)$ and let X be the $\mathcal{M}_f(D)$-valued process under P_{μ}. Then X is an $(L, \beta, \alpha; D)$-superprocess, where L corresponds to Y on D, $\beta(x) := c\gamma(x)$ and $\alpha(x) := \frac{1}{2}cm(x)$.

---

[46]This is the space of $\mathcal{M}_f(\widehat{D})$-valued paths, which are right-continuous with left limits in the weak topology. It is a separable completely metrizable topological space; with an appropriate 'Skorokhod-topology.'

(See [Engländer and Pinsky (1999)] for the proofs in our particular setting.)

The scaling limit in (1.64) is analogous to scaling limit constructions for Brownian motion and for other diffusions. Just like in those cases, the technical difficulty is compounded in proving the relative compactness (tightness) of $P_{\mu_n}^{(n)}$, $n \geq 1$.

Remark 1.26 ('Clock' in the critical case). In the particular case when $\beta \equiv 0$, an alternative approximation yields the same superprocess: in the nth approximation step one considers critical branching diffusions (the motion component corresponds to L and the branching is critical binary, i.e. either zero or two offspring with equal probabilities), but the branching rate is now $2n\alpha(x)$. So $\alpha(\cdot)$ in this case can also be thought of as the branching 'clock.'   ◇

1.15.3   Super-Brownian motion

When $L = \frac{1}{2}\Delta$, $\beta \equiv 0$ and $\alpha \equiv 1$, the corresponding measure-valued process is called a (standard, critical) super-Brownian motion.

According to the previous subsection, the following approximation scheme produces super-Brownian motion in the limit. Assume that in the nth approximation step each particle

- has mass $1/n$;
- branches at rate $2n$;
- while alive, performs Brownian motion;
- when dies, replaced by either zero or two particles (with equal probabilities) situated at the parent particle's final position.

And all these mechanisms are independent of each other.

Remark 1.27. Of course, slight variations are possible, depending on one's taste. For example the rate can be taken n instead of $2n$ but then one has to pick a mean one distribution with variance 2 for the branching law; if there is zero offspring with probability $2/3$ and three offspring with probability $1/3$, then we have such a distribution. Or, one can take a branching random walk instead of Brownian motion with step frequency n and step size $1/\sqrt{n}$, in which case, the random walk approximation for Brownian motion is 'built in' during the construction of the super-Brownian motion.   ◇

1.15.4   *More general branching*

The particular nonlinearity $\Psi(x, u) := \alpha(x)u^2(x)$, appearing in the cumulant equation (1.62) is convenient to work with. Not only we are dealing with a relatively tame semilinear equation, and so several ideas from the theory of linear PDEs are easy to adapt, but the corresponding superprocess has a continuous (in the weak-topology) version.

Even though in this book we will focus exclusively on the quadratic case, this review on superprocesses would not be complete without mentioning that (following E. Dynkin's work) superprocesses are also defined (and studied) for

$$\Psi(x, u) := \alpha(x)u^2(x) + \int_0^\infty [e^{-ku(x)} - 1 + ku(x)]\, n(x, dk)$$

in place of $\alpha(x)u^2$ in (1.62), via (1.61). This general branching term is usually referred to as *local branching*. Here n is a kernel from D to $[0, \infty)$, that is, $n(x, dk)$ is a measure on $[0, \infty)$ for each $x \in D$, while $n(\cdot, B)$ is a measurable function on D for every measurable $B \subset [0, \infty)$.

In particular, letting $\alpha \equiv 0$ and choosing an appropriate n, the nonlinearity takes the form $\Psi(x, u) = c(x)u^{1+p}, 0 < p < 1$, with some nonnegative, not identically zero function c. Even though this nonlinearity is as simple looking as the quadratic one (since we got rid of the integral term[47] in $\Psi(x, u)$), the path continuity (in the weak topology of measures) is no longer valid for the corresponding superprocess.

This is actually related to another difficulty: although the superprocess corresponding to $\Psi(x, u) = c(x)u^{1+p}, 0 < p < 1$, can still be constructed as the scaling limit of branching diffusions, the second moments of the branching mechanisms in those approximating processes are now unbounded. When the underlying motion is Brownian motion on $\mathbb{R}^d$, the corresponding superdiffusion is called the *infinite variance Dawson-Watanabe process*.

1.15.5   *Local and global behavior*

The notion of the 'extinction' of the $(L, \beta, \alpha; D)$-superprocess X can be approached in various ways, as the following definition shows. (One of them is the concept of local extinction, which we have met already, for branching diffusions.)

---

[47]Those familiar with the Lévy-Khinchine Theorem for infinitely divisible distributions, can suspect (justly), that the source of the somewhat mysterious integral term is in fact the infinite divisibility of $\langle g, X_t \rangle$ in (1.61).

Definition 1.17. Fix $0 \neq \mu \in \mathcal{M}_c(D)$. We say that

(i) X *exhibits local extinction* under P_μ if for every Borel set $B \subset\subset D$, there exists a random time τ_B such that
$$P_\mu(\tau_B < \infty) = 1 \quad \text{and} \quad P_\mu(X_t(B) = 0 \text{ for all } t \geq \tau_B) = 1.$$

(ii) X *exhibits weak local extinction* under P_μ if for every Borel set $B \subset\subset D$, $P_\mu(\lim_{t\to\infty} X_t(B) = 0) = 1$.

(iii) X *exhibits extinction* under P_μ if there exists a stopping time τ such that
$$P_\mu(\tau < \infty) = 1 \quad \text{and} \quad P_\mu(\|X_t\| = 0 \text{ for all } t \geq \tau) = 1.$$

(iv) X *exhibits weak extinction*[48] under P_μ if $P_\mu(\lim_{t\to\infty} \|X_t\| = 0) = 1$.

In [Pinsky (1996)] a criterion was obtained for the local extinction of X, namely, it was shown that X exhibits local extinction if and only if $\lambda_c = \lambda_c(L + \beta, D) \leq 0$. In particular, local extinction does not depend on the branching intensity α, but it does depend on L and β. (Note that, in regions where $\beta > 0$, β can be considered as mass creation, whereas in regions where $\beta < 0$, β can be considered as mass annihilation.) Since local extinction depends on the sign of $\lambda_c(L + \beta, D)$, heuristically, it depends on the competition between the outward speed of the L-particles and the spatially dependent mass creation β. The main tools of [Pinsky (1996)] are PDE techniques.

In [Engländer and Kyprianou (2004)], probabilistic (martingale and spine) arguments were used to show that $\lambda_c \leq 0$ implies weak local extinction, while $\lambda_c > 0$ implies local exponential growth. The following result[49] is Theorem 3 in [Engländer and Kyprianou (2004)].

Lemma 1.11 (Local extinction versus local exponential growth I).
Let P denote the law of the $(L, \beta, \alpha; D)$-superdiffusion X and let $\lambda_c := \lambda_c(L + \beta, D)$.

(i) Under P the process X exhibits local extinction if and only if $\lambda_c \leq 0$.

(ii) When $\lambda_c > 0$, it yields the 'right scaling exponent' in the sense that for any $\lambda < \lambda_c$ and any open $\emptyset \neq B \subset\subset D$, one has

$$P_\mu\left(\limsup_{t\uparrow\infty} e^{-\lambda t} X_t(B) = \infty\right) > 0, \text{ but}$$

$$P_\mu\left(\limsup_{t\uparrow\infty} e^{-\lambda_c t} X_t(B) < \infty\right) = 1.$$

---

[48] Alternatively, X 'extinguishes.'

[49] Cf. Lemma 2.1 later.

Putting things together, one realizes that part (ii) of Definition 1.17 is actually superfluous, and that

(a) *local extinction is in fact equivalent to weak local extinction,* and
(b) there is a dichotomy in the sense that the process either exhibits local extinction (when $\lambda_c \leq 0$), or there is local exponential growth with positive probability (when $\lambda_c > 0$).

On the other hand, extinction and weak extinction are different in general. The intuition behind this is that the total mass $\|X_t\|$ may stay positive but decay to zero, *while drifting out* (local extinction) and on its way obeying changing branching laws. (See Example 1.5 below.) This could not be achieved in a fixed compact region with fixed branching coefficients.

Similarly, without spatial motion (that is, for continuous state branching processes), the total mass cannot tend to zero without actual extinction, unless a usual assumption ('Grey-condition') is violated. (See again [Engländer, Ren and Song (2013)].)

Remark 1.28 (Discrete branching processes). For branching diffusions, an analogous result has been verified in [Engländer and Kyprianou (2004)], by using the same method — see Lemma 2.1 in the next chapter. (Note that for branching diffusions, weak (local) extinction and (local) extinction are obviously the same, because the local/total mass is an integer.) It was also noted that the growth rate of the total mass may exceed λ_c (see Remark 4 in [Engländer and Kyprianou (2004)]).   ◇

Let us consider now $D = \mathbb{R}^d$. Let T^β denote the semigroup corresponding to $L + \beta$, and let $\|T_t^\beta\|_{\infty,\infty}$ denote the L^∞-norm of the linear operator T_t^β for $t \geq 0$.

Definition 1.18 (L^∞-growth bound). Define

$$\lambda_\infty(L + \beta) := \lim_{t \to \infty} \frac{1}{t} \log \|T_t^\beta\|_{\infty,\infty}$$

$$= \lim_{t \to \infty} \frac{1}{t} \log \sup_{x \in \mathbb{R}^d} E_x \left[\exp \left(\int_0^t \beta(Y_s)\, ds \right) 1_{Y_t \in D} \right]. \quad (1.65)$$

(Here the diffusion Y under P_x corresponds to L on $\mathbb{R}^d$.) The existence of the limit can be demonstrated the same way as the analogous limit for λ_c, namely, by applying a well-known subadditivity argument.[50] We call $\lambda_\infty = \lambda_\infty(L + \beta)$ the *L^∞-growth bound.*

---

[50]Subadditivity means that $a_t := \log \|T_t^\beta\|_{\infty,\infty}$ satisfies $a_{t+s} \leq a_t + a_s$ for $t, s \geq 0$.

The probabilistic significance of λ_c and λ_∞ in the context of superprocesses is as follows. The quantity λ_∞ plays a crucial role in describing the behavior of the *total* mass of the superprocess, on a logarithmic scale, while λ_c describes the behavior of the *local* mass. As discussed above, the local mass cannot decay exponentially; the global mass can. That is, $\lambda_\infty(L+\beta) < 0$ may capture a logarithmic decay of the global mass, without actual extinction.

Note that from (1.25) and (1.65) it is obvious that $\lambda_\infty(L+\beta) \geq \lambda_c(L+\beta)$. In fact, $\lambda_\infty = \lambda_c$ and $\lambda_\infty > \lambda_c$ are both possible. For example, when L corresponds to a conservative diffusion, and β is constant, $\lambda_\infty(L+\beta) = \beta$, but $\lambda_c(L+\beta) = \lambda_c(L) + \beta$. So, when $\lambda_c(L) = 0$ ($\lambda_c(L) < 0$), we get $\lambda_\infty(L+\beta) = \lambda_c(L+\beta)$ ($\lambda_\infty(L+\beta) > \lambda_c(L+\beta)$).

Here we will only prove the basic result[51] that the global 'growth' rate for the $(L,\beta,\alpha;\mathbb{R}^d)$-superprocess, X, cannot exceed λ_∞, regardless of what α is. (When $\lambda_\infty < 0$, this, of course, means exponential decay in t, and in particular, weak extinction.)

In order to achieve this, we will need an assumption on the size of the mass creation term β, as follows. Let L correspond to the diffusion process Y on $\mathbb{R}^d$, and let $\{E_x\}_{x\in D}$ denote the corresponding expectations.

Definition 1.19 (Kato class). We say that β is in the *Kato class*[52] $\mathbf{K}(Y)$ if

$$\lim_{t\downarrow 0} \sup_{x\in\mathbb{R}^d} E_x\left(\int_0^t |\beta(Y_s)|\mathrm{d}s\right) = 0. \tag{1.66}$$

The Kato class assumption is significantly weaker than assuming the boundedness of β. However, it is sufficient to guarantee that $\lambda_\infty < \infty$ (see [Engländer, Ren and Song (2013)]). Note, that for superdiffusions, we have assumed that β is bounded from above.[53]

Theorem 1.17 (Over-scaling). *Let X be an $(L,\beta,\alpha;\mathbb{R}^d)$-superdiffusion. Assume that $\beta \in \mathbf{K}(Y)$. Then, for any $\lambda > \lambda_\infty$ and $\mu \in \mathcal{M}_f(\mathbb{R}^d)$,*

$$\mathbf{P}_\mu\left(\lim_{t\to\infty} e^{-\lambda t}\|X_t\| = 0\right) = 1. \tag{1.67}$$

In particular, if $\lambda_\infty < 0$, then X suffers weak extinction.

---

[51] See [Engländer, Ren and Song (2013)] for further results and examples concerning the weak extinction and global growth/decay of superdiffusions.

[52] Kato class is named after T. Kato; it plays an important role in 'Gauge Theory,' developed mostly by K.-L. Chung for Schrödinger operators and recently generalized for diffusion operators by Z.-Q. Chen and R. Song.

[53] As we noted earlier, this can be relaxed. In fact $\lambda_c < \infty$ would be sufficient, which in turn, is implied by $\lambda_\infty < \infty$.

Proof. By a standard Borel-Cantelli argument, it suffices to prove that with an appropriate choice of $T > 0$, and for any given $\epsilon > 0$,

$$\sum_n \mathbf{P}_\mu \left(\sup_{s \in [0,T]} e^{-\lambda(nT+s)} \|X_{nT+s}\| > \epsilon \right) < \infty. \tag{1.68}$$

Pick $\gamma \geq -\lambda$. Then

$$\mathbf{P}_\mu \left(\sup_{s \in [0,T]} e^{-\lambda(nT+s)} \|X_{nT+s}\| > \epsilon \right)$$

$$\leq \mathbf{P}_\mu \left(\sup_{s \in [0,T]} e^{\gamma(nT+s)} \|X_{nT+s}\| > \epsilon \cdot e^{(\lambda+\gamma)nT} \right). \tag{1.69}$$

Let $M_t^{(n)} := e^{\gamma(nT+t)} \|X_{nT+t}\|$ for $t \in [0,T]$. Pick a number $0 < a < 1$ and fix it. Let $\mathcal{F}_s^{(n)} := \sigma(X_{nT+r} : r \in [0,s])$. If we show that for a sufficiently small $T > 0$ and all $n \geq 1$, the process $\{M_t^{(n)}\}_{0 \leq t \leq T}$ satisfies that for all $0 < s < t < T$,

$$\mathbf{E}_\mu \left(M_t^{(n)} \mid \mathcal{F}_s^{(n)} \right) \geq a M_s^{(n)} \quad (\mathbf{P}_\mu\text{-a.s.}), \tag{1.70}$$

then, by using Lemma 1.2 along with the many-to-one formula (1.63) and the branching property, we can continue (1.69) with

$$\mathbf{P}_\mu \left(\sup_{s \in [0,T]} e^{-\lambda(nT+s)} \|X_{nT+s}\| > \epsilon \right) \leq \frac{1}{a\epsilon} e^{-(\lambda+\gamma)nT} \mathbf{E}_\mu \left[e^{\gamma(n+1)T} \|X_{(n+1)T}\| \right]$$

$$= \frac{1}{a\epsilon} e^{(\lambda+\gamma)T} e^{-\lambda(n+1)T} \mathbf{E}_\mu \|X_{(n+1)T}\|$$

$$\leq \frac{\|\mu\|}{a\epsilon} e^{(\lambda+\gamma)T} e^{-\lambda(n+1)T} \left\| T_{(n+1)T}^\beta 1 \right\|_\infty,$$

where T_t^β is as in (1.65). Since $\lambda > \lambda_\infty$ and since, by the definition of λ_∞,

$$\left\| T_{(n+1)T}^\beta 1 \right\|_\infty = \exp[\lambda_\infty (n+1)T + o(n)],$$

as $n \to \infty$, the summability (1.68) holds.

It remains to verify (1.70). Let $0 < s < t < T$. Using BMP at time $nT + s$,

$$\mathbf{E}_\mu \left[M_t^{(n)} \mid \mathcal{F}_s^{(n)} \right] = \mathbf{E}_{X_{nT+s}} e^{\gamma(nT+t)} \|X_{t-s}\|$$

$$= \left\langle \mathbf{E}_{\delta_x} e^{\gamma(nT+t)} \|X_{t-s}\|, \, X_{nT+s}(\mathrm{d}x) \right\rangle$$

$$= \left\langle \mathbf{E}_{\delta_x} e^{\gamma(t-s)} \|X_{t-s}\|, \, e^{\gamma(nT+s)} X_{nT+s}(\mathrm{d}x) \right\rangle. \tag{1.71}$$

We now determine T by using the Kato class assumption (1.66): pick $T > 0$ such that

$$\gamma t + E_x \int_0^t \beta(Y_s)\, ds \geq \log a,$$

for all $0 < t < T$ and all $x \in \mathbb{R}^d$. By Jensen's inequality,

$$\gamma t + \log E_x \exp\left(\int_0^t \beta(Y_s)\, ds\right) \geq \log a,$$

and thus

$$\mathbf{E}_{\delta_x} e^{\gamma t} \|X_t\| = e^{\gamma t} E_x \exp\left(\int_0^t \beta(Y_s)\, ds\right) \geq a$$

holds too, for all $0 < t < T$ and all $x \in \mathbb{R}^d$. Returning to (1.71), for $0 < s < t < T$, we have

$$\mathbf{E}_\mu\left[M_t^{(n)} \mid \mathcal{F}_s^{(n)}\right] \geq a\left\langle 1,\ e^{\gamma(nT+s)} X_{nT+s}\right\rangle = a M_s^{(n)},$$

$\mathbf{P}_\mu$-a.s., yielding (1.70). $\qquad\square$

Remark 1.29. We chose $D = \mathbb{R}^d$ for convenience. The result most probably can easily be extended for more general settings.

1.15.6   *Space-time H-transform; weighted superprocess*

Here we introduce a very useful transformation[54] of nonlinear operators/superprocesses, called H-transform. Recall from Subsection 1.7.5 that diffusion operators, in general, are not closed under Doob's h-transform, because the transformed operator has a 'potential' (zeroth order) term Lh/h. This difficulty vanishes with the semilinear operators we consider, if we define the transformation appropriately.

Consider the backward semilinear operator

$$\mathcal{A}(u) := \partial_s u + (L + \beta)u - \alpha u^2,$$

and let $0 < H \in C^{2,1,\eta}(D \times \mathbb{R}^+)$. (That is, $H \in C^{2,\eta}$ in space and $H \in C^{1,\eta}$ in time.) Analogously to Doob's h-transform for linear operators, introduce the new operator $\mathcal{A}^H(\cdot) := \frac{1}{H}\mathcal{A}(H\cdot)$. Then a direct computation gives

$$\mathcal{A}^H(u) = \frac{\partial_s H}{H} u + \partial_s u + Lu + a\frac{\nabla H}{H} \cdot \nabla u + \beta u + \frac{LH}{H} u - \alpha H u^2. \quad (1.72)$$

---

[54]See [Engländer and Winter (2006)] for more on H-transforms.

Note that the differential operator L is transformed into

$$L_0^H := Lu + a\frac{\nabla H}{H} \cdot \nabla,$$

while β and α transform into

$$\beta^H := \beta + \frac{LH}{H} + \frac{\partial_s H}{H}$$

and

$$\alpha^H := \alpha H,$$

respectively.

The transformation of operators described above has the following probabilistic impact. Let X be a $(L, \beta, \alpha; D)$-superdiffusion. We define a new process X^H by

$$X_t^H := H(\cdot, t)X_t \quad \left(\text{that is, } \frac{\mathrm{d}X_t^H}{\mathrm{d}X_t} = H(\cdot, t)\right), \quad t \geq 0. \qquad (1.73)$$

This way one obtains a new measure-valued process, which, in general,

(1) is not finite measure-valued, only locally finite measure-valued,
(2) is time-inhomogeneous.

Let $\mathcal{M}_{\mathrm{loc}}(D)$ denote the space of locally finite measures on D, equipped with the vague topology. As usual, $\mathcal{M}_c(D)$ denotes the space of finite measures on D with compact support. The connection between X^H and $\mathcal{A}^H$ is given by the following result.

Lemma 1.12 (Lemma 3 in [Engländer and Winter (2006)]). *The process X^H, defined by (1.73), is a superdiffusion corresponding to $\mathcal{A}^H$ on D (that is, an $(L_0^H, \beta^H, \alpha^H; D)$-superdiffusion) in the following sense:*

(i) X^H is an $\mathcal{M}_{loc}(D)$-valued (time-inhomogeneous) Markov process, $(X^H, \mathbf{P}_\mu^r; \mu \in \mathcal{M}_{loc}(D), r \geq 0)$, that is, a family $\{\mathbf{P}_\mu^r\}$ of probability measures where $\mathbf{P}_\mu^r$ is a probability on $C([r, \infty), \mathcal{M}_{loc}(D))$ and the family is indexed by $\mathcal{M}_c(D) \times [0, \infty)$, such that the following holds: for each $g \in C_c^+(D), \mu \in \mathcal{M}_c(D)$, and $r, t \geq 0$,

$$\mathbf{E}_\mu^r \left(\exp\langle -g, X_t^H\rangle\right) = \exp\langle -u(\cdot, r; t, g), \mu\rangle, \qquad (1.74)$$

where $u = u(\cdot, \cdot; t, g)$ is a particular non-negative solution to the backward equation $\mathcal{A}^H u = 0$ in $D \times (0, t)$, with $\lim_{r \uparrow t} u(\cdot, r; t, g) = g(\cdot)$.

(ii) To determine the solution u uniquely, use the equivalent forward equation along with the minimality of the solution: fix $t > 0$ and introduce the 'time-reversed' operator $\widehat{L}$ on $D \times (0, t)$ by

$$\widehat{L} := \frac{1}{2}\nabla \cdot \widehat{a}\nabla + \widehat{b} \cdot \nabla, \qquad (1.75)$$

where, for $r \in [0, t]$,
$$\widehat{a}(\cdot, r) := a(\cdot, t - r) \text{ and } \widehat{b}(\cdot, r) := \left(b + a \frac{\nabla H}{H} \right)(\cdot, t - r);$$
furthermore let
$$\widehat{\beta}(\cdot, r) := \beta^H(\cdot, t - r) \text{ and } \widehat{\alpha}(\cdot, r) := \alpha^H(\cdot, t - r).$$
Consider now v, the minimal non-negative solution to the forward equation
$$\partial_r v = \widehat{L} v + \widehat{\beta} v - \widehat{\alpha} v^2 \qquad \text{in } D \times (0, t),$$
$$\lim_{r \downarrow 0} v(\cdot, r; t, g) = g(\cdot). \tag{1.76}$$
Then
$$u(\cdot, r; t, g) = v(\cdot, t - r; t, g).$$

Example 1.3 (Transforming into critical process). If λ_c is the generalized principal eigenvalue of $L + \beta$ on D and $h > 0$ satisfies $(L + \beta - \lambda_c)h = 0$ on D (such a function always exists, provided λ_c is finite), and we define $H(x, t) := e^{-\lambda_c t} h(x)$, then $\beta^H \equiv 0$, which means that the $(L, \beta, \alpha; D)$-superdiffusion is transformed into a *critical* $(L_0^H, 0, \alpha^H; D)$-superdiffusion. (Here 'critical' refers to the branching.) At first sight this is surprising, since we started with a generic superdiffusion. The explanation is that we have paid a price for this simplicity, as the new 'clock' α^H is now time-dependent for $\lambda_c \neq 0$: $\alpha^H = \alpha h e^{-\lambda_c t}$.   ◇

Given a superdiffusion, H-transforms can be used to produce new superdiffusions that are weighted versions of the old one. Importantly, *the support process $t \mapsto supp(X_t)$ is invariant under H-transforms.*

In fact, as is clear from above, one way of defining a time-inhomogeneous superdiffusion is to start with a time-homogeneous one, and then to apply an H-transform. (One can, however, define them directly as well. Applying an H transform on a generic time-inhomogeneous superdiffusion is possible too, and it results in another superdiffusion — time-inhomogeneous in general.)

Example 1.4 (h-transform for superdiffusions). When H is temporarily constant, that is $H(x, t) = h(x)$ for $t \geq 0$, we speak about the h-transform of the superprocess — this is the case of re-weighting the superprocess by $h > 0$ as a spatial weight function. From an analytical point of view, the differential operator L is transformed into $L_0^h := L + a \frac{\nabla h}{h} \cdot \nabla$, while β and α transform into $\beta^h := \beta + \frac{Lh}{h}$ and $\alpha^h := \alpha h$, respectively; the time homogeneity of the coefficients is preserved. The new motion, an L_0^h-diffusion is obtained by a Girsanov transform from the old one.   ◇

The following example, which uses the h-transform technique along with Theorem 1.17, shows that weak extinction does not imply extinction.

Example 1.5 (Weak and also local extinction, but survival). Let $B, \epsilon >$ and consider the super-Brownian motion in $\mathbb{R}$ with $\beta(x) \equiv -B$ and $k(x) = \exp\left[\mp\sqrt{2(B+\epsilon)}x\right]$, that is, let X correspond to the semilinear elliptic operator $\mathcal{A}$ on $\mathbb{R}$, where

$$\mathcal{A}(u) := \frac{1}{2}\frac{d^2 u}{dx^2} - Bu - \exp\left[\mp\sqrt{2(B+\epsilon)}x\right]u^2.$$

By Theorem 1.17, X suffers weak extinction, and the total mass decays (at least) exponentially: for any $\delta > 0$,

$$\lim_{t\to 0} e^{(B-\delta)t}\|X_t\| = 0.$$

Also, clearly, $\lambda_c = -B$, yielding that X also exhibits local extinction.

Now we are going to show that, despite all the above, the process X survives with positive probability, that is

$$\mathbf{P}_\mu(\|X_t\| > 0, \ \forall \, t > 0) > 0,$$

for any nonzero $\mu \in \mathcal{M}_f(\mathbb{R}^d)$.

To see this, first notice that if, for a generic $(L, \beta, \alpha; D)$-superdiffusion, S_t is the event that $\|X_t\| > 0$, then

$$\mathbf{1}_{S_t} = \exp(-\lim_{n\to\infty} n\|X_t\|),$$

and so, by monotone convergence,

$$\mathbf{P}_\mu(S_t) = \lim_{n\to\infty} \mathbf{E}_\mu \exp(-n\|X_t\|).$$

Hence, the probability of survival can be expressed as

$$\lim_{t\to\infty} \mathbf{P}_\mu(S_t) = \lim_{t\to\infty} \lim_{n\to\infty} \mathbf{E}_\mu \exp(-n\|X_t\|)$$
$$= \lim_{t\to\infty} \lim_{n\to\infty} \exp\langle -u^{(n)}(\cdot, t), \mu\rangle, \tag{1.77}$$

where $u^{(n)}$ is the minimal nonnegative solution to

$$\left.\begin{aligned} u_t = Lu + \beta u - \alpha u^2 \quad &\text{on } D \times (0,\infty), \\ \lim_{t\downarrow 0} u(\cdot, t) = n. \end{aligned}\right\} \tag{1.78}$$

Returning to our specific example, a nonlinear h-transform with $h(x) := e^{\pm\sqrt{2(B+\epsilon)}x}$ transforms the operator $\mathcal{A}$ into $\mathcal{A}^h$, where

$$\mathcal{A}^h(u) := \frac{1}{h}\mathcal{A}(hu) = \frac{1}{2}\frac{d^2 u}{dx^2} \pm \sqrt{2(B+\epsilon)}\frac{du}{dx} + \epsilon u - u^2.$$

(Note that $h''/2 - (B + \epsilon)h = 0$.) Since the superprocess X^h corresponding to $\mathcal{A}^h$ is the same as the original process X, re-weighted by the function h (i.e. $X_t^h = hX_t$), survival (with positive probability) is invariant under h-transforms.

Applying (1.77) and (1.78) to X^h, it is easy to show (see e.g. [Engländer and Pinsky (1999)]) that X^h survives with positive probability[55]; the same is then true for X. ◇

1.16 Exercises

(1) Prove (A.1) of Appendix A.

(2) We have seen (see Appendix A) that the outer measure of Ω, as a subset of $\widehat{\Omega}$, is one. What is the outer measure of the set of discontinuous functions, $\widehat{\Omega} \setminus \Omega$, and why? (**Hint:** First think about this question: Which subsets of Ω belong to $\mathcal{B}'$?)

(3) Is Ω a measurable subset of $\widehat{\Omega}$ when, instead of all continuous functions, it stands for all *bounded* functions? How about all *increasing* functions? And, finally, how about all Lebesgue-measurable functions? (Consult Appendix A.)

(4) (**Feller vs. strong Markov**) Show that the following deterministic process is not Feller, even though it is strong Markovian:
  (a) For $x \geq 0$, let $P_x(X_t = x, \forall t \geq 0) = 1$;
  (b) for $x < 0$, let $P_x(X_t = x - t, \forall t \geq 0) = 1$.

(5) (**BM as a process of independent stationary increments**) Show that our Gaussian definition of Brownian motion (Definition 1.3) is equivalent to the following, alternative definition: Brownian motion is a process B with continuous paths, such that
  (a) $B_0 = 0$,
  (b) $B_t - B_s$ is a mean zero normal variable with variance $t - s$, for all $0 \leq s < t$,
  (c) B has independent increments.

(6) (**Level sets of BM**) Let B be standard one-dimensional Brownian motion and let $Z := \{t \geq 0 \mid B_t = 0\}$ be the set of zeros. Prove that Z has Lebesgue measure zero almost surely with respect to the Wiener measure. Your proof should also work for the level set $Z_a := \{t \geq 0 \mid$

---

[55] In fact, this can be done whenever the superdiffusion has a conservative motion component and constant branching mechanism, which is supercritical.

$B_t = a\}$, $a \in \mathbb{R}$.

(7) Prove that for any given time $t > 0$, t is almost surely not a local maximum of the one-dimensional Brownian motion.

(8) Show that W (d-dimensional standard Brownian motion) is a Feller process.

(9) Let τ be a stopping time with respect to the filtration $\{\mathcal{F}_t; t \geq 0\}$. Show that $\mathcal{F}_\tau$ is indeed a σ-algebra.

(10) Let L satisfy Assumption 1.1 on $\mathbb{R}^d$ and let $X = (X^{(1)}, X^{(2)}, ..., X^{(d)})$ denote the corresponding diffusion process. Prove that

$$\lim_{t \to 0} \frac{1}{t} E_x(X_t^{(i)} - X_0^{(i)}) = b_i(x), \ 1 \leq i \leq d$$

and

$$\lim_{t \to 0} \frac{1}{t} E_x \left[\left(X_t^{(i)} - X_0^{(i)} \right) \left(X_t^{(j)} - X_0^{(j)} \right) \right] = a_{ij}(x), \ 1 \leq i, j \leq d,$$

and interpret these as $b(x)$ being the local *infinitesimal mean*, and $a(x)$ being the local *infinitesimal covariance matrix* (variance in one-dimension).

(11) Let $T_t(f)(x) := E_x f(Y_t)$ for f bounded measurable on D, where Y is a diffusion on D. Prove that $T_{t+s} = T_t \circ T_s$ for $t, s \geq 0$. (Hint: Use the Markov property of Y.)

(12) With the setting of the previous problem, prove that $T_t 1 = 1$ if and only if Y is conservative on D.

(13) Prove that the martingale change of measure in (1.21) preserves the Markov property, using the multiplicative functional property and the Chapman-Kolmogorov equation. Is it true for martingale changes of measure in general, that the process under Q is Markov *if and only if* the density process is a multiplicative functional?

(14) Prove that if $n \geq 1$ and X_n denotes the size of the nth generation in a branching process, then its generating function h satisfies

$$E z^{X_n} = h(h(...(z)...)),$$

where on the right-hand side one has precisely the nth iterate of the function h.

(15) In Theorem 1.13, we 'doubled the rate' of a Poisson process by using a change of measure. What is the change of measure that changes the rate function g to kg, where $k \geq 3$ is an integer?

(16) Use a coupling argument to show that if Z and $\widehat{Z}$ are branching diffusions on the same domain D, with the same motion component Y and with branching rates β and $\widehat{\beta}$, respectively, such that $\beta \leq \widehat{\beta}$, then for

any given $t > 0$ and $B \subset D$, the random variable $Z_t(B)$ is stochastically smaller than $\widehat{Z}_t(B)$.

(17) Prove that property (2) of the generalized principal eigenvalue follows from properties (1) and (3).

(18) What is the generalized principal eigenvalue of Δ on $\mathbb{R}^d$?
(Hint: Replace $\mathbb{R}^d$ by a ball of radius $R > 0$ and express the generalized principal eigenvalue in terms of that of Δ on the ball of unit radius. Then use monotonicity in the domain.)

(19) Let $b \in \mathbb{R}$. Prove that $\lambda_c = -\frac{b^2}{2}$ for

$$L := \frac{1}{2}\frac{\mathrm{d}^2}{\mathrm{d}x^2} + b\frac{\mathrm{d}}{\mathrm{d}x} \text{ on } \mathbb{R}.$$

What does this tell you about the 'escape rate from compacts' for a Brownian motion with constant drift? (Hint: With an appropriate choice of $h > 0$, consider the operator $L^h(f) := (1/h)L(hf)$, which has zero drift part, but has a nonzero potential (zeroth order) term. Use that the generalized principal eigenvalue is invariant under h-transforms.)

(20) Generalize the result of the previous exercise for the operator

$$\frac{1}{2}\Delta + b \cdot \nabla \text{ on } \mathbb{R}^d,$$

and for multidimensional Brownian motion with constant drift, where $b \in \mathbb{R}^d$.

1.17   Notes

The material presented in this chapter up to Section 1.6 is standard and can be found in several textbooks. This is partially true for Section 1.7 too, but we used the notion of the 'generalized martingale problem' which has been introduced in [Pinsky (1995)]. In presenting Sections 1.10 and 1.11, we also followed [Pinsky (1995)]. These sections concern the 'criticality theory' of second order elliptic differential operators. The theory was developed by B. Simon, and later by M. Murata in the 1980s, and was applicable to Schrödinger, and more generally, to self-adjoint operators. Starting in the late 1980s Y. Pinchover succeeded to extend the definitions for general, non-selfadjoint elliptic operators. In the 1990s Pinsky reformulated and reproved several of Pinchover's results in terms of the diffusion processes corresponding to the operators.

Useful monographs with material on branching random walk, branching Brownian motion and branching diffusion are [Asmussen and Hering (1983); Révész (1994)].

The idea behind the notion of superprocesses can be traced back to W. Feller, who observed in his 1951 paper on diffusion processes in genetics, that for large populations one can employ a model obtained from the Galton-Watson process, by rescaling and passing to the limit. The resulting *Feller diffusion* thus describes the scaling limit of the population mass. This is essentially the idea behind the notion of *continuous state branching processes*. They can be characterized as $[0, \infty)$-valued Markov processes, having paths which are right-continuous with left limits, and for which the corresponding probabilities $\{P_x, x \geq 0\}$ satisfy the branching property: the distribution of the process at time $t \geq 0$ under P_{x+y} is the convolution of its distribution under P_x and its distribution under P_y for $x, y \geq 0$.

The first person who studied continuous state branching processes was the Czech mathematician M. Jiřina in 1958 (he called them 'stochastic branching processes with continuous state space'). Roughly ten years later J. Lamperti discovered an important one-to-one correspondence between continuous-state branching processes and Lévy processes (processes with independent, stationary increments and càdlàg paths) with no negative jumps, stopped whenever reaching zero, via random time changes. (See Section 12.1, and in particular, Theorem 12.2 in [Kyprianou (2014)].) This correspondence can be considered as a scaling limit of a similar correspondence between Galton-Watson processes and compound Poisson processes stopped at hitting zero. (See Section 1.3.4 in [Kyprianou (2014)].)

When the *spatial motion* of the individuals is taken into account as well, one obtains a scaling limit which is now a measure-valued branching process, or superprocess. The latter name was coined by E. B. Dynkin in the 1980s. Dynkin's work (including a long sequence of joint papers with S. E. Kuznetsov) concerning superprocesses and their connection to nonlinear partial differential equations was ground breaking. These processes are also called *Dawson-Watanabe processes* after the fundamental work of S. Watanabe in the late 1960s and of D. Dawson in the late 1970s. Among the large number of contributions to the superprocess literature we just mention the 'historical calculus' of E. Perkins, the 'Brownian snake representation' of J.-F. LeGall, the 'look down construction' (a countable representation) of P. Donnelly and T. G. Kurtz, and the result of R. Durrett and E. Perkins showing that for $d \geq 2$, rescaled contact processes converge to super-Brownian motion. In addition, *interacting superprocesses* and *superprocesses in random media* have been studied, for example, by Z.-Q. Chen, D. Dawson, J-F. Delmas, A. Etheridge, K. Fleischmann, H. Gill, P. Mörters, L. Mytnik, Y. Ren, R. Song, P. Vogt and H. Wang.

Besides being connected to partial differential equations, superprocesses are also related to so-called *stochastic partial differential equations* (SPDE's) via their spatial densities, when the latter exists. This is the case for super-Brownian motion in one dimension, where the SPDE for the time t density $u(t, x)$ has the form

$$\dot{u} = \Delta u + \sqrt{u}\, \dot{W},$$

and $\dot{W}$ is the so-called 'space-time white noise.' (One can consider such an equation as a heat equation with a random 'noise' term.)

The spatial h-transform for superprocesses was introduced in [Engländer and Pinsky (1999)]; the space-time H-transform was introduced in [Engländer and Winter (2006)]. Note that these transformations are easy to define for general local branching, exactly the same way as one does it for quadratic branching.

Independently, A. Schied introduced a spatial re-weighting transformation, for a particular class of superprocesses and weight functions [Schied (1999)].

Chapter 2

The Spine Construction and the Strong Law of Large Numbers for branching diffusions

In this chapter we study a strictly dyadic branching diffusion Z correspond-
ing to the operator $Lu + \beta(u^2 - u)$ on $D \subseteq \mathbb{R}^d$ (where β is as in (1.35)).
Our main purpose is to demonstrate that, when $\lambda_c \in (0, \infty)$ and $L + \beta - \lambda_c$
possesses certain 'criticality properties,' the random measures $e^{-\lambda_c t} Z_t$ converge almost surely in the vague topology as $t \to \infty$. As before, λ_c denotes
the generalized principal eigenvalue for the operator $L + \beta$ on D.

The reason we are considering vague topology instead of the weak one,
is that we are investigating the *local* behavior of the process. As it turns
out, local and global behaviors are different in general.

As a major tool, the 'spine change of measure' is going to be introduced;
we believe it is of interest in its own right.

2.1 Setting

Let $D \subseteq \mathbb{R}^d$ be a non-empty domain, recall that $\mathcal{M}(D)$ denotes finite
discrete measures on D:

$$\mathcal{M}(D) := \left\{ \sum_i^n \delta_{x_i} : n \in \mathbb{N}, \ x_i \in D, \text{ for } 1 \leq i \leq n \right\},$$

and consider Y, a diffusion process with probabilities $\{\mathbb{P}_x, \ x \in D\}$ that
corresponds to an elliptic operator L satisfying Assumption 1.2. At this
point, we do not assume that Y is conservative, that is, the exit time from
D may be finite with positive probability.

Assuming (1.35), recall from Chapter 1 the definition of the strictly
dyadic (precisely two offspring) $(L, \beta; D)$-*branching diffusion* with spatially
dependent rate β. In accordance with Section 1.14.3, instead of the assumption that $\sup_D \beta < \infty$ we work with the much milder assumption

$\lambda_c(L+\beta, D) < \infty$. We start the process from a measure in $\mathcal{M}(D)$; at each time $t > 0$, the state of the process is denoted by Z_t, where

$$Z_t \in \mathcal{M}_{\mathrm{disc}}(D) := \left\{ \sum_i \delta_{x_i} : x_i \in D \text{ for all } i \geq 1 \right\},$$

and the sum may run from 1 to ∞, as in general, Z_t is only *locally* finite: if $B \subset\subset D$, then $Z_t(B) < \infty$.

Probabilities corresponding to Z will be denoted by $\{P_\mu : \mu \in \mathcal{M}(D)\}$, and expectations by E_μ. (Except for δ_x, which will be replaced by x, as before.)

2.2   Local extinction versus local exponential growth

The following lemma, which we give here without proof, complements Lemma 1.11, and states a basic *dichotomy* for the large time local behavior of branching diffusions. Just like in Lemma 1.11, the interesting fact about the local behavior is that it depends on the sign of λ_c only.

Lemma 2.1 (Local extinction versus local exponential growth II).
Assume that $\sup_D \beta < \infty$. Let $0 \neq \mu \in \mathcal{M}(D)$. Then

(i) *Z under P_μ exhibits local extinction if and only if there exists a function $h > 0$ satisfying $(L+\beta)h = 0$ on D, that is, if and only if $\lambda_c \leq 0$.*

(ii) *When $\lambda_c > 0$, for any $\lambda < \lambda_c$ and $\emptyset \neq B \subset\subset D$ open,*

$$P_\mu(\limsup_{t \uparrow \infty} e^{-\lambda t} Z_t(B) = \infty) > 0, \text{ but}$$
$$P_\mu(\limsup_{t \uparrow \infty} e^{-\lambda_c t} Z_t(B) < \infty) = 1.$$

In particular, local extinction/local exponential growth does not depend on the initial measure $0 \neq \mu \in \mathcal{M}(D)$.

Remark 2.1. For the proof of Lemma 2.1, see [Engländer and Kyprianou (2004)].[1] In that article it is assumed that β is upper bounded, whereas we only assume $\lambda_c < \infty$. The proofs of [Engländer and Kyprianou (2004)] go through for our case too.

---

[1] The paper followed [Pinsky (1996); Engländer and Pinsky (1999)]. In the latter paper only $\lambda_c < \infty$ is assumed.

2.3 Some motivation

Let us discuss some heuristics now that motivates the rest of this chapter. Intuitively, Lemma 2.1 states that if $\emptyset \neq B \subset\subset D$ open, then

(a) $\underline{\lambda_c \leq 0}$: 'mass eventually leaves B.' This happens even though the entire process may survive with positive probability.[2]

(b) $\underline{\lambda_c > 0}$: 'mass accumulates on B'. With positive probability, $Z_t(B)$ grows faster than any exponential rate $\lambda < \lambda_c$, but this local rate cannot exceed λ_c.

It is natural to ask the following questions:

Does in fact λ_c yield an *exact* local growth rate? That is: is it true that $\lim_{t\to\infty} e^{-\lambda_c t} Z_t$ exists in the vague topology, almost surely, and is the limit non-degenerate? Or should we perhaps modify the pure exponential scaling by a smaller order correction factor? Beyond that, can one identify the limit?

A large number of studies for both branching diffusions and superprocesses have addressed these questions, and we shall review these in the notes at the end of this chapter.

Let $\{T_t\}_{t\geq 0}$ denote the semigroup corresponding to $L + \beta$ on D. According to the 'many-to-one formula' (Lemma 1.6),

$$E_x \langle g, Z_t \rangle = T_t(g)(x) \tag{2.1}$$

for $x \in D$ and for all non-negative bounded measurable g's.

Since the process *in expectation* is determined by $\{T_t\}_{t\geq 0}$, trusting that the Law of Large Numbers holds true for branching processes, one should expect that the process itself grows like the linear kernel, too. If this is the case, and the ratio

$$\frac{\langle g, Z_t \rangle}{E_x \langle g, Z_t \rangle}$$

tends to a non-degenerate limit P_x-a.s., as $t \to \infty$, $x \in D$, then we say that SLLN holds for the process.

On the other hand, it is easy to see that T_t does not in general scale precisely with $e^{-\lambda_c t}$ but sometimes with $f(t) e^{-\lambda_c t}$ instead, where $\lim_{t\to\infty} f(t) = \infty$ but f is sub-exponential. (Take for example $L = \Delta/2$ and $\beta > 0$ constant on $\mathbb{R}^d$, then $f(t) = t^{d/2}$.) One can show that the growth is purely exponential if and only if $L + \beta$ is product-critical (recall Definition

---

[2]If the motion Y is conservative in D for example, then the process survives with probability one.

1.10). Proving SLLN seems to be significantly harder in the general case involving the sub-exponential term f.

2.4   The 'spine'

In this section, we introduce a very useful change of measure which is related to designating a particular line of descent, called the *spine* in the branching process. This technology has a number of versions, both for branching diffusions and for superprocesses; we will choose one that suits our setting. In a sense, we are combining two changes of measures, namely the Girsanov transform (1.21) and the Poisson rate doubling theorem (Theorem 1.13). (See more comments in the notes at the end of this chapter.)

2.4.1   *The spine change of measure*

Let Z, L and β be as before, and let $\{\mathcal{F}_t : t \geq 0\}$ be the canonical filtration generated by Z. Assume that $L + \beta - \lambda_c$ is critical on D, let the ground state[3] be $\phi > 0$, and note that

$$(L + \beta - \lambda_c)\phi = 0 \text{ on } D. \tag{2.2}$$

Since $L+\beta-\lambda_c$ is critical, by Proposition 1.8, ϕ is the unique (up to constant multiples) invariant positive function for the linear semigroup corresponding to $L + \beta - \lambda_c$.

Remark 2.2 (Recurrence and ergodicity). By invariance under h-transforms, the operator $(L + \beta - \lambda_c)^\phi$ on D is also critical, and thus it corresponds to a recurrent diffusion Y on D, with probabilities $\{\mathbb{P}_x^\phi, x \in D\}$. For future reference we note that if one assumes in addition that $L+\beta-\lambda_c$ is product-critical (we do not do it for now), then, by the invariance of product-criticality under h-transforms, Y is actually positive recurrent (ergodic) on D. ◇

Returning to (2.1), we note that even though ϕ is not necessarily bounded from above, the term $T_t(\phi)$ makes sense and (2.1) remains valid when g is replaced by ϕ, because ϕ can be approximated with a monotone increasing sequence of g's and the finiteness of the limit is guaranteed precisely by the invariance property of ϕ. By this invariance, $E_x e^{-\lambda_c t}\langle \phi, Z_t \rangle = e^{-\lambda_c t} T_t(\phi)(x) = \phi(x)$, which, together with BMP, is sufficient to deduce

---

[3]The uniqueness of ϕ is meant up to positive multiples; fix any representative.

that W^ϕ is a martingale where

$$0 \le W_t^\phi := e^{-\lambda_c t} \langle \phi, Z_t \rangle, \ t \ge 0.$$

Indeed, note that, by BMP applied at time t,

$$E_x \left(e^{-\lambda_c(t+s)} \langle \phi, Z_{t+s} \rangle \mid \mathcal{F}_t \right) = e^{-\lambda_c t} E_{Z_t} \left(e^{-\lambda_c s} \langle \phi, Z_s \rangle \right) = e^{-\lambda_c t} \langle \phi, Z_t \rangle.$$

Being a non-negative martingale, P_x-almost sure convergence is guaranteed; the a.s. martingale limit $W_\infty^\phi := \lim_{t \to \infty} W_t^\phi$ will appear in Theorem 2.2, the main result of this chapter, which is a Strong Law of Large Numbers for branching processes.

Having a non-negative martingale at our disposal, we introduce a change of measure. For $x \in D$, normalize the martingale by its mean $\phi(x)$ and define a new law $\widetilde{P}_x$ by the change of measure

$$\left. \frac{\mathrm{d}\widetilde{P}_x}{\mathrm{d}P_x} \right|_{\mathcal{F}_t} = \frac{W_t^\phi}{\phi(x)}, \ t \ge 0. \tag{2.3}$$

The following important theorem describes the law $\widetilde{P}_x$ in an apparently very different way. (See also Exercise 3 at the end of this chapter.)

Theorem 2.1 (The spine construction). *The law of the spatial branching process constructed below in (i)–(iii) is exactly $\widetilde{P}_x$.*

(i) *A single particle, $Y = \{Y_t\}_{t \ge 0}$, referred to as the spine (or 'spine particle'), initially starts at x and moves as a diffusion process[4] corresponding to the h-transformed $(h = \phi)$ operator*

$$(L + \beta - \lambda_c)^\phi = L + a \frac{\nabla \phi}{\phi} \cdot \nabla \tag{2.4}$$

on D;

(ii) *the spine undergoes fission into two particles according to the accelerated rate $2\beta(Y)$, and whenever splits, out of the two offspring, one is selected randomly at the instant of fission, to continue the spine motion Y;*

(iii) *the remaining child gives rise to a copy of a P-branching diffusion started at its space-time 'point' of creation. (Those copies are independent of each other and of the spine.)*

---

[4]Which, according to our assumptions, is recurrent.

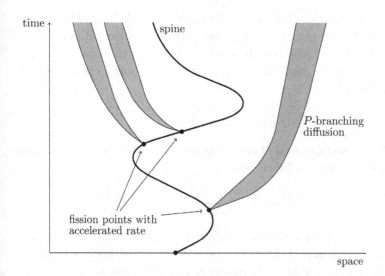

Fig. 2.1   The spine construction (courtesy of N. Sieben).

Remark 2.3 (Spine filtration). As obviously shown by Fig. 2.1, an equivalent description of the spine construction is as follows.

(i′) A single 'spine' particle, $Y = \{Y_t\}_{t \geq 0}$ initially starts at x and moves as a diffusion corresponding to (2.4);

(ii′) at rate $2\beta(Y)$, this path is augmented with space-time 'points';

(iii′) a copy of a P-branching diffusion emanates from each such space-time point. (Those copies are independent of each other and of the spine.)

Using the notion of *immigration*, $(Z, \widetilde{P}_x)$ has the same law as a process constructed in the following way: A $(Y, \mathbb{P}_x^\phi)$-diffusion is initiated, along which $(L, \beta; D)$-branching processes immigrate at space-time points $\{(Y_{\sigma_i}, \sigma_i) : i \geq 1\}$ where, given Y, $n = \{\{\sigma_i : i = 1, ..., n_t\} : t \geq 0\}$ is a Poisson process with law $\mathbb{L}^{2\beta(Y)}$.

Thinking of $(Z, \widetilde{P}_x)$ as being constructed in this way will have several advantages. It will also be convenient to define the canonical filtration of the spine *together with the birth process along the spine* by

$$\mathcal{G}_t := \sigma(Y_s, n_s : s \leq t).$$

This way we keep track of the spine, as well as the space-time points of immigration. ◇

Let us see now the proof of the spine construction.

Proof. First, one can show (left to the reader) that the spatial branching process described by the designated spine particle together with the immigrating P-branching trees, is a Markov process.[5] That the same is true for the process under the 'new law' $\widetilde{P}_x$ defined by (2.3), follows from the fact that M_t defined by

$$M_t(Z(\omega)) := \frac{e^{-\lambda_c t}\langle \phi, Z_t(\omega)\rangle}{\langle \phi, Z_0\rangle}$$

is a multiplicative functional of the measure-valued path. Indeed, equation (1.22) is satisfied for this functional:

$$\frac{e^{-\lambda_c(t+s)}\langle \phi, Z_{t+s}(\omega)\rangle}{\langle \phi, Z_0\rangle} = \frac{e^{-\lambda_c t}\langle \phi, Z_t(\omega)\rangle}{\langle \phi, Z_0\rangle} \cdot \frac{e^{-\lambda_c s}\langle \phi, Z_{t+s}(\omega)\rangle}{\langle \phi, Z_t(\omega)\rangle}.$$

Since we are trying to prove the equality of the laws of two Markov processes, it is enough to prove that their one-dimensional distributions (when starting both from the same δ_x measure) coincide. Moreover, since the one-dimensional distributions are determined by their Laplace transforms, it is sufficient to check the equality of the Laplace transforms.

To this end, let $g \in C_b^+(D)$ and let u_g denote the minimal non-negative solution to the initial-value problem

$$\left.\begin{array}{c} u_t = Lu + \beta(u^2 - u) \quad \text{on } D \times (0, \infty), \\ \lim_{t\downarrow 0} u(\cdot, t) = g(\cdot). \end{array}\right\} \tag{2.5}$$

Then one has to verify that

$$v(x,t) := \widetilde{E}_x\left(e^{-\langle g, Z_t\rangle}\right) = \mathbb{E}_x^\phi \mathbb{L}^{2\beta(Y)}\left(e^{-g(Y_t)}\prod_{k=1}^{n_t} u_g(Y_{\sigma_k}, t - \sigma_k)\right), \tag{2.6}$$

where $\mathbb{L}^{2\beta(Y)}$ denotes the law of the Poisson process with spatially varying rate 2β along a given path of Y, and $\mathbb{E}_x^\phi$ denotes the expectation for Y, starting at x and corresponding to the operator L^ϕ; the random times $\sigma_1, ..., \sigma_{n_t}$ are the Poisson times along the path Y up to time t. (The product after the double expectation expresses the requirement that the 'leaves' formed by the copies of the original process are independent of the 'spine' Y and of each other.)

Note that if $\sigma_1 < t$, then, after σ_1 the process branches into two independent copies of Z for the remaining $t - \sigma_1$ time. Let Y denote the first particle up to σ_1 and let $Z^{Y_{\sigma_1}}$ and $\widehat{Z}^{Y_{\sigma_1}}$ denote the two independent

---

[5]Imagine that the spine particle is black, while the others are blue. So at time $t > 0$ we know which one the spine particle is.

branches emanating from the position of Y at σ_1. Finally, given Y_{σ_1}, define the expression

$$M_t\left(Z^{Y_{\sigma_1}}\right) := \frac{e^{-\lambda_c t}\left\langle \phi, Z_t^{Y_{\sigma_1}}(\omega)\right\rangle}{\phi(Y_{\sigma_1})},$$

and let $M_t\left(\widehat{Z}^{Y_{\sigma_1}}\right)$ be defined the same way. (Cf. the definition of M_t at the beginning of this proof.) Then, by the strong BMP applied at the stopping time σ_1, and by (2.3),

$$\begin{aligned}
v(x,t) &= \widetilde{E}_x\left(e^{\langle -g, Z_t\rangle}; \sigma_1 > t\right) + \widetilde{E}_x\left(e^{\langle -g, Z_t\rangle}; \sigma_1 \le t\right)\\
&= \widetilde{E}_x\left(e^{-g(Y_t)}; \sigma_1 > t\right)\\
&\quad + \widetilde{E}_x E_{Y_{\sigma_1}}\left[\frac{\phi(Y_{\sigma_1})}{\phi(x)} e^{-\lambda \sigma_1}\left(M_{t-\sigma_1}(Z^{Y_{\sigma_1}}) + M_{t-\sigma_1}(\widehat{Z}^{Y_{\sigma_1}})\right)\right.\\
&\qquad \left. \cdot \exp\left(\left\langle -g, Z_{t-\sigma_1}^{Y_{\sigma_1}} + \widehat{Z}_{t-\sigma_1}^{Y_{\sigma_1}}\right\rangle\right); \sigma_1 \le t\right].
\end{aligned} \tag{2.7}$$

Turning from $\widetilde{E}_x$ to E_x and then to $E_x \times L^{\beta(Y)}$, this leads to the equation

$$\begin{aligned}
v(x,t) &= \left(E_x \times L^{\beta(Y)}\right)\left(1_{\{\sigma_1 > t\}} \frac{\phi(Y_t)}{\phi(x)} e^{-\lambda t - g(Y_t)}\right.\\
&\quad \left. + 1_{\{\sigma_1 \le t\}} \frac{\phi(Y_{\sigma_1})}{\phi(x)} e^{-\lambda \sigma_1} 2v(Y_{\sigma_1}, t - \sigma_1) u_g(Y_{\sigma_1}, t - \sigma_1)\right). \tag{2.8}
\end{aligned}$$

Plugging in the exponential density for σ_1, one obtains that

$$\begin{aligned}
v(x,t) &= E_x\left(e^{-\int_0^t \beta(Y_s)ds} \frac{\phi(Y_t)}{\phi(x)} e^{-\lambda t - g(Y_t)}\right. \tag{2.9}\\
&\quad \left. + \int_0^t \beta(Y_s) e^{-\int_0^s \beta(Y_u)du} \frac{\phi(Y_s)}{\phi(x)} e^{-\lambda s} 2v(Y_s, t - s) u_g(Y_s, t - s)\,ds\right).
\end{aligned}$$

Now, if $w(x,t)$ is the right-hand side of (2.6), then the Girsanov transform (1.21) along with the Poisson rate doubling theorem (Theorem 1.13) imply that w solves the functional equation (2.9) as well. To see this, first condition on σ_1 and get

$$\begin{aligned}
w(x,t) &= E_x^\phi L^{\beta(Y)}(1_{\{\sigma_1 > t\}} e^{-g(Y_t)} e^{-\int_0^t \beta(Y_s)ds} \tag{2.10}\\
&\quad + 1_{\{\sigma_1 \le t\}} e^{-\int_0^{\sigma_1} \beta(Y_s)ds} 2w(Y_{\sigma_1}, t - \sigma_1) u_g(Y_{\sigma_1}, t - \sigma_1)).
\end{aligned}$$

(Here the terms $e^{-\int_0^t \beta(Y_s)ds}$ and the factor 2 are consequences of the Poisson rate doubling theorem.) Then use the Girsanov transform, along with (2.2), to change the measure $\mathbb{P}_x^\phi$ to $\mathbb{P}_x$ (with $h := 1/\phi$ and writing $-\beta + \lambda_c$ in place of β in the density in (1.21)), and obtain that in fact, (2.10) is the same equation as (2.8), except that now w replaces v.[6] This yields (2.9) for w instead of v.

Let $\Pi := |v - w|$; our goal is to show that $\Pi \equiv 0$. Using the boundedness of the functions β, ϕ, u_g, Gronwall's inequality (Lemma 1.1) finishes the proof the same way as in the proof of Lemma 1.6.   □

2.4.2   The 'spine decomposition' of the martingale W^ϕ

Recall from Remark 2.3 the notion of the spine filtration $\{\mathcal{G}_t\}_{t \geq 0}$. The 'spine construction' of $(Z, \widetilde{P}_x)$ enables us to write that under $\widetilde{P}_x$, the random variable W_t^ϕ has the same law as

$$e^{-\lambda_c t} \phi(Y_t) + \sum_{i=1}^{n_t} e^{-\lambda_c \sigma_i} W_i,$$

for $t \geq 0$, where Y is the spine particle under $\mathbb{P}_x^\phi$, and conditional on $\mathcal{G}_t$, W_i is an independent copy of the martingale W_t^ϕ, started from position Y_{σ_i}, and run for a period of time $t - \sigma_i$, and σ_i is the ith fission time along the spine for $i = 1, \ldots, n_t$.

The fact that particles away from the spine are governed by the original law P, along with the equation $E_x(W_t^\phi) = \phi(x)$, yield the 'spine decomposition' of the conditional expectation:

$$\widetilde{E}_x \left(W_t^\phi \mid \mathcal{G}_t \right) = e^{-\lambda_c t} \phi(Y_t) + \sum_{i=1}^{n_t} e^{-\lambda_c \sigma_i} \phi(Y_{\sigma_i}), \qquad (2.11)$$

where Y is as above.

2.5   The Strong law

In the rest of the chapter and without further reference, we will always assume that the operator $L + \beta - \lambda_c$ on D is *product-critical*, that is $\langle \phi, \widetilde{\phi} \rangle < \infty$, and in this case we pick the representatives ϕ and $\widetilde{\phi}$ with the normalization $\langle \phi, \widetilde{\phi} \rangle = 1$.

We are going to show now the almost sure convergence in the vague topology of the exponentially discounted process, for a class of operators.

As usual, $C_c^+(D)$ denotes the space of non-negative, continuous and

---

[6]To be more precise, on the event $\{\sigma_1 > t\}$, one uses the Girsanov density up to t, whereas on $\{\sigma_1 \leq t\}$, one uses the Girsanov density up to σ_1, conditionally on σ_1.

compactly supported functions on D. Our Strong Law of Large Numbers[7] for the *local* mass of branching diffusions is as follows.

Theorem 2.2 (SLLN). *Assume that $L + \beta \in \mathcal{P}_p^*(D)$ for some $p \in (1,2]$ and $\langle \beta \phi^p, \widetilde{\phi} \rangle < \infty$. Let $x \in D$. Then $E_x\left(W_\infty^\phi\right) = \phi(x)$, and P_x-a.s.,*

$$\lim_{t \uparrow \infty} e^{-\lambda_c t} \langle g, Z_t \rangle = \langle g, \widetilde{\phi} \rangle W_\infty^\phi, \ g \in C_c^+(D). \tag{2.12}$$

If, in fact, $\sup_D \beta < \infty$, then the restriction $p \in (1,2]$ can be relaxed to $p > 1$.

Before turning to the proof of Theorem 2.2, we discuss some related issues in the next section. We conclude this section with the following Weak Law of Large Numbers.

We now change the class $\mathcal{P}_p^*(D)$ to the larger class $\mathcal{P}_p(D)$ and get $L^1(P_x)$-convergence instead of a.s. convergence (hence the use of the word 'weak'). It is important to point out however, that the class $\mathcal{P}_p^*(D)$ is already quite rich — see the next chapter, where we verify that key examples from the literature are in fact in $\mathcal{P}_p^*(D)$ and thus obey the SLLN.

Theorem 2.3 (WLLN). *Suppose that $L + \beta \in \mathcal{P}_p(D)$ for some $p \in (1,2]$ and $\langle \beta \phi^p, \widetilde{\phi} \rangle < \infty$. Then for all $x \in D$, $E_x(W_\infty^\phi) = \phi(x)$, and (2.12) holds in the $L^1(P_x)$-sense.*

Similarly to SLLN, if $\sup_D \beta < \infty$ then the restriction $p \in (1,2]$ can be relaxed to $p > 1$.

The proof of this theorem is deferred to subsection 2.5.4.

2.5.1   *The L^p-convergence of the martingale*

The a.s. convergence of the martingale W^ϕ is trivial (as it is non-negative) and it does not provide sufficient information. What we are interested in is whether L^p-convergence holds as well. The following result answers this question.

Lemma 2.2. *Assume that $L + \beta$ belongs to $\mathcal{P}_p(D)$ and that $\langle \beta \phi^p, \widetilde{\phi} \rangle < \infty$ for some $p \in (1,2]$. Then, for $x \in D$, W^ϕ is an $L^p(P_x)$-convergent martingale. If, in fact, $\sup_D \beta < \infty$, then the same conclusion holds assuming $p > 1$ only.*

---

[7]The reader will be asked in one of the exercises at the end of this chapter to explain why this name is justified.

To prove this lemma, we will use the result of Section 1.14.9 on ergodicity (Lemma 1.10), the spine decomposition (2.11), and the following, trivial inequality:

$$(u + v)^q \leq u^q + v^q \text{ for } u, v > 0, \text{ when } q \in (0, 1]. \tag{2.13}$$

Proof. Pick p so that $q = p - 1 \in (0, 1]$, $\langle \phi^p, \tilde{\phi} \rangle < \infty$ and $\langle \beta \phi^p, \tilde{\phi} \rangle < \infty$. (If $K := \sup_D \beta < \infty$ and we assume $p > 1$ only, then

$$\langle \phi^p, \tilde{\phi} \rangle = \langle \phi^q, \phi \tilde{\phi} \rangle < \infty$$

implies

$$\langle \phi^r, \tilde{\phi} \rangle = \langle \phi^{r-1}, \phi \tilde{\phi} \rangle < \infty, \forall r \in (0, p),$$

and

$$\langle \beta \phi^r, \tilde{\phi} \rangle \leq K \langle \phi^{r-1}, \phi \tilde{\phi} \rangle < \infty, \forall r \in (0, p),$$

and so we can assume that in fact $p \in (1, 2]$.)

The conditional Jensen's inequality (Theorem 1.3), along with (2.11) and (2.13), yield

$$
\begin{aligned}
\phi(x)^{-1} E_x \left[(W_t^\phi)^p \right] &= \tilde{E}_x \left[(W_t^\phi)^q \right] = \tilde{E}_x \left\{ \tilde{E} \left[(W_t^\phi)^q \mid \mathcal{G}_t \right] \right\} \\
&\leq \tilde{E}_x \left\{ \left[\tilde{E} \left(W_t^\phi \mid \mathcal{G}_t \right) \right]^q \right\} \\
&\leq \mathbb{E}_x^\phi \mathbb{L}^{2\beta(Y)} \left(e^{-\lambda_c q t} \phi(Y_t)^q + \sum_{i=1}^{n_t} e^{-\lambda_c q \sigma_i} \phi(Y_{\sigma_i})^q \right) \\
&= e^{-\lambda_c q t} \mathbb{E}_x^\phi [\phi(Y_t)^q] + \mathbb{E}_x^\phi \left[\int_0^t e^{-\lambda_c q s} 2\beta(Y_s) \phi(Y_s)^q \, ds \right].
\end{aligned}
$$

Now let us dub the two summands on the right-hand side the *spine term*, $A(x, t)$, and the *sum term*, $B(x, t)$, respectively.

Using the positive recurrence of Y and Lemma 1.10,

$$\lim_{t \uparrow \infty} e^{\lambda_c q t} A(t, x) = \lim_{t \uparrow \infty} \mathbb{E}_x^\phi (\phi(Y_t)^q) = \langle \phi^p, \tilde{\phi} \rangle < \infty,$$

for all $x \in D$. For the sum term, using that $\langle \beta \phi^p, \tilde{\phi} \rangle = \langle \beta \phi^q, \phi \tilde{\phi} \rangle < \infty$ and Lemma 1.10 again, we conclude that, for $x \in D$,

$$\lim_{s \uparrow \infty} \mathbb{E}_x^\phi (\beta(Y_s) \phi(Y_s)^q) = \langle \beta \phi^p, \tilde{\phi} \rangle < \infty,$$

and so, $\lim_{t \uparrow \infty} B(t, x) < \infty$. By Doob's inequality (Theorem 1.2), W^ϕ is an L^p-convergent, uniformly integrable martingale. $\square$

2.5.2   *Proof of Theorem 2.2 along lattice times*

The statement that $E_x(W_\infty^\phi) = \phi(x)$ as well as the one concerning the $p > 1$ case, follow from Lemma 2.2 and the first paragraph of its proof, respectively.

The rest of the proof will be based on the following key lemma.

Lemma 2.3. *Fix $\delta > 0$ and let $B \subset\subset D$. Define*

$$U_t = e^{-\lambda_c t}\langle \phi|_B, Z_t \rangle,$$

where $\phi|_B := \phi \mathbf{1}_B$. Then for any non-decreasing sequence $(m_n)_{n\geq 1}$,

$$\lim_{n\uparrow\infty} |U_{(m_n+n)\delta} - E(U_{(m_n+n)\delta} \mid \mathcal{F}_{n\delta})| = 0, \quad P_x\text{-a.s.}$$

Proof.   We will suppress the dependence on n in our notation below and simply write m instead m_n. Suppose that $\{Z_i : i = 1, ..., N_{n\delta}\}$ describes the configuration of particles at time $n\delta$. Note the decomposition

$$U_{(m+n)\delta} = \sum_{i=1}^{N_{n\delta}} e^{-n\delta\lambda_c} U_{m\delta}^{(i)}, \tag{2.14}$$

where given $\mathcal{F}_{n\delta}$, the elements of the collection $\{U_{m\delta}^{(i)} : i = 1, ..., N_{n\delta}\}$ are mutually independent, and the ith one is equal to $U_{m\delta}$ under P_{Z_i}, $i = 1, ..., N_{n\delta}$, in distribution.

By the Borel-Cantelli lemma, it is sufficient to prove that for $x \in D$ and for all $\epsilon > 0$,

$$\sum_{n\geq 1} P_x \left(|U_{(m+n)\delta} - E_x(U_{(m+n)\delta} \mid \mathcal{F}_{n\delta})| > \epsilon \right) < \infty.$$

To this end, use first the Markov inequality:

$$P_x \left(|U_{(m+n)\delta} - E_x(U_{(m+n)\delta} \mid \mathcal{F}_{n\delta})| > \epsilon \right)$$
$$\leq \frac{1}{\epsilon^p} E_x \left(|U_{(m+n)\delta} - E(U_{(m+n)\delta} \mid \mathcal{F}_{n\delta})|^p \right).$$

Recall the Biggins inequality (1.4) and the conditional Jensen inequality (Theorem 1.3), and note that for each $n \geq 1$, $|\sum_{i=1}^n u_i|^p \leq n^{p-1} \sum_{i=1}^n (|u_i|^p)$ and, in particular, $|u + v|^p \leq 2^{p-1}(|u|^p + |v|^p)$.

Notice that

$$U_{s+t} - E_x(U_{s+t} \mid \mathcal{F}_t) = \sum_{i=1}^{N_t} e^{-\lambda_c t} \left(U_s^{(i)} - E_x(U_s^{(i)} \mid \mathcal{F}_t) \right),$$

where conditional on $\mathcal{F}_t$, $Z_i := U_s^{(i)} - E_x(U_s^{(i)} \mid \mathcal{F}_t)$ are independent and Z_i has expectation zero. Hence, by (1.4) and the conditional Jensen inequality (Theorem 1.3),

$$E_x \left(|U_{s+t} - E_x(U_{s+t} \mid \mathcal{F}_t)|^p \mid \mathcal{F}_t \right)$$

$$\leq 2^p e^{-p\lambda_c t} \sum_{i=1}^{N_t} E_x \left(|U_s^{(i)} - E_x(U_s^{(i)} \mid \mathcal{F}_t)|^p \mid \mathcal{F}_t \right)$$

$$\leq 2^p e^{-p\lambda_c t} \sum_{i=1}^{N_t} E_x \left(2^{p-1} \left(|U_s^{(i)}|^p + |E_x(U_s^{(i)} \mid \mathcal{F}_t)|^p \right) \mid \mathcal{F}_t \right)$$

$$\leq 2^p e^{-p\lambda_c t} \sum_{i=1}^{N_t} 2^{p-1} E_x \left(|U_s^{(i)}|^p + E_x(|U_s^{(i)}|^p \mid \mathcal{F}_t) \mid \mathcal{F}_t \right)$$

$$\leq 2^{2p} e^{-p\lambda_c t} \sum_{i=1}^{N_t} E_x \left(|U_s^{(i)}|^p \mid \mathcal{F}_t \right).$$

Then, as a consequence of the previous estimate, we have that

$$\sum_{n \geq 1} E_x \left(|U_{(m+n)\delta} - E_x(U_{(m+n)\delta} \mid \mathcal{F}_{n\delta})|^p \right)$$

$$\leq 2^{2p} \sum_{n \geq 1} e^{-\lambda_c n\delta p} E_x \left(\sum_{i=1}^{N_{n\delta}} E_{\delta Z_i}[(U_{m\delta})^p] \right). \tag{2.15}$$

Recalling the definition of the 'spine term' $A(x,t)$ and the 'sum term' $B(x,t)$ from the proof of Lemma 2.2, and trivially noting that $U_t \leq W_t^\phi$, one has

$$\sum_{n \geq 1} E_x \left(|U_{(m+n)\delta} - E(U_{(m+n)\delta} \mid \mathcal{F}_{n\delta})|^p \right)$$

$$\leq 2^{2p} \sum_{n \geq 1} e^{-\lambda_c n\delta p} E_x \left(\sum_{i=1}^{N_{n\delta}} E_{\delta Z_i}[(W_{m\delta}^\phi)^p] \right)$$

$$\leq 2^{2p} \sum_{n \geq 1} E_x \left(\sum_{i=1}^{N_{n\delta}} e^{-p\lambda_c n\delta} \phi(Z_i)(A(Z_i, m\delta) + B(Z_i, m\delta)) \right)$$

$$= 2^{2p} \sum_{n \geq 1} \phi(x) e^{-q\lambda_c \delta n} \mathbb{E}_x^\phi \left(A(Y_{n\delta}, m\delta) + B(Y_{n\delta}, m\delta) \right), \tag{2.16}$$

where we have used the many-to-one formula (2.1) and the spine decomposition (2.11). Recall that the spine Y is a positive recurrent (ergodic) diffusion under $\mathbb{P}_x^\phi$. We have

$$\mathbb{E}_x^\phi [A(Y_{n\delta}, m\delta)] = e^{-\lambda_c qm\delta} \mathbb{E}_x^\phi (\phi(Y_{(m+n)\delta})^q). \tag{2.17}$$

Denote $m_\infty := \lim_{n \to \infty} m_n \in (0, \infty]$. According to Lemma 1.10, the right-hand side of (2.17) converges to $e^{-q\lambda_c m_\infty \delta} \langle \phi^p, \widetilde{\phi} \rangle$ (which will be zero if $m_\infty = \infty$) as $n \uparrow \infty$. Recall the assumption that $\langle \beta \phi^p, \widetilde{\phi} \rangle < \infty$. Just like before, we have that

$$\mathbb{E}_x^\phi [B(Y_{n\delta}, m\delta)] = 2 \int_0^{m\delta} e^{-\lambda_c qs} \mathbb{E}_x^\phi (\beta(Y_{s+n\delta}) \phi(Y_{s+n\delta})^q) ds,$$

and so

$$\lim_{n \to \infty} \mathbb{E}_x^\phi [B(Y_{n\delta}, m\delta)] = 2 \int_0^{m_\infty \delta} e^{-\lambda_c qs} \langle \beta \phi^p, \widetilde{\phi} \rangle ds < \infty.$$

These facts guarantee the finiteness of the last sum in (2.16), completing the Borel-Cantelli argument. $\qquad\square$

We now complete the proof of Theorem 2.2 along lattice times. Assume that $L + \beta \in \mathcal{P}_p^*$ for some $p > 1$. Recall now that $I(B) := \int_B \phi(y) \widetilde{\phi}(y) dy < 1$. In using $\{Z_i : i = 1, ..., N_t\}$ to describe the configuration of particles in the process at time $t > 0$, we are suppressing t in the notation. Note that, similarly to (2.14),

$E(U_{t+s} \mid \mathcal{F}_t)$

$$= \sum_{i=1}^{N_t} e^{-\lambda_c t} \phi(Z_i) p(Z_i, B, s) = \sum_{i=1}^{N_t} e^{-\lambda_c (t+s)} \int_B \phi(y) q(Z_i, y, s) \, dy$$

$$= \sum_{i=1}^{N_t} e^{-\lambda_c t} \phi(Z_i) \, I(B) + \sum_{i=1}^{N_t} e^{-\lambda_c (t+s)} \int_B \phi(y) Q(Z_i, y, s) \, dy$$

$$= I(B) W_t^\phi + \sum_{i=1}^{N_t} e^{-\lambda_c (t+s)} \int_B \phi(y) Q(Z_i, y, s) \, dy =: I(B) W_t^\phi + \Theta(t, s).$$

Let us replace now t by $n\delta$ and s by $m_n \delta$, where

$$m_n := \zeta(a_{n\delta})/\delta,$$

and a, ζ are the functions[8] appearing in the definition of $\mathcal{P}_p^*$. (Although we do not need it yet, we note that, according to (iv) in Definition 1.14, one has $m_n \leq Kn$, where $K > 0$ does not depend on δ.) Then

$$E(U_{(n+m_n)\delta} \mid \mathcal{F}_{n\delta}) = I(B) W_{n\delta}^\phi + \Theta(n\delta, m_n \delta).$$

Define the events

$$A_n := \{\text{supp}(Z_{n\delta}) \not\subset D_{a_{n\delta}}\}, \ n \geq 1.$$

---

[8]Note that x, B are fixed. Thus, according to our earlier comment, it is not necessary to indicate the dependency of ζ and α on B or the dependency of ζ and a on x.

Using the second part of Definition 1.14(iv) along with the choice of m_n and that $I(B) < 1$, we have

$$|\Theta(n\delta, m_n\delta)| \leq \sum_{i=1}^{N_{n\delta}} e^{-\lambda_c n\delta}\phi(Z_i)e^{-\lambda_c m_n\delta}\alpha_{m_n\delta} + |\Theta(n\delta, m_n\delta)|\mathbf{1}_{A_n}$$

$$= e^{-\lambda_c m_n\delta}\alpha_{m_n\delta} W_{n\delta}^{\phi} + |\Theta(n\delta, m_n\delta)|\mathbf{1}_{A_n}.$$

Since, according to Definition 1.14(iii), $\lim_{n\to\infty} \mathbf{1}_{A_n} = 0$ P_x-a.s., one has that

$$\limsup_{n\uparrow\infty} |\Theta(n\delta, m_n\delta)| \leq \lim_{n\uparrow\infty} e^{-\lambda_c m_n\delta}\alpha_{m_n\delta} W_{n\delta}^{\phi} = 0 \quad P_x\text{-a.s.},$$

and so

$$\lim_{n\uparrow\infty} \left| E_x(U_{(n+m_n)\delta} \mid \mathcal{F}_{n\delta}) - \langle\phi|_B, \widetilde{\phi}\,\mathrm{d}x\rangle W_{\infty}^{\phi}\right| = 0 \quad P_x\text{-a.s.} \qquad (2.18)$$

Now the result for lattice times follows by standard arguments, using the fact that $\mathrm{Span}\{\phi|_B, \ B \subset\subset D\}$ is dense in C_c^+, along with Lemma 2.3. $\square$

2.5.3 *Replacing lattice times with continuous time*

The following lemma upgrades convergence along lattice times to continuous time, and thus enables us to conclude the convergence in Theorem 2.2 — see the remark after the lemma.

Lemma 2.4 (Lattice to continuum). *Assume that* $\langle\phi^p, \widetilde{\phi}\rangle < \infty$ *for some $p > 1$. Assume furthermore that for all $\delta > 0$, $g \in C_c^+(D), x \in D$,*

$$\lim_{n\uparrow\infty} e^{-\lambda_c n\delta}\langle g, Z_{n\delta}\rangle = \langle g, \widetilde{\phi}\rangle W_{\infty}^{\phi}, \qquad P_x\text{-a.s.} \qquad (2.19)$$

Then (2.19) also holds for all $g \in C_c^+(D)$ and $x \in D$, with $n\delta$ and $\lim_{n\uparrow\infty}$ replaced by t and $\lim_{t\uparrow\infty}$, respectively.

Remark 2.4. Recall that we assumed that $\zeta(a_t) = \mathcal{O}(t)$ as $t \to \infty$, and so referring to the previous subsection, $m_n = \zeta(a_{n\delta})/\delta \leq Kn$ with some $K > 0$ which does not depend on δ. In fact, by possibly further increasing the function a, we can actually take $\zeta(a_t) = Kt$ and $m_n = Kn$. Then, from the previous subsection we already know that

$$\lim_{n\uparrow\infty} e^{-\lambda_c(K+1)n\delta}\langle g, Z_{(K+1)n\delta}\rangle = \langle g, \widetilde{\phi}\rangle W_{\infty}^{\phi} \qquad P_x\text{-a.s.}$$

Thus the assumption in Lemma 2.4 is indeed satisfied (write $\delta' := \delta(K+1)$ to see this). $\diamond$

Proof.   Let $B \subset\subset D$. For each $x \in D$ and $\epsilon > 0$, define

$$B^\epsilon(x) := \{y \in B : \phi(y) > (1 + \epsilon)^{-1}\phi(x)\}.$$

Note in particular that $x \in B^\epsilon(x)$ if and only if $x \in B$. Next define for each $\delta > 0$

$$\Xi_B^{\delta,\epsilon}(x) := \mathbf{1}_{\{\mathrm{supp}(Z_t) \subset B^\epsilon(x) \text{ for all } t \in [0,\delta]\}},$$

where $Z_0 = x$, and let $\xi_B^{\delta,\epsilon}(x) := E_x(\Xi_B^{\delta,\epsilon}(x))$. Note that

$$\lim_{\delta \downarrow 0} \xi_B^{\delta,\epsilon} = \mathbf{1}_B.$$

The crucial lower estimate is that for $t \in [n\delta, (n+1)\delta]$,

$$e^{-\lambda_c t}\langle \phi|_B, Z_t \rangle \geq \frac{e^{-\lambda_c \delta}}{(1 + \epsilon)} \sum_{i=1}^{N_{n\delta}} e^{-\lambda_c n\delta}\phi(Z_i)\,\Xi_i, \quad P_x\text{-a.s.},$$

where, given $\mathcal{F}_{n\delta}$, the random variables $\{\Xi_i : i = 1, ..., N_{n\delta}\}$ are independent and Ξ_i is equal in distribution to $\Xi_B^{\delta,\epsilon}(x)$ with $x = Z_i$ for $i = 1, ..., N_{n\delta}$, respectively. Note that the sum on the right-hand side is of the form of the decomposition in (2.14), where now the role of $U_{(m+n)\delta}$ is played by the right-hand side above and the role of $U_{m\delta}^{(i)}$ is played by

$$\phi(Z_i)\,\Xi_i \cdot e^{-\lambda_c \delta}.$$

Similar L^p-type estimates to those found in Lemma 2.3 show us that an estimate of the type of (2.15) is still valid in the setting here and hence

$$\sum_{n \geq 1} E_x \left(|U_{(m+n)\delta} - E(U_{(m+n)\delta} \mid \mathcal{F}_{n\delta})|^p \right)$$

$$\leq 2^{2p} \sum_{n \geq 1} e^{-\lambda_c n\delta p} E_x \left(\sum_{i=1}^{N_{n\delta}} \phi(Z_i)^p \xi_B^{\delta,\epsilon}(Z_i) \right).$$

Recall $q = p - 1$, and continue the upper estimate by

$$\leq 2^{2p} \sum_{n \geq 1} e^{-\lambda_c n\delta p} E_x \langle \phi^p, Z_{n\delta} \rangle = 2^{2p} \sum_{n \geq 1} e^{-\lambda_c n\delta q} \mathbb{E}_x^\phi(\phi(Y_{n\delta})^q) < \infty,$$

where the equality follows by equation (2.1), and the finiteness of the final sum follows from that of $\langle \phi^p, \widetilde{\phi} \rangle$ and the ergodicity of $\mathbb{P}_x^\phi$, in accordance with Lemma 1.10.

By the Borel-Cantelli Lemma we deduce that

$$\lim_{n \uparrow \infty} \left| \sum_{i=1}^{N_{n\delta}} e^{-\lambda_c n\delta}\phi(Z_i)\Xi_i - e^{-\lambda_c n\delta}\langle \phi\xi_B^{\delta,\epsilon}, Z_{n\delta} \rangle \right| = 0, \quad P_x\text{-a.s.},$$

and hence, using (2.19),

$$\liminf_{t \uparrow \infty} U_t \geq \frac{e^{-\lambda_c \delta}}{(1+\epsilon)} \langle \phi \xi_B^{\delta, \epsilon}, \widetilde{\phi} \rangle W_\infty^\phi.$$

Since $\xi_B^{\delta, \epsilon} \in [0, 1]$, taking $\delta \downarrow 0$, by dominated convergence we have that $\langle \phi \xi_B^{\delta, \epsilon}, \widetilde{\phi} \rangle \to \langle \phi|_B, \widetilde{\phi} \rangle$ in the lower estimate above; hence subsequently taking $\epsilon \downarrow 0$ gives us

$$\liminf_{t \uparrow \infty} U_t \geq \langle \phi|_B, \widetilde{\phi} \rangle W_\infty^\phi, \quad P_x\text{-a.s.} \tag{2.20}$$

Although this estimate was computed for $B \subset\subset D$, this restriction is not essential. Indeed, let $B \subseteq D$ (not necessarily bounded), and take a sequence of compactly embedded domains $\{B_n : n \geq 1\}$, with $B_n \uparrow B$. Now (2.20) is still valid, because for each $n \geq 1$,

$$\liminf_{t \uparrow \infty} U_t \geq \liminf_{t \uparrow \infty} e^{-\lambda_c t} \langle \phi|_{B_n}, Z_t \rangle \geq \langle \phi|_{B_n}, \widetilde{\phi} \rangle W_\infty^\phi,$$

and we can let $n \to \infty$.

After having a tight lower estimate for the liminf for arbitrary Borel $B \subseteq D$, we now handle the limsup, also for arbitrary Borel $B \subseteq D$. Using the normalization $\langle \phi, \widetilde{\phi} \rangle = 1$, one has, P_x-a.s.:

$$\limsup_{t \uparrow \infty} U_t = W_\infty^\phi - \liminf_{t \uparrow \infty} e^{-\lambda_c t} \langle \phi|_{D \setminus B}, Z_t \rangle \leq \langle \phi|_B, \widetilde{\phi} \rangle W_\infty^\phi;$$

hence (2.20) holds true with equality and with lim instead of lim inf.

Finally, just like for lattice times previously, a standard approximation argument shows that $\phi|_B$ can be replaced by an arbitrary test function $g \in C_c^+(D)$. $\qquad\square$

2.5.4 *Proof of the Weak Law (Theorem 2.3)*

Proof. The last part of the theorem is merely a consequence of the second paragraph of the proof of Lemma 2.2. Given $g \in C_c^+(D), s \geq 0$, define the function $h_s(x) := \mathbb{E}_x^\phi[g(Y_s)], x \in D$, and note that

$$\sup_{x \in D; s \geq 0} h_s(x) < \infty. \tag{2.21}$$

Now define $U_t[g] = e^{-\lambda_c t} \langle g\phi, Z_t \rangle$ and observe that, just as in Theorem 2.2, one has

$$U_{t+s}[g] = \sum_{i=1}^{N_t} e^{-\lambda_c t} U_s^{(i)}[g],$$

where by (2.1),

$$E_x(U_s^{(i)}[g] \mid \mathcal{F}_t) = \phi(Z_i(t))h_s(Z_i(t)).$$

Next, note from the Markov property at t and the proof[9] of Theorem 2.2 (along lattice times) that for fixed $s > 0$ and $x \in D$,

$$\lim_{t\uparrow\infty} E_x\left(|U_{t+s}[g] - E_x(U_{t+s}[g] \mid \mathcal{F}_t)|^p\right) = 0,$$

and hence, by the monotonicity of L^p-norms,

$$\lim_{t\uparrow\infty} E_x\left(|U_{t+s}[g] - E_x(U_{t+s}[g] \mid \mathcal{F}_t)|\right) = 0. \tag{2.22}$$

Next, making use of the many-to-one formula (2.1) and the spine construction (Theorem 2.1), we have that

$$E_x\left|E_x(U_{t+s}[g] \mid \mathcal{F}_t) - \langle \phi g, \widetilde{\phi}\rangle W_t^\phi\right|$$

$$\leq E_x\left(\sum_{i=1}^{N_t} e^{-\lambda_c t}\phi(Z_i(t))|h_s(Z_i(t)) - \langle \phi g, \widetilde{\phi}\rangle|\right)$$

$$= \phi(x)\mathbb{E}_x^\phi|h_s(Y_t) - \langle \phi g, \widetilde{\phi}\rangle|.$$

(Recall that Y corresponds to $\mathbb{P}_x^\phi$.) Hence taking limits as $t \uparrow \infty$, and using the ergodicity of the spine Y along with Lemma 1.10, as well as (2.21), we have

$$\lim_{t\uparrow\infty} E_x\left|E_x(U_{t+s}[g] \mid \mathcal{F}_t) - \langle \phi g, \widetilde{\phi}\rangle W_t^\phi\right| \leq \phi(x)\langle|h_s - \langle \phi g, \widetilde{\phi}\rangle|, \phi\widetilde{\phi}\rangle.$$

Finally, noting that $\lim_{s\uparrow\infty} h_s(x) = \langle \phi g, \widetilde{\phi}\rangle$, we have by dominated convergence and (2.21) that

$$\lim_{s\uparrow\infty}\lim_{t\uparrow\infty} E_x\left|E_x(U_{t+s}[g] \mid \mathcal{F}_t) - \langle \phi g, \widetilde{\phi}\rangle W_t^\phi\right| \leq \phi(x)\langle\lim_{s\uparrow\infty} |h_s - \langle \phi g, \widetilde{\phi}\rangle|, \phi\widetilde{\phi}\rangle = 0. \tag{2.23}$$

Now recall from Lemma 2.2 that $\lim_{t\to\infty} W_t^\phi = W_\infty^\phi$ in L^p, and hence

$$\lim_{t\uparrow\infty} E_x\left(\left|W_t^\phi - W_\infty^\phi\right|\right) = 0. \tag{2.24}$$

The proof becomes complete by an application of the triangle inequality along with (2.22),(2.23),(2.24) and taking $g = \kappa/\phi$ for any $\kappa \in C_c^+(D)$. $\square$

---

[9]Note that even though U_t is defined differently, we still have martingale differences and the key upper estimate of $U_t \leq \text{const} \cdot W_t^\phi$ still holds.

2.6 Exercises

(1) Give a rigorous proof for the fact that the spatial branching process described in the spine construction (by the designated spine particle along with the immigrating original branching trees) is a Markov process.

(2) Give an example of two stochastic processes with the same one-dimensional distributions, such that the corresponding laws are different. Give such an example when the first process is a Markov process.

(3) Was it important in the spine construction (Theorem 2.1) that we worked with λ_c and the ground state ϕ corresponding to it, or could we replace them with any $\lambda > \lambda_c$ and $\phi > 0$, respectively, which satisfy that $(L + \beta - \lambda)\phi = 0$ on D? (Recall, that, according to general criticality theory, for any $\lambda > \lambda_c$, there exists a twice differentiable $\phi > 0$, such that $(L + \beta - \lambda)\phi = 0$ on D.)

(4) Generalize Theorem 2.1 for the case when, instead of dyadic branching, one considers a given offspring distribution $\{p_0, p_1, ..., p_r\}$, $r > 0$.

(5) Generalize also the spine decomposition (2.11).

(6) Explain why it is justified to call Theorem 2.2 'Strong Law of Large Numbers.' **Hint:** Rewrite the statement as

$$\lim_{t \to \infty} \frac{\langle g, Z_t \rangle}{E_x \langle g, Z_t \rangle} \cdot \frac{E_x \langle g, Z_t \rangle}{\phi(x) e^{\lambda_c t} \langle g, \widetilde{\phi} \rangle} = \frac{W_\infty^\phi}{\phi(x)}, \quad P_x\text{-a.s.}$$

What can you say about the second fraction? Are the terms in $\langle g, Z_t \rangle = \sum_{i=1}^{N_t} g(Z_t^i)$ independent? (Here N_t is the number of particles alive at t and Z_t^i is the ith one.) Are we talking about a 'classical' or an 'array-type' SLLN?

(7) Referring to the Notes of this chapter (see penultimate paragraph), what measurability problems may possibly arise, when one tries to give a pathwise construction of the spine with immigration in the superprocess case?

(8) **(Local spine construction)** This, somewhat more difficult, exercise consists of several parts:

   (a) Derive the 'local version' of the spine construction, given below, by modifying the proof of the global version.

   (b) Give an interpretation of the decomposition using immigration.

   (c) Check that the change of measure preserves the Markov property.

   (d) Check that the right-hand side of (2.25) is a mean one martingale. When will absolute continuity (of the new law $\widetilde{P}_\mu$ with respect to P_μ) hold in (2.25) up to $t = \infty$? (Answer: when $\lambda_c > 0$.)

Let Z be a branching diffusion on $D \subset \mathbb{R}^d$ with motion process $(Y, \mathbb{P}_x)$ corresponding to the operator L and branching rate β, satisfying our usual assumptions.

Let $B \subset\subset D$ be a smooth subdomain. Assume that μ is a finite measure with $\operatorname{supp}\mu \subset B$, that is $\mu = \sum_i \delta_{x_i}$ where $\{x_i\}$ is a finite set of points in B.

By general theory, if λ_c denotes the principal eigenvalue of $L + \beta$ on B, then there exists a unique (up to constant multiples) positive harmonic Dirichlet-eigenfunction on $\overline{B}$, that is a function ϕ such that $\phi > 0$ and $(L + \beta - \lambda_c)\phi = 0$ in B, and satisfying that $\lim_{x \to x_0} \phi(x) = 0$, $x_0 \in \partial B$. Let $\{\mathcal{G}_t\}_{t \geq 0}$ denote the natural filtration of Y, and recall that if τ^B denotes the first exit time of Y from B, then under the change of measure

$$\left.\frac{\mathrm{d}\mathbb{P}_x^\phi}{\mathrm{d}\mathbb{P}_x}\right|_{\mathcal{G}_t} = \frac{\phi(Y_{t \wedge \tau^B})}{\phi(x)} \exp\left\{-\int_0^{t \wedge \tau^B} (\lambda_c - \beta(Y_s))\,\mathrm{d}s\right\}$$

the process $(Y, \mathbb{P}_x^\phi)$ corresponds to the h-transformed ($h = \phi$) generator $(L + \beta - \lambda_c)^\phi = L + a\phi^{-1}\nabla\phi \cdot \nabla$. In fact, it is ergodic (positive recurrent) on B, and in particular, it never hits ∂B.

Let $\{\mathcal{F}_t\}_{t \geq 0}$ denote the natural filtration of Z, and let Z^B denote the branching process with killing at ∂B, that is let Z^B be obtained from Z by removing particles upon reaching ∂B. Define $\widetilde{P}_\mu$ by the martingale change of measure (check that the right-hand side is indeed a martingale!)

$$\left.\frac{\mathrm{d}\widetilde{P}_\mu}{\mathrm{d}P_\mu}\right|_{\mathcal{F}_t} = e^{-\lambda_c t}\frac{\langle \phi, Z_t^B \rangle}{\langle \phi, \mu \rangle}. \tag{2.25}$$

Next, given a non-negative bounded continuous function $\gamma(t)$, $t \geq 0$, $(n, \mathbb{L}^\gamma)$ will denote the Poisson process, where[10]

$$n = \{\{\sigma_i : i = 1, ..., n_t\} : t \geq 0\}$$

has instantaneous rate $\gamma(t)$.

Finally, for $g \in C_b^+(D)$, let u_g denote the minimal non-negative solution to $\dot{u} = Lu + \beta u^2 - \beta u$ on $D \times (0, \infty)$ with $\lim_{t \downarrow 0} u(\cdot, t) = g(\cdot)$. Then, for $t \geq 0$ and $g \in C_b^+(D)$, the Laplace-transform of the 'new

---

[10]That is, n_t is the number of events σ_i up to t.

law' satisfies

$$\widetilde{E}_\mu \left(e^{-\langle g, Z_t \rangle} \right)$$

$$= \sum_i \frac{\phi(x_i)}{\langle \phi, \mu \rangle} \left\{ \mathbb{E}^\phi_{x_i} \mathbb{L}^{2\beta(Y)} \left(e^{-g(Y_t)} \prod_{k=1}^{n_t} u_g(Y_{\sigma_k}, t - \sigma_k) \right) \prod_{j \neq i} u_g(x_j, t) \right\}.$$

2.7   Notes

Local extinction and its connection to the generalized principal eigenvalue were studied in [Pinsky (1996)], [Engländer and Pinsky (1999)] (for superprocesses) and [Engländer and Kyprianou (2004)] (for branching diffusions).

The results of this chapter were proved in [Engländer, Harris and Kyprianou (2010)], which was motivated by

- a cluster of articles due to Asmussen and Hering, dating from the mid 1970s,

- the more recent work concerning analogous results for superdiffusions of [Engländer and Turaev (2002); Engländer and Winter (2006)].

In the former, the study of growth of typed branching processes on compact domains of the type space was popularized by Asmussen and Hering, long before a revival in this field appeared in the superprocess community.

In the late 1970s the two authors wrote a series of papers concerning weak and strong laws of large numbers for a reasonably general class of branching processes, including branching diffusions. See [Asmussen and Hering (1976a)] and [Asmussen and Hering (1976b)]. Their achievements relevant to our context were as follows.

- They showed (see Section 3 in [Asmussen and Hering (1976a)]) that, when D is a *bounded* one-dimensional[11] interval, for branching diffusions and for a special class of operators $L + \beta$, the rescaled process $\{\exp\{-\lambda_c t\} Z_t : t \geq 0\}$ converges in the vague topology, almost surely.

- For the same class of $L + \beta$, when D is unbounded, they proved the existence of the limit *in probability* of $\exp\{-\lambda_c t\} Z_t$ as $t \uparrow \infty$ (in the vague topology).

The special class of operators alluded to above are called 'positively regular' by those authors. The latter is a subclass of our class $\mathcal{P}^*_p(D)$. They actually proved the convergence of $e^{-\lambda_c t} \langle Z_t, g \rangle$ for all $0 \leq g \in L^1(\widetilde{\phi}(x) \mathrm{d}x)$. Their method is robust in the sense that it extends to many other types of branching processes; discrete time, discrete space, etc.

---

[11]Though they remark that 'For greater clarity we therefore restrict ourselves to this case. However, all results and proofs of this and the following section can be formulated with n-dimensional diffusions, and we shall do this in the more comprehensive framework of a future publication.'

Our Lemma 2.4 is based on the idea to be found in Lemma 8 of [Asmussen and Hering (1976a)].

Interestingly, preceding all work of Asmussen and Hering is the single article [Watanabe (1967)]. Watanabe demonstrates[12] that when a suitable Fourier analysis is available with respect to the operator $L + \beta$, by spectrally expanding any $g \in C_c^+(D)$, one can show that $\{\langle g, Z_t \rangle : t \geq 0\}$ is almost surely asymptotically equivalent to its mean, yielding the classic Strong Law of Large Numbers for strictly dyadic branching Brownian motion in $\mathbb{R}^d$: when $L = \Delta/2$ and $\beta > 0$ is a constant,

$$\lim\nolimits_{t \uparrow \infty} t^{d/2} e^{-\beta t} Z_t(B) = (2\pi)^{d/2} |B| \times N_\mu,$$

where B is any Borel set ($|B|$ is its Lebesgue measure) and $N_\mu > 0$ is a random variable depending on the initial configuration $\mu \in \mathcal{M}(\mathbb{R}^d)$.

Notice, however, that $\Delta/2 + \beta \notin \mathcal{P}_1(D)$, while [Engländer and Turaev (2002); Engländer and Winter (2006); Chen and Shiozawa (2007)] all assume that the operator is in $\mathcal{P}_1(D)$. For a result on supercritical super-Brownian motion, analogous to Watanabe's theorem, see [Engländer (2007a)].

In our context, being in the 'positively regular' class of Asmussen and Hering means that

(A) $\lambda_c > 0$ (in [Asmussen and Hering (1976a)] this property is called 'supercriticality'),

(B) ϕ is bounded from above,

(C) $\langle \widetilde{\phi}, 1 \rangle < \infty$.

Obviously, **(B)**–**(C)** is stronger than the assumption $\langle \phi, \widetilde{\phi} \rangle < \infty$ (product-criticality).

Secondly, $\{T_t\}_{t \geq 0}$, the semigroup corresponding to $L + \beta$ ('expectation semigroup') satisfies the following condition. If η is a non-negative, bounded measurable function on $\mathbb{R}^d$, then

(D)

$$T_t(\eta)(x) = \langle \eta, \widetilde{\phi} \rangle \, \phi(x) \left[e^{\lambda_c t} + o\left(e^{\lambda_c t}\right) \right] \quad \text{as } t \uparrow \infty, \text{ uniformly in } \eta.$$

Let $\overline{S}_t := \exp\{-\lambda_c t\} T_t$ (which corresponds to $L + \beta - \lambda_c$), and $\{T_t\}_{t \geq 0}$ be the semigroup defined by $T_t(f) := \overline{S}_t^\phi(f) = \frac{1}{\phi} \overline{S}_t(\phi f)$, for all $0 \leq f$ measurable with ϕf being bounded. Then $\{T_t\}_{t \geq 0}$ corresponds to the h-transformed ($h = \phi$) operator $(L + \beta - \lambda_c)^\phi = L + a\phi^{-1} \nabla \phi \cdot \nabla$ and to a positive recurrent diffusion. Next, assuming that ϕ is bounded, it is easy to check that the following condition would suffice for **(D)** to hold:

$$\limsup_{t \uparrow \infty} \sup_{x \in D} \sup_{\|g\| \leq 1} \left| T_t(g) - \langle g, \phi \widetilde{\phi} \rangle \right| = 0, \tag{2.26}$$

---

[12]A glitch in the proof was later fixed by Biggins in [Biggins (1992)].

where $\|\cdot\|$ denotes sup-norm. However this is not true in most cases on unbounded domains (or even on bounded domains with general unbounded coefficients) because of the requirement on the uniformity in x. (See our examples in the next chapter — neither of the examples on $\mathbb{R}^d$ satisfy (2.26).)

We note that later, in their book [Asmussen and Hering (1983)], the above authors gave a short chapter about SLLN in unbounded domains. In the notes they explain that they wanted to show examples when their regularity assumptions do not hold. They only treat two cases though: branching Brownian motion and one-dimensional branching Ornstein-Uhlenbeck process.

More recently, in [Chen and Shiozawa (2007)], almost sure limits were established for a class of Markov branching processes,[13] using mostly functional analytic methods.

Related to their analytical (as opposed to probabilistic) approach is the difference that, because of the L^2-approach, their setting had to be restricted to symmetric operators, unlike in the results of this chapter (see Example 13 of the next chapter).

In fact, even within the symmetric case, our milder spectral assumptions include e.g. Examples 10 and 11 of the next chapter, which do not satisfy the assumptions in [Chen and Shiozawa (2007)]: Example 10 does not satisfy the assumption that $\sup \phi < \infty$; in Example 11, since β is constant, $\beta \notin K_\infty(Y)$.[14]

Turning to superprocesses, one sees considerably fewer results of this kind in the literature (see the references [Dawson (1993); Dynkin (1991, 1994); Etheridge (2000)] for superprocesses in general). Some recent and general work in this area are [Engländer and Turaev (2002); Engländer and Winter (2006); Engländer (2007a)], and [Chen, Ren and Wang (2008)].

In [Engländer and Turaev (2002)] it was proved that (in the vague topology) $\{\exp\{-\lambda_c t\}X_t : t \geq 0\}$ converges in law where X is the $(L, \beta, \alpha, \mathbb{R}^d)$-superprocess satisfying that $L + \beta \in \mathcal{P}_1(D)$ and that $\alpha\phi$ is bounded from above. (An additional requirement was that $\langle \phi, \mu \rangle < \infty$ where $\mu = X_0$ is the deterministic starting measure.) Later, the convergence in law was upgraded to convergence in probability and instead of $\mathbb{R}^d$, a general Euclidean domain $D \subseteq \mathbb{R}^d$ was considered. (See [Engländer and Winter (2006)].)

The *general dichotomy conjecture* (for branching diffusions or superprocesses) is that either

(a) $\lambda_c \leq 0$ and local extinction holds; or

(b) $\lambda_c > 0$ and local SLLN holds.

We have seen that this statement has been verified under various assumptions on the operator; proving or disproving it in full generality is still an open problem.

---

[13] A similar limit theorem for superprocesses has been obtained in [Chen, Ren and Wang (2008)].

[14] The class $K_\infty(Y)$ depends on the motion process Y, and is defined in [Chen and Shiozawa (2007)] with the help of Kato classes; it contains rapidly decaying functions.

As far as the spine technique[15] is concerned, it has been introduced to the literature by [Lyons, Pemantle and Peres (1995)] and by the references given there (in particular [Kallenberg (1977)]) and involves a change of measure inducing a 'spine' decomposition. In [Lyons, Pemantle and Peres (1995)] however, the spine is non-spatial, and it is thus simply size-biasing.

A simple use of spines to get right-most particle speeds is given in [Harris and Harris (2009)]; while some more recent applications of the spine method can be found in [Harris and Roberts (2012, 2013a)]. The paper [Harris and Roberts (2013b)] is a more recent extension of spine ideas but with more than one spine; a somewhat spine-related, recent work is [Harris, Hesse and Kyprianou (2013)]. See also [Athreya (2000); Kyprianou (2004); Engländer and Kyprianou (2004); Hardy and Harris (2009)].

For yet further references, see for example [Evans (1993); Etheridge (2000)] as well as the discussion in [Engländer and Kyprianou (2004)].

For spine techniques with superprocesses, the initial impetus came from the so-called 'Evans immortal particle' [Evans (1993)]. The immigration in such results is continuous in time, and the contribution of the immigration up to time t is expressed via a time integral up to t. (See e.g. [Engländer and Pinsky (1999)] for more elaboration.)

We mention also [Salisbury and Verzani (1999)], where conditioning the so-called *exit measure* of the superprocess to hit a number of specified points on the boundary of a domain was investigated. The authors used spine techniques, but in a more general sense, as the change of measure they employ is given by a martingale which need not arise from a single harmonic function.

The reader may also want to take a look at Section 4.1 of [Engländer (2007b)] which explains the construction and the probabilistic meaning of the quantities appearing in the formulas in the spine construction discussed there.

We note, that, as far as spine/skeleton 'constructions' for superprocesses are concerned (e.g. in [Engländer and Pinsky (1999)]), these are usually hardly actual constructions. Although one is tempted to interpret the equality of certain Laplace-transforms as equality of processes, the 'construction' of the spine/skeleton with immigration in the superprocess case is prevented by measurability issues. This shortcoming has recently[16] been fixed by Kyprianou et al., by discovering a truly *pathwise* approach. See Section 5, and in particular, Theorem 5.2 in [Kyprianou, Liu, Murillo-Salas and Ren (2012)]; see also [Berestycki, Kyprianou and Murillo-Salas (2011)]. In the latter work, the so-called *Dynkin-Kuznetsov N-measure* plays an important role.

Finally, in [Eckhoff, Kyprianou and Winkel (2014)] the authors use a skeleton decomposition to derive the Strong Law of Large Numbers for a wide class of superdiffusions from the corresponding result for branching diffusions.

---

[15]'Spine' and 'backbone/skeleton' decompositions are different. In the latter, one considers a whole branching process as a distinguished object and not just one particular path. A typical application is conditioning a supercritical spatial branching process on survival.

[16]Although some of the ideas can be traced back to Salisbury's work.

Chapter 3

Examples of The Strong Law

In this chapter we provide examples which satisfy all the assumptions we had in the previous chapter, and thus, according to Theorem 2.2, obey the SLLN.[1]

We begin with discussing the important particular case when the domain is *bounded*.

Example 3.1 (Bounded domain). First note that when D is bounded, an important subset of $\mathcal{P}_p(D)$, $p > 1$ is formed by the operators $L + \beta$ which are uniformly elliptic on D with bounded coefficients which are smooth up to the boundary of D and with $\lambda_c > 0$. That is, in this case $L + \beta - \lambda_c$ is critical (see [Pinsky (1995)], Section 4.7), and since ϕ and $\widetilde{\phi}$ are Dirichlet eigenfunctions (that is, zero at ∂D), it is even product-p-critical for all $p > 1$. WLLN (Theorem 2.3) thus applies.

Although in this case Y is not conservative in D, in fact even SLLN (Theorem 2.2) is applicable whenever (iv^*) can be strengthened to the following uniform convergence on D:

$$\lim_{t \to \infty} \sup_{z \in D, y \in B} \left| \frac{p(z, y, \zeta(t))}{\phi \widetilde{\phi}(y)} - 1 \right| = 0. \tag{3.1}$$

(Note that [Asmussen and Hering (1976a)] has a similar global uniformity assumption — see the paragraph after (2.26).) Indeed, then the proof of Theorem 2.2 can be simplified, because the function a is not actually needed: D_{a_n} can be replaced by D for all $n \geq 1$.

As far as (3.1) is concerned, it is often relatively easy to check. For example, assume that $d = 1$ (the method can be extended for radially symmetric settings too) and so let $D = (r, s)$. Then the drift term of the

---

[1]Note that those examples do not fall into the setting in [Asmussen and Hering (1976a,b)] and two of them are not covered by [Chen and Shiozawa (2007)] either.

spine is $b + a(\log \phi)'$. Now, if this is negative and bounded away from zero at $s - \epsilon < x < s$ and positive and bounded away from zero at $r < x < r + \epsilon$ with some $\epsilon \in (0, s - r)$, then (3.1) can be verified by a method similar to the one in Example 3.4 (the last example in this section). The above condition on the drift is not hard to check in a concrete example. It helps to keep in mind that, since ϕ satisfies the zero Dirichlet boundary condition at r and s, therefore $\lim_{x \to y} \log \phi(x) = -\infty$ for $y = r, s$.

If we relax the regularity assumptions on $L + \beta$ then for example ϕ is not necessarily upper bounded, and so we are leaving the family of operators handled in [Asmussen and Hering (1976b)] (see the four paragraphs preceding (2.26)); nevertheless our method still works as long as $L + \beta \in \mathcal{P}_p^*(D)$, $p > 1$ (for the SLLN) or $L + \beta \in \mathcal{P}_p(D)$, $p > 1$ (for the WLLN).

The next two examples are related to multidimensional Ornstein-Uhlenbeck (OU) processes.

Example 3.2 (OU process with quadratic branching rate). This model has been introduced and extensively studied in [Harris (2000)].

Let $\sigma, \mu, a, b > 0$ and consider

$$L := \frac{1}{2}\sigma^2 \Delta - \mu x \cdot \nabla \text{ on } \mathbb{R}^d$$

corresponding to an (inward) OU process, the equilibrium distribution of which is given by the normal density

$$\pi(x) = \left(\frac{\mu}{\pi \sigma^2}\right)^{d/2} \exp\left\{-\frac{\mu}{\sigma^2} x^2\right\}.$$

Let $\beta(x) := b x^2 + a$. Since L corresponds to a recurrent diffusion, it follows by Proposition 1.10 that $\lambda_c > 0$.

Assume that $\mu > \sigma\sqrt{2b}$, and define the shorthands

$$\gamma^{\pm} := \frac{1}{2\sigma^2}\left(\mu \pm \sqrt{\mu^2 - 2b\sigma^2}\right);$$

$$c^- := \left(1 - (2b\sigma^2/\mu^2)\right)^{\frac{d}{8}}, \quad c^+ := c^- \left(\mu/(\pi\sigma^2)\right)^{\frac{d}{2}}.$$

It is then easy to check that

$$\lambda_c = \sigma^2 \gamma^- + a, \quad \phi(x) = c^- \exp\{\gamma^- x^2\} \text{ and } \widetilde{\phi}(x) = c^+ \exp\{-\gamma^+ x^2\}. \tag{3.2}$$

Indeed, (3.2) follows from the fact that (as we will see right below) $L + \beta - \lambda_c$ with $\lambda_c := \sigma^2 \gamma^- + a$ can be h-transformed into an operator corresponding to a (positive) recurrent diffusion. Since such an operator has to be critical, thus, by h-transform invariance, $L + \beta - \lambda_c$ is a critical operator itself, possessing a unique ground state, up to constant multiples.

Indeed, the spine, corresponding to the h-transformed operator ($h = \phi$) is also an (inward) Ornstein-Uhlenbeck process with parameter $\alpha := \mu - 2\gamma^-\sigma^2 = \sqrt{\mu^2 - 2b\sigma^2}$ where

$$(L + \beta - \lambda_c)^\phi = L + \sigma^2 \frac{\nabla\phi}{\phi} \cdot \nabla = \frac{1}{2}\sigma^2\Delta - \alpha x \cdot \nabla \text{ on } \mathbb{R}^d,$$

and for transition density one has

$$p(x,y,t) = \left(\frac{\alpha}{\pi\sigma^2 \left(1 - e^{-2(\alpha/\sigma^2)t}\right)} \right)^{d/2} \exp\left[-\frac{\alpha \sum_{i=1}^{d}(y_i - x_i e^{-(\alpha/\sigma^2)t})^2}{\sigma^2(1 - e^{-2(\alpha/\sigma^2)t})} \right].$$

Let us check now that all necessary conditions are satisfied for Theorem 2.2 to hold. We see that the drift of the inward OU process causes the influence of any starting position to decrease exponentially with time. Indeed, one can take $\zeta(t) = (1 + \epsilon)(\sigma^2/2\alpha)\log t$ for any $\epsilon > 0$ for condition (iv^*) in Definition 1.14 to hold. Trivially, $\zeta(a_t) = \mathcal{O}(t)$ (in fact, only $\log t$ growth). Finally, to guarantee that condition (iii) in Definition 1.14 holds, one can pick $a_t = \sqrt{\lambda t/\gamma^+}$ for any $\lambda > \lambda_c$.

Remark 3.1. (i) This non-trivial model highlights the strength of our general result. In particular, it is known that a quadratic breeding rate is *critical* in the sense that a BBM Z with breeding rate $\beta(x) = \text{const} \cdot x^p$

- explodes in a finite time a.s.[2], when $p > 2$;
- explodes in the *expected* population size, even though the population size itself remains finite for all times a.s., when $p = 2$;
- the expected population size remains finite for all times, when $p < 2$.

In our case though, an inward OU process replaced Brownian motion. We have seen that a strong enough drift with $\mu > \sigma\sqrt{2b}$ could balance the high breeding, whereas any weaker drift would have led to a dramatically different behavior.

(ii) In order to calculate the *expected* growth of the support, one can utilize the Many-to-one formula (2.1), and obtain, that in expectation, the support of the process grows like $\sqrt{\lambda_c t/\gamma^+}$ as $t \to \infty$. ◇

Example 3.3 (Outward OU process with constant branching rate). Let $\sigma, \mu, b > 0$ and consider

$$L := \frac{1}{2}\sigma^2\Delta + \mu x \cdot \nabla \text{ on } \mathbb{R}^d,$$

---

[2]That is, there exists a finite random time T such that $\lim_{t \to T} \|Z_t\| = \infty$ a.s.

corresponding to an 'outward' Ornstein-Uhlenbeck process Y, and let $\beta(\cdot) \equiv b$. As the spatial motion has no affect on branching, the global population grows like $e^{\beta t}$, this being the 'typical' behavior. The local growth, on the other hand, is smaller. Indeed, it is well known (and easy to check) that $\lambda_c(L) = -\mu$. Hence, $\lambda_c = b - \mu < b$, it being associated with the *local*, as opposed to *global*, growth rate.

The corresponding ground state is $\phi(x) = \text{const} \cdot \exp\{-(\mu/\sigma^2)x^2\}$, and, despite the highly transient nature of Y, the h-transformed motion ($h = \phi$) of the spine is positive recurrent. It is in fact an *inward* OU process, corresponding to the operator

$$(L + \beta - \lambda_c)^\phi = L + \sigma^2 \frac{\nabla\phi}{\phi} \cdot \nabla = \frac{1}{2}\sigma^2\Delta - \mu x \cdot \nabla \text{ on } \mathbb{R}^d,$$

with equilibrium density $\phi\widetilde{\phi}(x) \propto \exp\{-(\mu/\sigma^2)x^2\}$.

Let us check now that the conditions required for Theorem 2.2 to hold are satisfied. After some expectation calculations similar to those alluded to at Example 3.2, one finds that an upper bound on the spread of the process is roughly the same as for an individual outward OU particle. In other words, one can take $a_t := \exp\{(1 + \delta)(\mu/\sigma^2)t\}$ for any $\delta > 0$. Finally, let $\zeta(t) = (1+\epsilon)(\sigma^2/\mu)\log t$ for any $\epsilon > 0$. Then $\zeta(a_t) = (1+\epsilon)(1+\delta)t = \mathcal{O}(t)$.

Remark 3.2. Intuitively, the spine's motion is the one that 'maximizes the *local* growth rate' at λ_c. (Here it is Y 'conditioned to keep returning to the origin.')   ◇

In the next example the motion process is a recurrent Brownian motion, and the branching only takes place in the vicinity of the origin.

Example 3.4 (BBM with $0 \not\equiv \beta \in C_c^+(\mathbb{R}^d)$ for $d = 1, 2$). Consider the $(\frac{1}{2}\Delta + \beta)$-branching diffusion where $\beta \in C_c^+(\mathbb{R}^d)$ and $\beta \not\equiv 0$ for $d = 1, 2$. Since Brownian motion is recurrent in dimensions $d = 1, 2$, it follows that $\lambda_c > 0$ and in fact, the operator $\frac{1}{2}\Delta + \beta - \lambda_c$ is product-critical, and even product-p-critical for all $p > 1$ (see Example 22 in [Engländer and Turaev (2002)]).

We begin by showing how to find a ζ that satisfies (iv^*) in Definition 1.14. We do it for $d = 1$; the case $d = 2$ is similar, and is left to the reader.

Let $b > 0$ be so large that $\text{supp}(\beta) \subset [-b, b]$ and let $M := \max_{\mathbb{R}} \beta$. Recall that $p(t, x, y)$ denotes the (ergodic) kernel corresponding to $(\frac{1}{2}\Delta + \beta - \lambda_c)^\phi$. In this example P will denote the corresponding probability. By comparison with the constant branching rate case, it is evident that $a_t := \sqrt{2M} \cdot t$ is an appropriate choice, because a BBM with constant rate M has velocity

$\sqrt{2M}$. (Recall Proposition 1.16.) Therefore we have to find a ζ which satisfies that for any fixed ball B,

$$\lim_{t \to \infty} \sup_{|z| \le t} \left| \frac{p(z, B, \zeta(t))}{\int_B \phi \tilde{\phi}(y) \, dy} - 1 \right| = 0,$$

together with the condition that $\zeta(a_t) = \zeta(\sqrt{2M} \cdot t) = \mathcal{O}(t)$ as $n \to \infty$.

An easy computation (see again Example 22 in [Engländer and Turaev (2002)]) shows that on $\mathbb{R} \setminus [-b, b]$,

$$\left(\frac{1}{2}\Delta + \beta - \lambda_c \right)^\phi = \frac{1}{2}\Delta - \text{sgn}(x) \cdot \sqrt{2\lambda_c} \frac{d}{dx},$$

where $\text{sgn}(x) := x/|x|$, $x \ne 0$. Fix an ϵ and let $\tau_{\pm b}$ and τ_0 denote the first hitting time (by a single Brownian particle) of $[-b, b]$ and of 0, respectively. We first show that as $t \to \infty$,

$$\sup_{b < |x| \le t} P_x \left[\tau_{\pm b} > \frac{t(1 + \epsilon)}{\sqrt{2\lambda_c}} \right] \to 0. \tag{3.3}$$

Obviously, it suffices to show that for example

$$\lim_{t \to \infty} \mathbf{P}_t \left[\tau_0 > \frac{t(1 + \epsilon)}{\sqrt{2\lambda_c}} \right] = 0,$$

where $\mathbf{P}$ corresponds to $\frac{1}{2}\Delta - \sqrt{2\lambda_c} \frac{d}{dx}$ on $[0, \infty)$. Indeed, if $\mathcal{W}$ denotes standard Brownian motion starting at the origin, under probability Q, then

$$\mathbf{P}_t \left[\tau_0 > \frac{t(1 + \epsilon)}{\sqrt{2\lambda_c}} \right] \le \mathbf{P}_t \left[Y_{\frac{t(1+\epsilon)}{\sqrt{2\lambda_c}}} > 0 \right]$$

$$= Q \left[t - \sqrt{2\lambda_c} \frac{t(1 + \epsilon)}{\sqrt{2\lambda_c}} + \mathcal{W}_{\frac{t(1+\epsilon)}{\sqrt{2\lambda_c}}} > 0 \right]$$

$$= Q \left[\mathcal{W}_{\frac{t(1+\epsilon)}{\sqrt{2\lambda_c}}} > \epsilon t \right] \to 0$$

(the last term tends to zero by the SLLN for $\mathcal{W}$).

Define the shorthand $I(B) := \int_B \phi \tilde{\phi}(y) \, dy$. We now claim that $\zeta(t) := \frac{t(1+2\epsilon)}{\sqrt{2\lambda_c}}$ satisfies

$$\lim_{t \to \infty} \sup_{|z| \le t} \left| \frac{p(z, B, \zeta(t))}{I(B)} - 1 \right| = 0.$$

(The condition $\zeta(a_t) = \mathcal{O}(t)$ is obviously satisfied.) By the positive recurrence of the motion corresponding to $p(t, x, y)$, it suffices to verify that ζ satisfies

$$\lim_{t \to \infty} \sup_{b < |z| \le t} \left| \frac{p(z, B, \zeta(t))}{I(B)} - 1 \right| = 0.$$

Let, for example $b < x \le t$. By the strong Markov property at τ_b (the hitting time of b) and by (3.3),

$$\frac{p(x, B, \zeta(t))}{I(B)} = \frac{p\left(b, B, \zeta(t) - \frac{t(1+\epsilon)}{\sqrt{2\lambda_c}}\right)}{I(B)} P_x \left[\tau_b \le \frac{t(1 + \epsilon)}{\sqrt{2\lambda_c}} \right] + o(1),$$

uniformly in $b < x \le t$.

Finally, again because of the positive recurrence,

$$\lim_{t \to \infty} p\left(b, B, \zeta(t) - \frac{t(1 + \epsilon)}{\sqrt{2\lambda_c}}\right) = I(B),$$

noting that

$$\lim_{t \to \infty} \left[\zeta(t) - \frac{t(1 + \epsilon)}{\sqrt{2\lambda_c}} \right] = \lim_{t \to \infty} \frac{t\epsilon}{\sqrt{2\lambda_c}} = \infty.$$

This completes the proof of our claim about ζ.

Finally, for the sake of concreteness, we present a simple example for a *non-symmetric* operator[3] that satisfies our assumptions.

Example 3.5 (Non-symmetric operator). We now slightly modify the setting of Example 3.4.

In Example 3.4 set $d = 2$. Now *add a drift* $b(x, y)$ as follows. Let $b = (b_1, b_2)^T$, where $b_1(x, y) := m(x)n(y)$ and $b_2(x, y) := p(x)q(y)$. If m, n, p, q are smooth, compactly supported functions, then so is b, and the same argument as the one in Example 3.4 shows that the conditions are satisfied. Nonetheless, if $m(x)n'(y)$ is not equal to $p'(x)q(y)$ for all (x, y), that is, if $(m/p')(x) \ne (q/n')(y)$, then the operator is not symmetric, because then b is not a gradient vector.

Hence, whenever q/n' is not a constant or m/p' is not a constant, this setting constitutes a non-symmetric example for Theorem 2.2.

---

[3]The reader not familiar with symmetric operators can find more background in Section 4.10 of [Pinsky (1995)], for example. But reviewing our Section 1.10 is enough here.

3.1 Exercises

(1) In Example 3.2, prove that in expectation, the support of the process grows like $\sqrt{\lambda_c t / \gamma^+}$ as $t \to \infty$.
(2) In Example 3.4, complete the calculations for $d = 2$.
(3) Give a detailed proof of (3.2).

3.2 Notes

A strong law for a generalization of the model of Example 3.2 can be found in [Harris (2000)], where the convergence is proved using a martingale expansion for continuous functions $g \in L^2(\pi)$ (rather than compactly supported g). Almost sure asymptotic growth rates (and a.s. support) for the same model have been studied in [Git, Harris and Harris (2007)].

Having seen the specific examples above, we discuss next some *heuristic* computations — by seeing them, the reader may get a better idea as to how one can find such examples.

3.2.1 *Local versus global growth*

A natural question to ask is whether there is any quick way to guess, in a given setting, that the local and global rates are different. A simple approach is to look at the *expected* growth rates, as follows.

From (2.1), we have

$$E_x Z_t(B) = e^{\lambda_c t} \frac{\phi(x)}{\phi(y)} p(t, x, B),$$

for B Borel. When $B = D$, by ergodicity,

$$e^{-\lambda_c t} E_x \langle 1, Z_t \rangle = \phi(x) \int_D \frac{p(t, x, y)}{\phi(y)} \, dy \to \phi(x) \langle \widetilde{\phi}, 1 \rangle, \quad \text{as } t \to \infty,$$

provided $\langle \widetilde{\phi}, 1 \rangle < \infty$. (Recall Lemma 1.10.) Hence, if $\langle \widetilde{\phi}, 1 \rangle < \infty$, then the global population growth is the same as the local population growth (in expectation), whereas, if $\langle \widetilde{\phi}, 1 \rangle = \infty$ the global growth rate exceeds the local growth rate. The latter is the case in Example 3.3 too, since in that setting $\widetilde{\phi} \equiv 1$.

A much more thorough investigation of the question regarding 'local versus global growth' can be found in [Engländer, Ren and Song (2013)], albeit for superprocesses only. As already mentioned at the end of Chapter 1 (recall (1.65) and the paragraphs afterward), it has been shown, that, under quite general conditions, the global growth rate is given by the L^∞-growth bound (denoted by $\lambda_\infty = \lambda_\infty(L + \beta)$) of the semigroup corresponding to the operator $L + \beta$. Methods to distinguish between the cases $\lambda_\infty = \lambda_c$ and $\lambda_\infty > \lambda_c$ are also given in [Engländer, Ren and Song (2013)].

To generalize [Engländer, Ren and Song (2013)] so that it includes discrete branching diffusions too, however, is still to be achieved.

3.2.2 *Heuristics for a and ζ*

One may also wonder how one can find the functions a and ζ as in Definition 1.14(iii)–(iv). Let us see a quite straightforward method (due to S. Harris), which, despite being heuristic, is often efficient. Fix $x \in D$. Using the Borel-Cantelli lemma, if one can pick a deterministic, increasing function a such that, for all $\delta > 0$,

$$\sum_{n=1}^{\infty} P_x \left(\mathrm{supp}(Z_{n\delta}) \not\subset D_{a_{n\delta}} \right) < \infty, \tag{3.4}$$

then the function a is an appropriate choice, whenever also $\zeta(a_t) = \mathcal{O}(t)$ holds. Furthermore, since $P_x(Z_t(B) > 0) \leq E_x Z_t(B)$, for $t > 0$ and B Borel, thus instead of (3.4), we may simply check that

$$\sum_{n=1}^{\infty} E_x Z_{n\delta}(D^c_{a_{n\delta}}) < \infty, \tag{3.5}$$

which is a much easier task.

Next, recall from Definition 1.14 that $q(t, x, y)$ is the transition kernel corresponding to $L + \beta$ on D. If, for example, $D = \mathbb{R}^d$ and $D_t = B_t$, then using that

$$q(t, x, y) = e^{\lambda_c t} \phi(x) \frac{p(t, x, y)}{\phi(y)},$$

one has

$$E_x Z_{n\delta}(D^c_{a_{n\delta}}) = \int_{|y| > a_{n\delta}} q(t, x, y) \, \mathrm{d}y = e^{\lambda_c t} \cdot \int_{|y| > a_{n\delta}} \phi(x) \frac{p(t, x, y)}{\phi(y)} \, \mathrm{d}y.$$

Hence, (3.5) holds, whenever we can choose a_t such that, for some $\epsilon > 0$,

$$\int_{|y| > a_t} \frac{p(x, y, t)}{\phi(y)} \, \mathrm{d}y \leq e^{-(\lambda_c + \epsilon)t}, \, \forall t > 0. \tag{3.6}$$

How to find a function a satisfying (3.6) though? Let $F(\alpha) :=$ $\int_{|y| > \alpha} \widetilde{\phi}(y) \, \mathrm{d}y$. Since $\widetilde{\phi} > 0$, therefore $F \downarrow$, and we can define the inverse $G := F^{-1}$. Now, *if* the convergence $\lim_{t \to \infty} p(t, x, y) = \phi(y)\widetilde{\phi}(y)$ actually implies that the integral in (3.6) is 'close' to $F(a_t)$ for large times, then a natural candidate for the function a is given by $a_t = G(e^{-\lambda t})$, where $\lambda > \lambda_c$. This formal computation will be justified in situations when, loosely speaking, we have a 'nicely decaying' $\tilde{\phi}$, and the spine's transition density $p(t, x, y)$ converges to its equilibrium $\phi(y)\tilde{\phi}(y)$ 'sufficiently quickly' even for 'very large' y.

In order to gain some further insight, and to be able to have an educated guess for the function ζ too, we now use some ideas from the well-known *Freidlin-Wentzell large deviations theory of stochastic processes*. If our ergodic spine starts at a very *large* position, it will tend to move back toward the origin, albeit taking a potentially *large* time. Hence, according to the theory alluded to above, the spine particle will 'closely' follow the path of a deterministic particle with the same drift function.[4]

To be a bit more concrete, consider, just like in some of the previous examples, the operator $L = \frac{1}{2}\sigma^2(x)\Delta + \mu(x) \cdot \nabla$, and consider also the function $f_t = f(t)$ on $[0, \infty)$, solving the *deterministic* ordinary differential equation, corresponding to the h-transformed ($h = \phi$) operator:

$$\dot{f}_t = \mu^\phi(f_t), \tag{3.7}$$

where

$$\mu^\phi := \mu + \sigma^2 \, \nabla(\log \phi).$$

If we can solve (3.7) (with a generic initial condition), then we can use f to *guess* for a suitable form for ζ, as follows. Let us try to find out heuristically, how far away the spine particle may start in order that it both returns to the vicinity of the origin and then 'ergodizes' towards its invariant measure before large time t. To achieve this, the Freidlin-Wentzell theory suggests to approximate the path of the spine particle by 'its most probable realization,' given by the deterministic function f. This means that the spine's position at $\zeta(t)$ will be close to $f(\zeta(t))$. We want that quantity to be 'slightly smaller' than t.

For instance when $d = 1$, one should[5] set $\zeta(t)$ 'slightly larger' than $f^{-1}(t)$. (f^{-1} denotes the inverse of f.)

It turns out that these are precisely the heuristics, for both a and ζ, that yield some of the examples in this chapter.

---

[4]What this means is that the particle's motion is considered a small random perturbation of a dynamical system, and to 'significantly' deviate from the solution of the dynamical system is a 'large deviation' event.

[5]Assuming that $f \downarrow$ and the initial condition is positive, for example.

Chapter 4

The Strong Law for a type of self-interaction; the center of mass

In this chapter our goal is to obtain a strong (a.s.) limit theorem for a spatial branching system, but with a new feature: the particles, besides moving and branching, now also *interact* with each other.

To this end, we introduce a branching Brownian motion (BBM) with 'attraction' or 'repulsion' between the particles.

4.1 Model

Consider a dyadic (i.e. precisely two offspring replaces the parent) BBM in $\mathbb{R}^d$ with unit time branching and with the following interaction between particles: if Z denotes the process and Z_t^i is the ith particle, then Z_t^i 'feels' the drift

$$\frac{1}{n_t} \sum_{1 \leq j \leq n_t} \gamma \cdot \left(Z_t^j - \cdot \right),$$

where $\gamma \neq 0$, that is, at least intuitively,[1] the particle's motion is an $L^{(i)}$-diffusion, where

$$L^{(i)} := \frac{1}{2}\Delta + \frac{1}{n_t} \sum_{1 \leq j \leq n_t} \gamma \cdot \left(Z_t^j - x \right) \cdot \nabla. \tag{4.1}$$

(Here and in the sequel, n_t is a shorthand for $2^{\lfloor t \rfloor}$, where $\lfloor t \rfloor$ is the integer part of t.) If $\gamma > 0$, then this means *attraction*, if $\gamma < 0$, then it means *repulsion*.

To provide a rigorous construction, we define the process by induction as follows. Z_0 is a single particle at the origin. In the time interval $[m, m+1)$ we define a system of 2^m *interacting diffusions*, starting at the position of

---

[1]The drift depends on time and on the other particles' position too.

their parents at the end of the previous step (at time $m-0$) by the following system of stochastic differential equations:

$$dZ_t^i = dW_t^{m,i} + \frac{\gamma}{2^m} \sum_{1 \le j \le 2^m} (Z_t^j - Z_t^i)\, dt; \quad i = 1, 2, \ldots, 2^m, \qquad (4.2)$$

where $W^{m,i}, i = 1, 2, \ldots, 2^m; m = 0, 1, \ldots$ are independent Brownian motions.

The reason our interactive model is actually well-defined, is clear if we recall Theorem 1.9. Notice that the 2^m interacting diffusions on $[m, m+1)$ can be considered as a single $2^m d$-dimensional Brownian motion with linear (and therefore Lipschitz) drift $\mathbf{b} : \mathbb{R}^{2^m d} \to \mathbb{R}^{2^m d}$:

$$\mathbf{b}\big(x_1, x_2, \ldots, x_d, x_{1+d}, x_{2+d}, \ldots, x_{2d}, \ldots, x_{1+(2^m-1)d}, x_{2+(2^m-1)d}, \ldots, x_{2^m d}\big)$$
$$:= \gamma(\beta_1, \beta_2, \ldots, \beta_{2^m d})^T, \qquad (4.3)$$

where

$$\beta_k := 2^{-m} \left(x_{\widehat{k}} + x_{\widehat{k}+d} + \ldots + x_{\widehat{k}+(2^m-1)d} \right) - x_k, \quad 1 \le k \le 2^m d, \qquad (4.4)$$

and $\widehat{k} \equiv k \pmod{d}$, $1 \le \widehat{k} \le d$. By Theorem 1.9, this yields existence and uniqueness for our model.

Remark 4.1 (Weakly interacting particles). If there were no branching and the interval $[m, m+1)$ were extended to $[0, \infty)$, then for $\gamma > 0$ the interaction (4.2) would describe the *ferromagnetic Curie-Weiss model*, a model of weakly interacting stochastic particles, appearing in the microscopic statistical description of a spatially homogeneous gas in a granular medium. It is known that as $m \to \infty$, a Law of Large Numbers, the *McKean-Vlasov limit* holds and the normalized empirical measure

$$\rho_m(t) := 2^{-m} \sum_{i=1}^{2^m} \delta_{Z_t^i}$$

tends to a probability measure-valued solution of

$$\frac{\partial}{\partial t}\rho = \frac{1}{2}\Delta\rho + \frac{\gamma}{2}\nabla \cdot \left(\rho \nabla f^{(\rho)}\right),$$

where $f^{(\rho)}(x) := \int_{\mathbb{R}^d} |x - y|^2 \rho(dy)$.

In fact, besides the Curie-Weiss model, some other (non-linear) kinds of interactions between stochastic particles have interpretations in gas kinetics too. Branching is not present in any of these models. (See pp. 23–24 in [Feng and Kurtz (2006)] and the references therein for all the above.)   ◇

Convention: Before proceeding, we point out that in this chapter an Ornstein-Uhlenbeck process may refer to *either inward or outward* O-U processes. Similarly, we will use the phrase 'branching Ornstein-Uhlenbeck process' in both senses.

We are interested in the long time behavior of Z, and also whether we can say something about the number of particles in a given compact set for n large ('local asymptotics').

In order to answer these questions, we will first show that Z asymptotically becomes a branching Ornstein-Uhlenbeck process (inward for attraction and outward for repulsion), however

(1) the origin is shifted to a random point which has d-dimensional normal distribution $\mathcal{N}(0, 2\mathbf{I}_d)$, and

(2) the Ornstein-Uhlenbeck particles are not independent but constitute a system with a degree of freedom which is less than their number by precisely one.

The main result will concern the local behavior of the system: we will prove a scaling limit theorem (Theorem 4.1) for the local mass in the attractive ($\gamma > 0$) case, and formulate and motivate a conjecture (Conjecture 4.1) for the repulsive ($\gamma < 0$) case.

Finally, we remind the reader the notation $\langle g, Z_t \rangle = \langle Z_t, g \rangle := \sum_{i=1}^{n_t} g(Z_t^i)$, which we will frequently use.

4.2   The mass center stabilizes

Notice that

$$\frac{1}{n_t} \sum_{1 \leq j \leq n_t} \left(Z_t^j - Z_t^i \right) = \overline{Z}_t - Z_t^i, \tag{4.5}$$

and so *the net attraction pulls the particle towards the center of mass* (net repulsion pushes it away from the center of mass).

Since the interaction is in fact through the center of mass, it is important to analyze how it behaves for large times.

Lemma 4.1 (Mass center stabilizes). *The mass center performs a Brownian motion, slowed down by a factor 2^m in the unit time interval $[m, m+1)$, $m = 0, 1, 2, ...$; in particular, it is a Markov process. Furthermore, there exists a random variable $N \sim \mathcal{N}(\mathbf{0}, 2\mathbf{I}_d)$ such that $\lim_{t \to \infty} \overline{Z}_t = N$ a.s.*

Proof.   Let $m \in \mathbb{N}$. For $t \in [m, m+1)$ there are 2^m particles moving around and particle Z_t^i's $(1 \leq i \leq 2^m)$ motion is governed by the stochastic differential equation

$$dZ_t^i = dW_t^{m,i} + \gamma(\overline{Z}_t - Z_t^i)dt.$$

Since $2^m \overline{Z}_t = \sum_{i=1}^{2^m} Z_t^i$, we obtain that

$$d\overline{Z}_t = 2^{-m} \sum_{i=1}^{2^m} dZ_t^i = 2^{-m} \sum_{i=1}^{2^m} dW_t^{m,i} + \frac{\gamma}{2^m} \left(2^m \overline{Z}_t - \sum_{i=1}^{2^m} Z_t^i \right) dt$$

$$= 2^{-m} \sum_{i=1}^{2^m} dW_t^{m,i}.$$

Since, for intermediate times, we have

$$\overline{Z}_{m+\tau} = \overline{Z}_m + 2^{-m} \bigoplus_{i=1}^{2^m} W_\tau^{m,i} =: \overline{Z}_m + 2^{-m/2} B^{(m)}(\tau), \qquad (4.6)$$

where $0 \leq \tau < 1$, and $B^{(m)}$ is a Brownian motion on $[m, m+1)$, using induction, we obtain that[2]

$$\overline{Z}_t = B^{(0)}(1) \oplus \frac{1}{\sqrt{2}} B^{(1)}(1) \oplus \cdots \oplus \frac{1}{2^{k/2}} B^{(k)}(1) \oplus$$

$$\cdots \oplus \frac{1}{\sqrt{2^{\lfloor t \rfloor - 1}}} B^{(\lfloor t \rfloor - 1)}(1) \oplus \frac{1}{\sqrt{n_t}} B^{(\lfloor t \rfloor)}(\tau), \qquad (4.7)$$

where $\tau := t - \lfloor t \rfloor$.

Next, observe, that by Brownian scaling, the random variables

$$W^{(m)}(\cdot) := 2^{-m/2} B^{(m)}(2^m \cdot), \; m \geq 1$$

are (independent) Brownian motions, implying that

$$\overline{Z}_t = W^{(0)}(1) \oplus W^{(1)} \left(\frac{1}{2} \right) \oplus \cdots \oplus W^{(\lfloor t \rfloor - 1)} \left(\frac{1}{2^{\lfloor t \rfloor - 1}} \right) \oplus W^{(\lfloor t \rfloor)} \left(\frac{\tau}{n_t} \right),$$

To see that $\overline{Z}$ is a Markov process, let $\{\mathcal{F}_t\}_{t \geq 0}$ and $\{\mathcal{G}_t\}_{t \geq 0}$ be the canonical filtrations for Z and $\overline{Z}$, respectively. Since $\mathcal{G}_s \subset \mathcal{F}_s$, it is enough to check the Markov property with $\mathcal{G}_s$ replaced by $\mathcal{F}_s$.

Assume first $0 \leq s < t$, $\lfloor s \rfloor = \lfloor t \rfloor =: m$. Then the distribution of $\overline{Z}_t$, conditional on $\mathcal{F}_s$, is the same as conditional on Z_s, because Z itself is a Markov process. But the distribution of $\overline{Z}_t$ only depends on Z_s through $\overline{Z}_s$, as

$$\overline{Z}_t \stackrel{d}{=} \overline{Z}_s \oplus W^{(2^m)} \left(\frac{t-s}{2^m} \right), \qquad (4.8)$$

---

[2]It is easy to check that, as the notation suggests, the summands are independent.

whatever Z_s is. That is, $P(\overline{Z}_t \in \cdot \mid \mathcal{F}_s) = P(\overline{Z}_t \in \cdot \mid Z_s) = P(\overline{Z}_t \in \cdot \mid \overline{Z}_s)$. Note that this is even true when $s \in \mathbb{N}$ and $t = s + 1$, because $\overline{Z}_t = \overline{Z}_{t-0}$.

Assume now that $s < \lfloor t \rfloor =: m$. Then the equation $P(\overline{Z}_t \in \cdot \mid \mathcal{F}_s) = P(\overline{Z}_t \in \cdot \mid \overline{Z}_s)$, is obtained by conditioning successively on $m, m-1, ..., \lfloor s \rfloor + 1, s$.

By the Markov property, applied at $t = 1, \frac{1}{2}, ..., \frac{1}{2^{\lfloor t \rfloor - 1}}$, in fact

$$\overline{Z}_t = \widehat{W}\left(1 + \frac{1}{2} + \cdots + \frac{1}{2^{\lfloor t \rfloor - 1}} + \frac{\tau}{n_t}\right),$$

where $\widehat{W}$ is a Brownian motion (the concatenation of the $W^{(i)}$'s), and since $\widehat{W}$ has a.s. continuous paths, $\lim_{t \to \infty} \overline{Z}_t = \widehat{W}(2)$, a.s. □

For another interpretation see the remark after Lemma 4.3.

We will also need the following fact later, the proof of which we leave to the reader as an easy exercise.

Lemma 4.2. *The coordinate processes of Z are independent one-dimensional interactive branching processes of the same type as Z.*

Remark 4.2 (Dambis-Dubins-Schwarz Theorem viewpoint). This remark is for the reader familiar with the *Dambis-Dubins-Schwarz Theorem*.[3] The first statement of Lemma 4.1 is in fact a manifestation of that theorem. In our case the increasing process is deterministic, and even piecewise linear. The variance is being reduced in every time step, since it is that of the average of more and more independent particle positions. ◇

4.3 Normality via decomposition

As before, we denote $m := \lfloor t \rfloor$. We will need the following decomposition result.

Lemma 4.3 (Decomposition). *Consider the $d \cdot n_t$-dimensional process $(Z_t^1, Z_t^2, ..., Z_t^{n_t})$. This process can be decomposed into two components: a d-dimensional Brownian motion and an independent $d(n_t - 1)$-dimensional Ornstein-Uhlenbeck process with parameter γ. More precisely, in the time interval $[m, m+1)$, each coordinate process (as a 2^m-dimensional process) can be decomposed into two components:*

---

[3]It states that every d-dimensional continuous martingale with independent coordinate processes is a time-changed Brownian motion, where the time change is determined by the 'increasing process' of the martingale.

- *a one-dimensional Brownian motion in the direction* $(1, 1, ..., 1)$
- *an independent* $(2^m - 1)$-*dimensional Ornstein-Uhlenbeck process with parameter* γ *in the ortho-complement of the vector* $(1, 1, ..., 1)$.

Furthermore, the vector $Z_t = (Z_t^1, Z_t^2, ..., Z_t^{n_t})$, *conditioned on* Z_s, *is* (dn_t)-*dimensional joint normal for all* $t > s \geq 0$.

Proof. By Lemma 4.2, we may assume that $d = 1$. Recall (4.3-4.4), and note that for $d = 1$, they simplify to

$$\mathbf{b}(x_1, x_2, ..., x_{2^m}) =: \gamma(\beta_1, \beta_2, ..., \beta_{2^m})^T,$$

where

$$\beta_k = 2^{-m}(x_1 + x_2 + ... + x_{2^m}) - x_k, \quad 1 \leq k \leq 2^m. \tag{4.9}$$

What this means is that defining the 2^m-dimensional process Z^* on the time interval $t \in [m, m+1)$ by

$$Z_t^* := (Z_t^1, Z_t^2, ..., Z_t^{2^m}),$$

Z^* is a Brownian motion with drift

$$\gamma\left[(\overline{Z}_t, \overline{Z}_t, ..., \overline{Z}_t) - (Z_t^1, Z_t^2, ..., Z_t^{2^m})\right] \in \mathbb{R}^{2^m},$$

starting at a random position. (*Warning*: the reader should not confuse this 'artificial' space with the 'true' state space of the process, which is now simply $\mathbb{R}$ with 2^m interacting particles in it. The significance of working with this 'artificial' space is given exactly by the fact that we can ignore the dependence of particles.)

Notice the important fact that by the definition of $\overline{Z}_t$, this drift is orthogonal to the vector[4] $\mathbf{v} := (1, 1, ..., 1) \in \mathbb{R}^{2^m}$, that is, the vector $(\overline{Z}_t, \overline{Z}_t, ..., \overline{Z}_t) \in \mathbb{R}^{2^m}$ is nothing but the orthogonal projection of $(Z_t^1, Z_t^2, ..., Z_t^{2^m})$ in the direction of $\mathbf{v}$.

Notice also that the one-dimensional process $(\overline{Z}_t, \overline{Z}_t, ..., \overline{Z}_t)$ on the line spanned by $\mathbf{v}$ is precisely Brownian motion. Indeed, although $\overline{Z}_t$ is a Brownian motion slowed down by factor 2^m, we have 2^m of them, and, by the Pythagoras Theorem along with Brownian scaling, this precisely cancels the slowdown out.

These observations immediately lead to the statement of the lemma concerning the decomposition.

Note that the concatenation of the first (Brownian) components in the decomposition constitutes a single Brownian motion on $[0, \infty)$. The reason

---

[4]For simplicity, we use row vectors in the rest of the proof.

is that there is no jump at the time of the fission since each particle splits into precisely offspring, leaving the center of mass unchanged.

We now prove the normality of $(Z_t^1, Z_t^2, ..., Z_t^{N_t})$ by induction.

(i) In the time interval $[0, 1)$ the statement is trivially true.

(ii) Assume now that it is true on the time interval $[0, m)$. Consider the time m position of the 2^{m-1} particles $(Z_m^1, Z_m^2, ..., Z_m^{2^{m-1}})$ 'directly before' the fission. At the instant of the fission we obtain the 2^m-dimensional vector

$$(Z_m^1, Z_m^1, Z_m^2, Z_m^2, ..., Z_m^{2^{m-1}}, Z_m^{2^{m-1}}),$$

which has the same distribution on the 2^{m-1}-dimensional subspace

$$S := \{x \in \mathbb{R}^{2^m} \mid x_1 = x_2, x_3 = x_4, ..., x_{2^m-1} = x_{2^m}\}$$

of $\mathbb{R}^{2^m}$ as the vector $\sqrt{2}(Z_m^1, Z_m^2, ..., Z_m^{2^{m-1}})$ on $\mathbb{R}^{2^{m-1}}$.

Since, by the induction hypothesis, $(Z_m^1, Z_m^2, ..., Z_m^{2^{m-1}})$ is normal, the vector formed by the particle positions 'right after' the fission is a 2^m-dimensional degenerate normal.[5]

The normality on the time interval $[m, m+1)$ now follows from the fact that the convolution of normals is normal, along with the Gaussian property of the Wiener and Ornstein-Uhlenbeck processes (applied to $(\overline{Z}_t, \overline{Z}_t, ..., \overline{Z}_t)$ and $(\overline{Z}_t, \overline{Z}_t, ..., \overline{Z}_t) - (Z_t^1, Z_t^2, ..., Z_t^{2^m})$, respectively).

That $Z_t = (Z_t^1, Z_t^2, ..., Z_t^{n_t})$, conditioned on Z_s ($0 \leq s < t$) is joint normal, follows exactly the same way as in the $s = 0$ case above. □

Remark 4.3 (Mass center stabilizes, via decomposition). Consider the Brownian component in the decomposition appearing in the previous proof. Since, on the other hand, this coordinate is $2^{m/2}\overline{Z}_t$, using Brownian scaling, one obtains a slightly different way of seeing that $\overline{Z}_t$ stabilizes at a position which is distributed as the time $1+2^{-1}+2^{-2}+...+2^{-m}+ ... = 2$ value of a Brownian motion. (The decomposition shows this for $d = 1$ and then it is immediately upgraded to general d by independence.) ◇

Corollary 4.1 (Asymptotics for finite subsystem). *Let $k \geq 1$ and consider the subsystem $(Z_t^1, Z_t^2, ..., Z_t^k)$, $t \geq m_0$ for $m_0 := \lfloor \log_2 k \rfloor + 1$. (This means that at time m_0 we pick k particles and at every fission replace the parent particle by randomly picking one of its two descendants.) Let the real numbers $c_1, ..., c_k$ satisfy*

$$\sum_{i=1}^{k} c_i = 0, \quad \sum_{i=1}^{k} c_i^2 = 1. \tag{4.10}$$

---

[5]The reader can easily visualize this for $m = 1$: the distribution of (Z_1^1, Z_1^1) is clearly $\sqrt{2}$ times the distribution of a Brownian particle at time 1, i.e. $\mathcal{N}(0, \sqrt{2})$ on the line $x_1 = x_2$.

Define $\Psi_t = \Psi_t^{(c_1,...,c_k)} := \sum_{i=1}^{k} c_i Z_t^i$ *and note that* Ψ_t *is invariant under the translations of the coordinate system. Let* $\mathcal{L}_t$ *denote its law.*

For every $k \geq 1$ *and* $c_1, ..., c_k$ *satisfying* (4.10), $\Psi^{(c_1,...,c_k)}$ *is the same d-dimensional Ornstein-Uhlenbeck process corresponding to the operator* $1/2\Delta - \gamma\nabla \cdot x$, *and in particular, for* $\gamma > 0$,

$$\lim_{t\to\infty} \mathcal{L}_t = \mathcal{N}\left(0, \frac{1}{2\gamma}\mathbf{I}_d\right).$$

For example, taking $c_1 = 1/\sqrt{2}, c_2 = -1/\sqrt{2}$, we obtain that when viewed from a tagged particle's position, any given other particle moves as $\sqrt{2}$ times the above Ornstein-Uhlenbeck process.

Proof. By independence (Lemma 4.2) it is enough to consider $d = 1$. For m fixed, consider the decomposition appearing in the proof of Lemma 4.3 and recall the notation there. By (4.10), whatever $m \geq m_0$ is, the 2^m-dimensional unit vector

$$(c_1, c_2, ..., c_k, 0, 0, ..., 0)$$

is orthogonal to the 2^m-dimensional vector $\mathbf{v}$. This means that $\Psi^{(c_1,...,c_k)}$ is a one-dimensional projection of the Ornstein-Uhlenbeck component of Z^*, and thus it is itself a one-dimensional Ornstein-Uhlenbeck process (with parameter γ) on the unit time interval.

Now, although as m grows, the Ornstein-Uhlenbeck components of Z^* are defined on larger and larger spaces ($S \subset \mathbb{R}^{2^m}$ is a 2^{m-1}-dimensional linear subspace), the projection onto the direction of $(c_1, c_2, ..., c_k, 0, 0, ..., 0)$ is always the same one-dimensional Ornstein-Uhlenbeck process, i.e. the different unit time 'pieces' of $\Psi^{(c_1,...,c_k)}$ obtained by those projections may be concatenated. □

4.4 The interacting system as viewed from the center of mass

Recall that by (4.6) the interaction has no effect on the motion of $\overline{Z}$. Let us see now how the interacting system looks like when viewed from $\overline{Z}$.

4.4.1   The description of a single particle

Using our usual notation, assume that $t \in [m, m+1)$ and let $\tau := t - \lfloor t \rfloor$. When viewed from $\overline{Z}$, the relocation[6] of a particle is governed by the stochastic differential equation

$$d(Z_t^1 - \overline{Z}_t) = dZ_t^1 - d\overline{Z}_t = dW_t^{m,1} - 2^{-m} \sum_{i=1}^{2^m} dW_t^{m,i} - \gamma(Z_t^1 - \overline{Z}_t)dt.$$

So if $Y^1 := Z^1 - \overline{Z}$, then

$$dY_t^1 = dW_t^{m,1} - 2^{-m} \sum_{i=1}^{2^m} dW_t^{m,i} - \gamma Y_t^1 dt.$$

Clearly,

$$W_\tau^{m,1} - 2^{-m} \bigoplus_{i=1}^{2^m} W_\tau^{m,i} = \bigoplus_{i=2}^{2^m} 2^{-m} W_\tau^{m,i} \oplus (1 - 2^{-m}) W_\tau^{m,1};$$

and, by a trivial computation, the right-hand side is a Brownian motion with mean zero and variance $(1 - 2^{-m})\tau \mathbf{I}_d := \sigma_m^2 \tau \mathbf{I}_d$. That is,

$$dY_t^1 = \sigma_m d\widetilde{W}_t^{m,1} - \gamma Y_t^1 dt,$$

where $\widetilde{W}^{m,1}$ is a standard Brownian motion.

By Remark 1.8, this means that on the time interval $[m, m+1)$, Y^1 corresponds to the Ornstein-Uhlenbeck operator

$$\frac{1}{2}\sigma_m^2 \Delta - \gamma x \cdot \nabla. \tag{4.11}$$

Since for m large σ_m is close to one, the relocation viewed from the center of mass is *asymptotically governed by an Ornstein-Uhlenbeck process corresponding to* $\frac{1}{2}\Delta - \gamma x \cdot \nabla$.

Remark 4.4 (Asymptotically vanishing correlation). Let us now investigate the correlation between the driving BM's. To this end, let $\widetilde{W}^{m,i,k}$ be the kth coordinate of the ith Brownian motion: $\widetilde{W}^{m,i} = (\widetilde{W}^{m,i,k}, k = 1, 2, ..., d)$ and $B^{m,i,k}$ be the kth coordinate of $W^{m,i}$. For $1 \le i \ne j \le 2^m$, we have

$$E\left[\sigma_m \widetilde{W}_\tau^{m,i,k} \cdot \sigma_m \widetilde{W}_\tau^{m,j,k}\right]$$

$$= E\left[\left(B_\tau^{m,i,k} - 2^{-m} \bigoplus_{r=1}^{2^m} B_\tau^{m,r,k}\right)\left(B_\tau^{m,j,k} - 2^{-m} \bigoplus_{r=1}^{2^m} B_\tau^{m,r,k}\right)\right]$$

$$= -2^{-m}\left[\mathrm{Var}\left(B_\tau^{m,i,k}\right) + \mathrm{Var}\left(B_\tau^{m,j,k}\right)\right] + 2^{-2m} \cdot 2^m \tau$$

$$= (2^{-m} - 2^{1-m})\tau = -2^{-m}\tau,$$

---

[6]I.e. the relocation between time m and time t.

that is, for $i \neq j$,

$$E\left[\widetilde{W}_\tau^{m,i,k}\,\widetilde{W}_\tau^{m,j,\ell}\right] = -\frac{\delta_{k\ell}}{2^m - 1}\,\tau. \tag{4.12}$$

Hence *the pairwise correlation tends to zero* as $t \to \infty$ (recall that $m = \lfloor t \rfloor$ and $\tau = t - m \in [0, 1)$).

And of course, for the variances we have

$$E\left[\widetilde{W}_\tau^{m,i,k}\,\widetilde{W}_\tau^{m,i,\ell}\right] = \delta_{k\ell} \cdot \tau, \text{ for } 1 \leq i \leq 2^m. \qquad \diamond \tag{4.13}$$

4.4.2   The description of the system; the 'degree of freedom'

Fix $m \geq 1$ and for $t \in [m, m + 1)$ let $Y_t := (Y_t^1, ..., Y_t^{2^m})^T$, where $()^T$ denotes transposed. (This is a vector of length 2^m where each component itself is a d-dimensional vector; one can actually view it as a $2^m \times d$ matrix too.) We then have

$$\mathrm{d}Y_t = \sigma_m \mathrm{d}\widetilde{W}_t^{(m)} - \gamma Y_t \mathrm{d}t,$$

where

$$\widetilde{W}^{(m)} = \left(\widetilde{W}^{m,1}, ..., \widetilde{W}^{m,2^m}\right)^T$$

and the random variables

$$\widetilde{W}_\tau^{m,i} = \sigma_m^{-1}\left(W_\tau^{m,i} - 2^{-m}\bigoplus_{j=1}^{2^m} W_\tau^{m,j}\right), \ i = 1, 2, ..., 2^m$$

are mean zero Brownian motions with correlation structure given by (4.12)–(4.13).

Just like as in the argument for existence and uniqueness (see the paragraph after equation (4.2)), we can consider Y as a single $2^m d$-dimensional diffusion. Each of its components is an Ornstein-Uhlenbeck process with asymptotically unit diffusion coefficient.

By independence, it is enough to consider the one-dimensional case, and so from now on, in this subsection we assume that $d = 1$.

Let us first describe the distribution of $\widetilde{W}_t^{(m)}$ for $t \geq 0$ fixed. Recall that $\{W_s^{m,i}, \ s \geq 0; \ i = 1, 2, ..., 2^m\}$ are independent Brownian motions. By definition, $\widetilde{W}_t^{(m)}$ is a 2^m-dimensional multivariate normal:

$$\widetilde{W}_t^{(m)} = \sigma_m^{-1} \cdot \begin{pmatrix} 1 - 2^{-m} & -2^{-m} & ... & -2^{-m} \\ -2^{-m} & 1 - 2^{-m} & ... & -2^{-m} \\ & \cdot & & \\ & \cdot & & \\ & \cdot & & \\ -2^{-m} & -2^{-m} & ... & 1 - 2^{-m} \end{pmatrix} W_t^{(m)} =: \sigma_m^{-1}\mathbf{A}^{(m)}W_t^{(m)},$$

where $W_t^{(m)} = (W_t^{m,1}, ..., W_t^{m,2^m})^T$, yielding

$$dY_t = \mathbf{A}^{(m)} dW_t^{(m)} - \gamma Y_t dt.$$

Since we are viewing the system from the center of mass, $\widetilde{W}_t^{(m)}$ is a *singular* multivariate normal and thus Y is a degenerate diffusion. The 'true' dimension of $\widetilde{W}_t^{(m)}$ is $r(\mathbf{A}^{(m)})$.

Lemma 4.4. $r(\mathbf{A}^{(m)}) = 2^m - 1$.

Proof.   We will simply write $\mathbf{A}$ instead of $\mathbf{A}^{(m)}$. Since the columns of $\mathbf{A}$ add up to zero, the matrix $\mathbf{A}$ is not of full rank: $r(\mathbf{A}) \leq 2^m - 1$. On the other hand,

$$2^m \mathbf{A} + \begin{pmatrix} 1\,1\,...\,1 \\ 1\,1\,...\,1 \\ \cdot \\ \cdot \\ \cdot \\ 1\,1\,...\,1 \end{pmatrix} = 2^m \mathbf{I},$$

where $\mathbf{I}$ is the 2^m-dimensional unit matrix, and so, by the subadditivity of the rank, $r(\mathbf{A}) + 1 = r(2^m \mathbf{A}) + 1 \geq 2^m$. $\square$

By Lemma 4.4, $\widetilde{W}_t^{(m)}$ is concentrated on S, and there the vector $\widetilde{W}_t^{(m)}$ has *non-singular* multivariate normal distribution.[7] What this means is that even though $\widetilde{W}^{m,1}, ..., \widetilde{W}^{m,2^m}$ are not independent, their 'degree of freedom' is $2^m - 1$, i.e. the 2^m-dimensional vector $\widetilde{W}_t^{(m)}$ is *determined by* $2^m - 1$ *independent components* (corresponding to $2^m - 1$ principal axes).

Remark 4.5 (Connection with Lemma 4.3). The reader has already been warned not to confuse the 'physical' state space with the 2^m-dimensional space (for $d = 1$) appearing in Lemma 4.3. Nevertheless, the statement about the $2^m - 1$ degrees of freedom in the 'physical' space and the statement that the O-U process appearing in the decomposition in Lemma 4.3 is $(2^m - 1)$-dimensional, describe the exact same phenomenon.◇

4.5   Asymptotic behavior

4.5.1   *Conditioning*

Our next purpose, quite naturally, is to 'put together' two facts:

[7]Recall that S is the $(2^m - 1)$-dimensional linear subspace given by the orthogonal complement of the vector $(1, 1, ..., 1)^T$.

(1) that $\overline{Z}_t$ tends to a random final position,
(2) the description of the system 'as viewed from $\overline{Z}_t$.'

The following lemma is the first step in this direction. It shows that the terminal position of the center does not affect the statistics of the relative system.

Lemma 4.5 (Independence). *Let $\mathcal{T}$ be the tail σ-algebra of $\overline{Z}$.*

(1) For $t \geq 0$, the random vector Y_t is independent of the path $\{\overline{Z}_s\}_{s \geq t}$.
(2) The process $Y = (Y_t; t \geq 0)$ is independent of $\mathcal{T}$.

Proof. In both parts we will refer to the following fact. Let $s \leq t$, $s \in [\widehat{m}, \widehat{m} + 1); t \in [m, m + 1)$ with $\widehat{m} \leq m$. Since the random variables $Z_t^1, Z_t^2, ..., Z_t^{2^m}$ are exchangeable, thus, denoting $\widehat{n} := 2^{\widehat{m}}, n := 2^m$, the vectors $\overline{Z}_t$ and $Z_s^1 - \overline{Z}_s$ are uncorrelated for $0 \leq s \leq t$. Indeed, by Lemma 4.2, we may assume that $d = 1$ and then

$$E[\overline{Z}_t \cdot (Z_s^1 - \overline{Z}_s)]$$

$$= E\left[\frac{Z_t^1 + Z_t^2 + ... + Z_t^n}{n} \cdot \left(Z_s^1 - \frac{Z_s^1 + Z_s^2 + ... + Z_s^{\widehat{n}}}{\widehat{n}}\right)\right]$$

$$= \frac{1}{n}E\left(Z_t^1 \cdot Z_s^1\right) + \frac{n-1}{n}E\left(Z_s^1 \cdot Z_t^2\right) - \frac{\widehat{n}}{n\widehat{n}}E\left(Z_t^1 \cdot Z_s^1\right)$$

$$- \frac{\widehat{n}(n-1)}{n\widehat{n}}E\left(Z_t^2 \cdot Z_s^1\right) = 0.$$

(Of course the index 1 can be replaced by i for any $1 \leq i \leq 2^m$.)

<u>Part (1)</u>: First, for any $t > 0$, the ($d \cdot 2^m$-dimensional) vector Y_t is independent of the (d-dimensional) vector $\overline{Z}_t$, because the $d(2^m+1)$-dimensional vector

$$(\overline{Z}_t, Z_t^1 - \overline{Z}_t, Z_t^2 - \overline{Z}_t, ..., Z_t^{2^m} - \overline{Z}_t)^T$$

is normal (since it is a linear transformation of the $d \cdot 2^m$-dimensional vector $(Z_t^1, Z_t^2, ..., Z_t^{2^m})^T$, which is normal by Lemma 4.3), and so it is sufficient to recall that $\overline{Z}_t$ and $Z_t^i - \overline{Z}_t$ are uncorrelated for $1 \leq i \leq 2^m$.

To complete the proof of (a), recall (4.6) and (4.7) and notice that the conditional distribution of $\{\overline{Z}_s\}_{s \geq t}$ given $\mathcal{F}_t$ only depends on its starting point $\overline{Z}_t$, as it is that of a Brownian path appropriately slowed down, whatever Y_t (or, equivalently, whatever $Z_t = Y_t + \overline{Z}_t$) is. Since, as we have seen, Y_t is independent of $\overline{Z}_t$, we are done.

Part (2): Let $A \in \mathcal{T}$. By Dynkin's π-λ-Lemma (Lemma 1.1), it is enough to show that $(Y_{t_1}, ..., Y_{t_k})$ is independent of A for $0 \le t_1 < ... < t_k$ and $k \ge 1$. Since $A \in \mathcal{T} \subset \sigma(\overline{Z}_s; s \ge t_k + 1)$, it is sufficient to show that $(Y_{t_1}, ..., Y_{t_k})$ is independent of $\{\overline{Z}_s\}_{s \ge t_k + 1}$.

To see this, similarly as in Part (1), notice that the conditional distribution of $\{\overline{Z}_s\}_{s \ge t_k + 1}$ given $\mathcal{F}_{t_k + 1}$ only depends on its starting point $\overline{Z}_{t_k + 1}$, as it is that of a Brownian path appropriately slowed down, whatever the vector $(Y_{t_1}, ..., Y_{t_k})$ is. If we show that $(Y_{t_1}, ..., Y_{t_k})$ is independent of $\overline{Z}_{t_k + 1}$, we are done.

To see why the latter is true, one just have to repeat the argument in (a), using again normality[8] and recalling that the vectors $\overline{Z}_t$ and $Z_s^i - \overline{Z}_s$ are uncorrelated. □

Remark 4.6 (Conditioning on the final position of $\overline{Z}$). Recall that $N := \lim_{t \to \infty} \overline{Z}_t$ exists and $N \sim \mathcal{N}(0, 2\mathbf{I}_d)$. Define the conditional laws

$$P^x(\cdot) := P(\cdot \mid N = x), \; x \in \mathbb{R}^d.$$

By Lemma 4.5, $P^x(Y_t \in \cdot) = P(Y_t \in \cdot)$ for almost all $x \in \mathbb{R}^d$. It then follows that the decomposition $Z_t = \overline{Z}_t \oplus Y_t$ as well as the result obtained for the distribution of Y in subsections 4.4.1 and 4.4.2 are true under P^x too, for almost all $x \in \mathbb{R}^d$. ◇

4.5.2 *Main result and a conjecture*

Here is a summary of what we have shown up to now:

(1) On the time interval $[m, m + 1)$, Y^1 is an Ornstein-Uhlenbeck process corresponding to the operator

$$\frac{1}{2}\sigma_m^2 \Delta - \gamma x \cdot \nabla;$$

(2) $\sigma_m \to 1$ as $m \to \infty$;

(3) There is asymptotically vanishing correlation between the driving Brownian motions;

(4) The process Y satisfies

$$dY_t = \mathbf{A}^{(m)} dW_t^{(m)} - \gamma Y_t dt,$$

where $\{W_s^{m,i}, \; s \ge 0; \; i = 1, 2, ..., 2^m\}$ are independent Brownian motions;

---

[8]We now need normality for finite dimensional distributions and not just for one-dimensional marginals, but this is still true by Lemma 4.3.

(5) $r(\mathbf{A}^{(m)}) = 2^m - 1$;

(6) The terminal position of the center of mass is independent of the relative motions; the relative motions are independent of the 'future' of the center of mass (Lemma 4.5).

We now go beyond these preliminary results and state a theorem (the main result of this chapter) and a conjecture on the local behavior of the system.

Of course, once we have the description of Y as in (1-6) above, we may attempt to put them together with Theorem 2.2 for the process Y. If the components of Y were independent and the branching rate were exponential, the theorem would be readily applicable. However, since the 2^m components of Y are not independent (their degree of freedom is $2^m - 1$, as expressed by (5) above) and since, unlike in the non-interacting case, we now have unit time branching, the method of the previous chapter has to be adapted to our setting. As we will see, this adaptation requires quite a bit of extra work.

Recall that one can consider Z_n as 'empirical measure,' that is, as an element of $\mathcal{M}_f(\mathbb{R}^d)$, by putting unit point mass at the site of each particle; with a slight abuse of notation we will write $Z_n(dy)$. Let $\{P^x, x \in \mathbb{R}^d\}$ be as in Remark 4.6. Our main result below says that in the attractive case, the *normalized empirical measure* has a limit as $n \to \infty$, P^x-a.s.

Theorem 4.1 (Scaling limit for the attractive case). *If $\gamma > 0$, then, as $n \to \infty$,*

$$2^{-n} Z_n(dy) \overset{w}{\Rightarrow} \left(\frac{\gamma}{\pi}\right)^{d/2} \exp\left(-\gamma|y - x|^2\right) dy, \qquad (4.14)$$

almost surely under P^x, for almost all $x \in \mathbb{R}^d$. Consequently,

$$2^{-n} E Z_n(dy) \overset{w}{\Rightarrow} f^\gamma(y) dy, \qquad (4.15)$$

where

$$f^\gamma(\cdot) = \left(\pi(4 + \gamma^{-1})\right)^{-d/2} \exp\left[\frac{-|\cdot|^2}{4 + \gamma^{-1}}\right].$$

Remark 4.7. Notice that f^γ, which is the limiting density of the intensity measure, is the density for $\mathcal{N}\left(0, \left(2 + \frac{1}{2\gamma}\right) \mathbf{I}_d\right)$. This is the convolution of two terms:

(1) $\mathcal{N}(0, 2\mathbf{I}_d)$, representing the randomness of the final position of the center of mass (cf. Lemma 4.1);

(2) $\mathcal{N}\left(\mathbf{0}, \left(\frac{1}{2\gamma}\right)\mathbf{I}_d\right)$, representing the final distribution of the mass scaled Ornstein-Uhlenbeck branching particle system around its center of mass (cf. (4.14)).

For strong attraction, the contribution of the second term is negligible. $\diamond$

Next, we state a conjecture, and provide some explanation.

Conjecture 4.1 (Dichotomy for the repulsive case). *Let* $\gamma < 0$.

(1) If $|\gamma| \geq \frac{\log 2}{d}$, *then* Z *suffers local extinction:*
$$Z_n(\mathrm{d}y) \overset{v}{\Rightarrow} 0, \quad a.s. \ under \ P.$$

(2) If $|\gamma| < \frac{\log 2}{d}$, *then*
$$2^{-n} e^{d|\gamma|n} Z_n(\mathrm{d}y) \overset{v}{\Rightarrow} \mathrm{d}y, \quad a.s. \ under \ P.$$

4.5.3 The intuition behind the conjecture

A heuristic picture behind the conjecture, and in particular behind the phase transition at $\log 2/d$, is given below.

Recall first the situation for ordinary (non-interacting) $(L, \beta; D)$-branching diffusions from subsection 1.15.5: either local extinction or local exponential growth takes place according to whether $\lambda_c \leq 0$ or $\lambda_c > 0$, where $\lambda_c = \lambda_c(L + \beta)$ is the generalized principle eigenvalue of $L + \beta$ on $\mathbb{R}^d$. In particular, for $\beta \equiv B > 0$, the criterion for local exponential growth becomes $B > |\lambda_c(L)|$, where $\lambda_c(L) \leq 0$ is the generalized principle eigenvalue of L. Since λ_c is also the 'exponential rate of escape from compacts' for the diffusion corresponding to L, the interpretation of the criterion in this case is that a large enough mass creation can compensate the fact that individual particles drift away from a given bounded set. (Note that if L corresponds to a recurrent diffusion, then $\lambda_c(L) = 0$.)

Now return to our interacting model. The situation is similar as before, with $\lambda_c = d\gamma$ for the outward Ornstein-Uhlenbeck process, taking into account that for unit time branching, the role of B is played by $\log 2$. The condition for local exponential growth should therefore be $\log 2 > d|\gamma|$.

The scaling $2^{-n} e^{d|\gamma|n}$ comes from a similar consideration, noting that in our unit time branching setting, 2^n replaces the term $e^{\beta t}$ appearing in the exponential branching case, while $e^{\lambda_c(L)t}$ becomes $e^{\lambda_c(L)n} = e^{d\gamma n}$.

Note that since the rescaled (vague) limit of $Z_n(\mathrm{d}y)$ is translation invariant (i.e. Lebesgue), the final position of the center of mass plays no role.

Although we will not prove Conjecture 4.1, we will discuss some of the technicalities in section 4.7.

4.6 Proof of Theorem 4.1

Fix $x \in \mathbb{R}^d$, and abbreviate

$$\nu^{(x)}(dy) := \left(\frac{\gamma}{\pi}\right)^{d/2} \exp\left(-\gamma|y - x|^2\right) dy.$$

Before proving (4.14), we note the following. Clearly,

$$2^{-n} Z_n(dx), \ \nu^{(x)} \in \mathcal{M}_1(\mathbb{R}^d).$$

Consequently, by a standard fact from functional analysis[9], the convergence $2^{-n} Z_n(dx) \overset{w}{\Rightarrow} \nu^{(x)}$ is equivalent to the statement that

$$\forall g \in \mathcal{E}: \ 2^{-n}\langle g, Z_n \rangle \to \langle g, \nu^{(x)} \rangle,$$

where $\mathcal{E}$ is any given family of bounded measurable functions with $\nu^{(x)}$-zero (Lebesgue-zero) sets of discontinuity, that is separating[10] for $\mathcal{M}_1(\mathbb{R}^d)$.

In fact, one can pick a *countable* $\mathcal{E}$, which, furthermore, consists of compactly supported functions. Such an $\mathcal{E}$ is given by the indicators of sets in $\mathcal{R}$.

Fix such a family $\mathcal{E}$. Since $\mathcal{E}$ is countable, in order to show (4.14), it is sufficient to prove that for almost all $x \in \mathbb{R}^d$,

$$P^x(2^{-n}\langle g, Z_n \rangle \to \langle g, \nu^{(x)} \rangle) = 1, \ g \in \mathcal{E}. \tag{4.16}$$

We will carry out the proof of (4.16) in several subsections.

4.6.1 *Putting Y and $\overline{Z}$ together*

Let $\widetilde{f}(\cdot) = \widetilde{f}^\gamma(\cdot) := (\frac{\gamma}{\pi})^{d/2} \exp\left(-\gamma|\cdot|^2\right)$, and note that $\widetilde{f}$ is the density for $\mathcal{N}(\mathbf{0}, (2\gamma)^{-1}\mathbf{I}_d)$.

We now assert that in order to show (4.16), it suffices to prove that for almost all x,

$$P^x(2^{-n}\langle g, Y_n \rangle \to \langle g, \widetilde{f} \rangle) = 1, \ g \in \mathcal{E}. \tag{4.17}$$

This is because

$$\lim_{n\to\infty} 2^{-n}\langle g, Z_n \rangle = \lim_{n\to\infty} 2^{-n}\langle g, Y_n + \overline{Z}_n \rangle = \lim_{n\to\infty} 2^{-n}\langle g(\cdot + \overline{Z}_n), Y_n \rangle = I + II,$$

---

[9]See Proposition 4.8.12 and the proof of Propositions 4.8.15 in [Breiman (1992)].
[10]This means that for $\mu \neq \widehat{\mu} \in \mathcal{M}_1(\mathbb{R}^d)$, there exists an $f \in \mathcal{E}$ with $\langle f, \mu \rangle \neq \langle f, \widehat{\mu} \rangle$.

where

$$I := \lim_{n \to \infty} 2^{-n} \langle g(\cdot + x), Y_n \rangle$$

and

$$II := \lim_{n \to \infty} 2^{-n} \langle g(\cdot + \overline{Z}_n) - g(\cdot + x), Y_n \rangle.$$

Now, (4.17) implies that for almost all x, $I = \langle g(\cdot + x), \widetilde{f}(\cdot) \rangle$ P^x-a.s., while the compact support of g, and Heine's Theorem yields that $II = 0$, P^x-a.s. Hence, $\lim_{n \to \infty} 2^{-n} \langle g, Z_n \rangle = \langle g(\cdot + x), \widetilde{f}(\cdot) \rangle = \langle g(\cdot), \widetilde{f}(\cdot - x) \rangle$, P^x-a.s., giving (4.16).

Next, let us see how (4.14) implies (4.15). Let g be continuous and bounded. Since $2^{-n} \langle Z_n, g \rangle \leq \|g\|_\infty$, it follows by bounded convergence that

$$\lim_{n \to \infty} E 2^{-n} \langle Z_n, g \rangle = \int_{\mathbb{R}^d} E^x \left(\lim_{n \to \infty} 2^{-n} \langle Z_n, g \rangle \right) Q(\mathrm{d}x)$$

$$= \int_{\mathbb{R}^d} \langle g(\cdot), \widetilde{f}(\cdot - x) \rangle Q(\mathrm{d}x),$$

where $Q \sim \mathcal{N}(0, 2\mathbf{I}_d)$. Now, if $\widehat{f} \sim \mathcal{N}(0, 2\mathbf{I}_d)$ then, since $f^\gamma \sim \mathcal{N}\left(0, \left(2 + \frac{1}{2\gamma}\right) \mathbf{I}_d\right)$, it follows that $f^\gamma = \widehat{f} * \widetilde{f}$ and

$$\int_{\mathbb{R}^d} \langle g(\cdot), \widetilde{f}(\cdot - x) \rangle Q(\mathrm{d}x) = \langle g(\cdot), f^\gamma \rangle,$$

yielding (4.15).

Next, notice that it is in fact sufficient to prove (4.17) *under P instead of P^x*. Indeed, by Lemma 4.5,

$$P^x \left(\lim_{n \to \infty} 2^{-n} \langle g, Y_n \rangle = \langle g, \widetilde{f} \rangle \right) = P \left(\lim_{n \to \infty} 2^{-n} \langle g, Y_n \rangle = \langle g, \widetilde{f} \rangle \mid N = x \right)$$

$$= P \left(\lim_{n \to \infty} 2^{-n} \langle g, Y_n \rangle = \langle g, \widetilde{f} \rangle \right).$$

Let us use the shorthand $U_n(\mathrm{d}y) := 2^{-n} Y_t(\mathrm{d}y)$; in general $U_t(\mathrm{d}y) := \frac{1}{n_t} Y_t(\mathrm{d}y)$. With this notation, our goal is to show that

$$P(\langle g, U_n \rangle \to \langle g, \widetilde{f} \rangle) = 1, \ g \in \mathcal{E}. \tag{4.18}$$

Now, as mentioned earlier, we may (and will) set $\mathcal{E} := \mathcal{I}$, where $\mathcal{I}$ is the family of indicators of sets in $\mathcal{R}$. Then, it remains to show that

$$P \left(U_n(B) \to \int_B \widetilde{f}(x) \mathrm{d}x \right) = 1, \ B \in \mathcal{R}. \tag{4.19}$$

4.6.2   Outline of the further steps

Notation 4.1. In the sequel $\{\mathcal{F}_t\}_{t\geq 0}$ will denote the canonical filtration for Y, rather than the canonical filtration for Z.

The following key lemma (Lemma 4.6), which is similar to Lemma 2.3, will play an important role. It will be derived using Lemma 4.7 and (4.27), where the latter will be derived with the help of Lemma 4.7 too. Then, Lemma 4.6 together with (4.29) will be used to complete the proof of (4.19) and hence, that of Theorem 4.1.

Lemma 4.6 (Key Lemma). *Let $B \subset \mathbb{R}^d$ be a bounded measurable set, and let $\{m_n\}_{n\geq 1}$ be any non-decreasing sequence. Then, P-a.s.,*

$$\lim_{n\to\infty} [U_{n+m_n}(B) - E(U_{n+m_n}(B) \mid \mathcal{F}_n)] = 0. \qquad (4.20)$$

4.6.3   Establishing the crucial estimate (4.27) and the key Lemma 4.6

Let Y_n^i denote the '*i*th' particle at time n, $i = 1, 2, ..., 2^n$. Since B is a fixed set, in the sequel we will simply write U_n instead of $U_n(B)$. Recall the time inhomogeneity (piecewise constant coefficients in time) of the underlying diffusion process and note that by the branching property, we have the *clumping decomposition*: for $n, m \geq 1$,

$$U_{n+m} = \sum_{i=1}^{2^n} 2^{-n} U_m^{(i)}, \qquad (4.21)$$

where given $\mathcal{F}_n$, each member in the collection $\{U_m^{(i)} : i = 1, ..., 2^n\}$ is defined similarly to U_m but with Y_m replaced by the time m configuration of the particles starting at Y_n^i, $i = 1, ..., 2^n$, respectively, and with motion component $\frac{1}{2}\sigma_{n+k}\Delta - \gamma x \cdot \nabla$ in the time interval $[k, k + 1)$.

4.6.3.1   The functions a and ζ

Next, we define two positive functions, a and ζ on $(1, \infty)$. Our motivation is the same as in the previous two chapters, where we have investigated the SLLN for a branching diffusion without interaction. Namely,

(i) The function a. will be related (via (4.24) below) to the *radial speed* of the particle system Y.

(ii) The function $\zeta(\cdot)$, will be related (via (4.25) below) to the *speed of ergodicity* of the underlying Ornstein-Uhlenbeck process.

For $t > 1$, define

$$a_t := C_0 \cdot \sqrt{t}, \tag{4.22}$$
$$\zeta(t) := C_1 \log t, \tag{4.23}$$

where C_0 and C_1 are positive (non-random) constants to be determined later. Note that

$$m_t := \zeta(a_t) = C_3 + C_4 \log t$$

with $C_3 = C_1 \log C_0 \in \mathbb{R}$ and $C_4 = C_1/2 > 0$. We will use the shorthand

$$\ell_n := 2^{\lfloor m_n \rfloor}.$$

Recall that $\widetilde{f}^\gamma$ is the density for $\mathcal{N}(0, (2\gamma)^{-1}\mathbf{I}_d)$ and let $q(x, y, t) = q^{(\gamma)}(x, y, t)$ and $q(x, dy, t) = q^{(\gamma)}(x, dy, t)$ denote the transition density and the transition kernel, respectively, corresponding to the operator $\frac{1}{2}\Delta - \gamma x \cdot \nabla$. We are going to show below that for sufficiently large C_0 and C_1, the following holds. For each given $x \in \mathbb{R}^d$ and $B \subset \mathbb{R}^d$ non-empty bounded measurable set,

$$P\left(\exists n_0, \forall n_0 < n \in \mathbb{N} : \operatorname{supp}(Y_n) \subset B_{a_n}\right) = 1, \text{ and} \tag{4.24}$$

$$\lim_{t \to \infty} \sup_{z \in B_t, y \in B} \left| \frac{q(z, y, \zeta(t))}{\widetilde{f}^\gamma(y)} - 1 \right| = 0. \tag{4.25}$$

For (4.24), recall that in Example 3.2 of the previous chapter, similar calculations have been carried out for the case when the underlying diffusion is an Ornstein-Uhlenbeck process and the breeding is quadratic. It is important to recall that in that example the estimates followed from *expectation* calculations, and thus they can be mimicked in our case for the Ornstein-Uhlenbeck process performed by the particles in Y (which corresponds to the operator $\frac{1}{2}\sigma_m \Delta - \gamma x \cdot \nabla$ on $[m, m+1)$, $m \geq 1$), despite the fact that the particle motions are now correlated. These expectation calculations lead to the estimate that the growth rate of the support of Y satisfies (4.24) with a sufficiently large $C_0 = C_0(\gamma)$. The same example shows that (4.25) holds with a sufficiently large $C_1 = C_1(\gamma)$.

Remark 4.8. Denote by $\nu = \nu^\gamma \in \mathcal{M}_1(\mathbb{R}^d)$ the normal distribution $\mathcal{N}(0, (2\gamma)^{-1}\mathbf{I}_d)$. Let $B \subset \mathbb{R}^d$ be a non-empty bounded measurable set. Taking $t = a_n$ in (4.25) and recalling that $\zeta(a_n) = m_n$, one obtains that

$$\lim_{n \to \infty} \sup_{z \in B_{a_n}, y \in B} \left| \frac{q(z, y, m_n)}{\widetilde{f}^\gamma(y)} - 1 \right| = 0.$$

Since $\widetilde{f}^\gamma$ is bounded, this implies that for any bounded measurable set $B \subset \mathbb{R}^d$,

$$\lim_{n \to \infty} \sup_{z \in B_{a_n}} [q(z, B, m_n) - \nu(B)] = 0. \tag{4.26}$$

We will use (4.26) in Subsection 4.6.4. ◇

4.6.3.2   *Covariance estimates*

Let $\{Y_{m_n}^{i,j}, \ j = 1, ..., \ell_n\}$ be the descendants of $Y_n{}^i$ at time $m_n + n$. So $Y_{m_n}^{1,j}$ and $Y_{m_n}^{2,k}$ are respectively the jth and kth particles at time $m_n + n$ of the trees emanating from the first and second particles at time n. It will be useful to control the covariance between $\mathbf{1}_B(Y_{m_n}^{1,j})$ and $\mathbf{1}_B(Y_{m_n}^{2,k})$, where B is a non-empty, bounded open set. To this end, we will need the following lemma, the proof of which is relegated to Section 4.8 in order to minimize the interruption in the main flow of the argument.

Lemma 4.7. *Let $B \subset \mathbb{R}^d$ be a bounded measurable set.*

(a) *There exists a non-random constant $K(B)$ such that if $C = C(B, \gamma) := \frac{3}{\gamma}|B|^2 K(B)$, then*

$$P\Big[\forall n \text{ large enough and } \forall \xi, \tilde{\xi} \in \Pi_n, \ \xi \neq \tilde{\xi}:$$

$$\big|P(\xi_{m_n}, \tilde{\xi}_{m_n} \in B \mid \mathcal{F}_n) - P(\xi_{m_n} \in B \mid \mathcal{F}_n)P(\tilde{\xi}_{m_n} \in B \mid \mathcal{F}_n)\big| \leq \frac{Cn}{2^n}\Big]$$

$$= 1,$$

where Π_n denotes the collection of those ℓ_n particles, which start at some time-n location of their parents and run for (an additional) time m_n.

(b) *Let $C = C(B) := \nu(B) - (\nu(B))^2$. Then*

$$P\Big[\lim_{n \to \infty} \sup_{\xi \in \Pi_n} \big|\mathrm{Var}\left(\mathbf{1}_{\{\xi_{m_n} \in B\}} \mid \mathcal{F}_n\right) - C\big| = 0\Big] = 1.$$

Remark 4.9. In the sequel, instead of writing ξ_{m_n} and $\tilde{\xi}_{m_n}$, we will use the notation $Y_{m_n}^{i_1,j}$ and $Y_{m_n}^{i_2,k}$ with $1 \leq i_1, i_2 \leq n; 1 \leq j, k \leq \ell_n$ satisfying that $i_1 \neq i_2$ or $j \neq k$.   ◇

4.6.3.3   *The crucial estimate (4.27)*

Let $B \subset \mathbb{R}^d$ be a bounded measurable set and $C = C(B, \gamma)$ as in Lemma 4.7. Define

$$\mathcal{Z}_i := \frac{1}{\ell_n} \sum_{j=1}^{\ell_n} \big[\mathbf{1}_B(Y_{m_n}^{i,j}) - P(Y_{m_n}^{i,j} \in B \mid Y_n^i)\big], \quad i = 1, 2, ..., 2^n.$$

With the help of Lemma 4.7, we will establish the following crucial estimate, the proof of which is provided in Section 4.8.

Claim 4.1. *There exists a non-random constant $\widehat{C}(B, \gamma) > 0$ such that the following event holds P-almost surely:*

$$\sum_{1 \le i \ne j \le 2^n} E\left[\mathcal{Z}_i \mathcal{Z}_j \mid \mathcal{F}_n\right] \le \widehat{C}(B, \gamma) n \ell_n \sum_{i=1}^{2^n} E\left[\mathcal{Z}_i^2 \mid \mathcal{F}_n\right], \text{ for all large } n \in \mathbb{N}.$$

(4.27)

The significance of Claim 4.1 is as follows.

Claim 4.2. *Lemma 4.7 together with the estimate (4.27) implies Lemma 4.6.*

Proof of Claim 4.2. Assume that (4.27) holds. By the clumping decomposition under (4.21),

$$U_{n+m_n} - E(U_{n+m_n} \mid \mathcal{F}_n) = \sum_{i=1}^{2^n} 2^{-n} \left(U_{m_n}^{(i)} - E(U_{m_n}^{(i)} \mid \mathcal{F}_n) \right).$$

Since $U_{m_n}^{(i)} = \ell_n^{-1} \sum_{j=1}^{\ell_n} \mathbf{1}_B(Y_{m_n}^{i,j})$, therefore

$$U_{m_n}^{(i)} - E(U_{m_n}^{(i)} \mid \mathcal{F}_n) = U_{m_n}^{(i)} - E(U_{m_n}^{(i)} \mid Y_n^i) = \mathcal{Z}_i.$$

Hence,

$$E\left([U_{n+m_n} - E(U_{n+m_n} \mid \mathcal{F}_n)]^2 \mid \mathcal{F}_n \right)$$

$$= E\left(\left[\sum_{i=1}^{2^n} 2^{-n} \left(U_{m_n}^{(i)} - E(U_{m_n}^{(i)} \mid \mathcal{F}_n) \right) \right]^2 \mid \mathcal{F}_n \right)$$

$$= E\left(\left[\sum_{i=1}^{2^n} 2^{-n} \mathcal{Z}_i \right]^2 \mid \mathcal{F}_n \right)$$

$$= 2^{-2n} \left[\sum_{i=1}^{2^n} E\left(\mathcal{Z}_i^2 \mid \mathcal{F}_n \right) + \sum_{1 \le i \ne j \le 2^n} E\left[\mathcal{Z}_i \mathcal{Z}_j \mid \mathcal{F}_n \right] \right].$$

By (4.27), P-almost surely, this can be upper estimated for large n's by

$$2^{-2n} \left[(Cn\ell_n + 1) \sum_{i=1}^{2^n} E\left(\mathcal{Z}_i^2 \mid \mathcal{F}_n \right) \right] \le 2^{-2n} \left[C' n\ell_n \sum_{i=1}^{2^n} E\left(\mathcal{Z}_i^2 \mid \mathcal{F}_n \right) \right],$$

where $\widehat{C}(B, \gamma) < C'$. Now note that by Lemma 4.7,

$$\ell_n^2 E[\mathcal{Z}_1^2 \mid \mathcal{F}_n]$$

$$= \sum_{j,k=1}^{\ell_n} \left\{ P(Y_{m_n}^{1,j}, Y_{m_n}^{1,k} \in B \mid \mathcal{F}_n) - P(Y_{m_n}^{1,j} \in B \mid \mathcal{F}_n) P(Y_{m_n}^{1,k} \in B \mid \mathcal{F}_n) \right\}$$

$$= (\ell_n^2 - \ell_n) \left\{ P(Y_{m_n}^{1,1}, Y_{m_n}^{1,2} \in B \mid \mathcal{F}_n) - P(Y_{m_n}^{1,1} \in B \mid \mathcal{F}_n) P(Y_{m_n}^{1,2} \in B \mid \mathcal{F}_n) \right\}$$

$$+ \ell_n \mathrm{Var}\left(\mathbf{1}_{\{Y_{m_n}^{1,1} \in B\}} \mid \mathcal{F}_n \right) = \mathcal{O}(n2^{-n}\ell_n^2) + \mathcal{O}(\ell_n).$$

(Here the first term corresponds to the $k \neq j$ case and the second term corresponds to the $k = j$ case.)

Since, by Lemma 4.7, this estimate remains *uniformly* valid when the index 1 is replaced by anything between 1 and 2^n, therefore,

$$\ell_n^2 \sum_{i=1}^{2^n} E[\mathcal{Z}_i^2 \mid \mathcal{F}_n] = \mathcal{O}(n\ell_n^2) + \mathcal{O}(2^n \ell_n) = \mathcal{O}(2^n \ell_n) \text{ a.s.}$$

(Recall that $m_n = C_3 + C_4 \log n$.) Thus,

$$\sum_{i=1}^{2^n} E[\mathcal{Z}_i^2 \mid \mathcal{F}_n] = \mathcal{O}(2^n/\ell_n) \text{ a.s.}$$

It then follows that, P-almost surely, for large n's,

$$E\left([U_{n+m_n} - E(U_{n+m_n} \mid \mathcal{F}_n)]^2 \mid \mathcal{F}_n \right) \leq C'' \cdot n2^{-n}.$$

The summability immediately implies Lemma 4.6; nevertheless, since conditional probabilities are involved, one needs a conditional version of Borel-Cantelli, as follows. First, we have that P-almost surely,

$$\sum_{n=1}^{\infty} E\left([U_{n+m_n} - E(U_{n+m_n} \mid \mathcal{F}_n)]^2 \mid \mathcal{F}_n \right) < \infty$$

Then, by the (conditional) Markov inequality, for any $\delta > 0$, P-almost surely,

$$\sum_{n=1}^{\infty} P\left(|U_{n+m_n} - E(U_{n+m_n} \mid \mathcal{F}_n)| > \delta \mid \mathcal{F}_n \right) < \infty.$$

Finally, by a well-known conditional version of the Borel-Cantelli lemma (see e.g. Theorem 1 in [Chen (1978)]), it follows that

$$P\left(|U_{n+m_n} - E(U_{n+m_n} \mid \mathcal{F}_n)| > \delta \text{ occurs finitely often} \right) = 1,$$

which implies the result in Lemma 4.6. □

Remark 4.10 (No spine argument needed). In the proof of Theorem 2.2, this part of the analysis was more complicated, because the upper estimate there involved the analogous term U_s, which, unlike here, was not upper bounded. This is why we had to proceed with the spine change of measure and with further calculations. That part of the work is saved now. Notice that the martingale by which the change of measure was defined in the previous chapter, now becomes identically one: $2^{-n}\langle 1, Y_n \rangle = 1$. (Because now 2^{-n} plays the role of $e^{-\lambda_c t}$ and the function 1 plays the role of the positive $(L + \beta - \lambda_c)$-harmonic function ϕ.)   ◇

4.6.4   *The rest of the proof*

Recall the definition of ν and $\mathcal{R}$, and that our goal is to show that for any $B \in \mathcal{R}$,

$$P\left(\lim_{n\to\infty} U_n(B) = \nu(B) \right) = 1. \tag{4.28}$$

Let us fix $B \in \mathcal{R}$ for the rest of the subsection, and simply write U_t instead of $U_t(B)$.

Next, recall the limit in (4.26), but note that the underlying diffusion is only asymptotically Ornstein-Uhlenbeck[11], that is $\sigma_n^2 = 1 - 2^{-n}$, and so the transition kernels q_n defined by

$$q_n(x, \mathrm{d}y, k) := P(Y_k^1 \in \mathrm{d}y \mid Y_n^1 = x), \ k \geq n,$$

are slightly different from q. Note also the decomposition

$$E\left(U_{n+m_n} \mid \mathcal{F}_n \right) = \sum_{i=1}^{2^n} 2^{-n} E(U_{m_n}^{(i)} \mid \mathcal{F}_n) = 2^{-n} \sum_{i=1}^{2^n} q_n(Y_n^i, B, n + m_n).$$

In addition, recall the following facts.

(1) If $A_n := \{\mathrm{supp}(Y_n) \not\subset B_{a_n}\}$, then $\lim_{n\to\infty} 1_{A_n} = 0$, P-a.s.;
(2) $m_t = \zeta(a_t) = C_3 + C_4 \log t$;
(3) Lemma 4.6.

From these it follows that the limit

$$\lim_{n\to\infty} \sup_{x_0 \in B_{a_n}} |q_n(x_0, B, n + m_n) - \nu(B)| = 0, \tag{4.29}$$

which we will verify below, implies (4.28) with U_n replaced by U_{n+m_n}.

---

[11]Unlike in the non-interacting setting of Theorem 2.2, where we had $\sigma_n \equiv 1$.

Remark 4.11 (n and $N(n)$). Notice that (4.28) must then also hold P-a.s. for U_n, and even for U_{t_n} with *any given* sequence $t_n \uparrow \infty$ replacing n. Indeed, define the sequence $N(n)$ by the equation

$$N(n) + m_{N(n)} = t_n.$$

Clearly, $N(n) = \Theta(t_n)$, and in particular $\lim_{n \to \infty} N(n) = \infty$. Now, it is easy to see that in the proof of Theorem 4.1, including the remainder of this chapter, all the arguments go through when replacing n by $N(n)$, yielding thus (4.28) with U_n replaced by $U_{N(n)+m_{N(n)}} = U_{t_n}$. In those arguments it never plays any role that n is actually an integer.　　　◇

(We preferred to provide Remark 4.11 instead of presenting the proof with $N(n)$ replacing n everywhere, and to avoid notation even more difficult to follow[12].)

In light of Remark 4.11, we need to show (4.29). To achieve this goal, first recall that on the time interval $[l, l+1)$, $Y = Y^1$ corresponds to the d-dimensional Ornstein-Uhlenbeck operator

$$\frac{1}{2}\sigma_l^2 \Delta - \gamma x \cdot \nabla,$$

where $\sigma_l^2 = 1 - 2^{-l}$, $l \in \mathbb{N}$. That is, if $\sigma^{(n)}(\cdot)$ is defined by $\sigma^{(n)}(s) := \sigma_{n+l}$ for $s \in [l, l+1)$, then, given $\mathcal{F}_n$ and with a Brownian motion W, one has (recalling the representation (1.15)) that

$$Y_{m_n} - E(Y_{m_n} \mid \mathcal{F}_n) = Y_{m_n} - e^{-\gamma m_n} Y_0 = \int_0^{m_n} \sigma^{(n)}(s) e^{\gamma(s-m_n)} I_d \cdot dW_s$$

$$= \int_0^{m_n} e^{\gamma(s-m_n)} I_d \cdot dW_s - \int_0^{m_n} [1 - \sigma^{(n)}(s)] e^{\gamma(s-m_n)} I_d \cdot dW_s.$$

We now proceed to show that

$$\lim_{n \to \infty} P\left(\left| \int_0^{m_n} [1 - \sigma^{(n)}(s)] e^{\gamma(s-m_n)} I_d \cdot dW_s \right| > \epsilon \right) = 0. \qquad (4.30)$$

To show (4.30), we assume $d = 1$ for convenience; to upgrade the argument to $d > 1$ is trivial, using that

$$\left| \int_0^{m_n} [1 - \sigma^{(n)}(s)] e^{\gamma(s-m_n)} I_d \cdot dW_s \right|^2 = \sum_{i=1}^d \left| \int_0^{m_n} [1 - \sigma^{(n)}(s)] e^{\gamma(s-m_n)} dW_s^{(i)} \right|^2,$$

where $W^{(i)}$ is the ith coordinate of W.

---

[12]For example, one should replace 2^n with $2^{\lfloor N(n) \rfloor}$ or $n_{N(n)}$ everywhere.

Fix $\epsilon > 0$. By the Chebyshev inequality and the Itô-isometry (Proposition 1.5),

$$P\left(\left|\int_0^{m_n}[1-\sigma^{(n)}(s)]e^{\gamma(s-m_n)}\,dW_s\right| > \epsilon\right)$$

$$\leq \epsilon^{-2}E\left[\left(\int_0^{m_n}[1-\sigma^{(n)}(s)]e^{\gamma(s-m_n)}\,dW_s\right)^2\right]$$

$$= \epsilon^{-2}\int_0^{m_n}[1-\sigma^{(n)}(s)]^2 e^{2\gamma(s-m_n)}\,ds.$$

Now,

$$[1-\sigma^{(n)}(s)]^2 \leq [1-\sigma_n]^2 = (1-\sqrt{1-2^{-n}})^2 = \left(\frac{2^{-n}}{1+\sqrt{1-2^{-n}}}\right)^2 \leq 2^{-2n}.$$

Hence,

$$P\left(\left|\int_0^{m_n}[1-\sigma^{(n)}(s)]e^{\gamma(s-m_n)}\,dW_s\right| > \epsilon\right) \leq \epsilon^{-2}\int_0^{m_n}2^{-2n}e^{2\gamma(s-m_n)}\,ds.$$

Since $e^{-m_n} = e^{-C_3}n^{-C_4}$, we obtain that

$$\epsilon^{-2}\int_0^{m_n}2^{-2n}e^{2\gamma(s-m_n)}\,ds = \epsilon^{-2}e^{-2\gamma C_3}2^{-2n}n^{-2\gamma C_4}\int_0^{m_n}e^{2\gamma s}\,ds$$

$$= \epsilon^{-2}e^{-2\gamma C_3}2^{-2n}n^{-2\gamma C_4}\cdot\frac{e^{2\gamma C_3}n^{2\gamma C_4}-1}{2\gamma} \to 0, \text{ as } n \to \infty.$$

Therefore, (4.30) holds. We have

$$q_n(x_0,B,n+m_n) = P(Y_{n+m_n}^1 \in B \mid Y_n^1 = x_0)$$

$$= P\left(\int_0^{m_n}\sigma^{(n)}(s)e^{\gamma(s-m_n)}I_d\cdot dW_s \in B - x_0 e^{-\gamma m_n}\right),$$

and

$$q(x_0,B,m_n) = P\left(\int_0^{m_n}e^{\gamma(s-m_n)}I_d\cdot dW_s \in B - x_0 e^{-\gamma m_n}\right).$$

As before, it is easy to see, that it is sufficient to check the $d = 1$ case, and this is how we proceed now.

For estimating $q_n(x_0,B,n+m_n)$ let us use the inequality

$$\mathring{A}^\epsilon \subset A+b \subset A^\epsilon, \text{ for } A \subset \mathbb{R}^d, b \in \mathbb{R}^d, |b| < \epsilon, \epsilon > 0.$$

So, for any $\epsilon > 0$,

$$q_n(x_0, B, n + m_n)$$

$$= P\left(\int_0^{m_n} e^{\gamma(s-m_n)} \, \mathrm{d}W_s - \int_0^{m_n} [1 - \sigma^n(s)] e^{\gamma(s-m_n)} \, \mathrm{d}W_s \in B - x_0 e^{-\gamma m_n}\right)$$

$$= P\left(\int_0^{m_n} e^{\gamma(s-m_n)} \, \mathrm{d}W_s \in B - x_0 e^{-\gamma m_n} + \int_0^{m_n} [1 - \sigma^n(s)] e^{\gamma(s-m_n)} \, \mathrm{d}W_s\right)$$

$$\leq P\left(\int_0^{m_n} e^{\gamma(s-m_n)} \, \mathrm{d}W_s \in B^\epsilon - x_0 e^{-\gamma m_n}\right)$$

$$+ P\left(\left|\int_0^{m_n} [1 - \sigma^{(n)}(s)] e^{\gamma(s-m_n)} \, \mathrm{d}W_s\right| > \epsilon\right)$$

$$= q(x_0, B^\epsilon, m_n) + P\left(\left|\int_0^{m} [1 - \sigma^{(n)}(s)] e^{\gamma(s-m_n)} \, \mathrm{d}W_s\right| > \epsilon\right).$$

Taking $\limsup_{n\to\infty} \sup_{x_0 \in B_{a_n}}$, the second term vanishes by (4.30) and the first term becomes $\nu(B^\epsilon)$ by (4.26).

The lower estimate is similar:

$$q_n(x_0, B, n + m_n)$$

$$\geq P\left(\int_0^{m_n} e^{\gamma(s-m_n)} \, \mathrm{d}W_s \in \dot{B}^\epsilon - x_0 e^{-\gamma m_n}\right)$$

$$- P\left(\left|\int_0^{m_n} [1 - \sigma^{(n)}(s)] e^{\gamma(s-m_n)} \, \mathrm{d}W_s\right| > \epsilon\right)$$

$$= q(x_0, \dot{B}^\epsilon, m_n) - P\left(\left|\int_0^{m_n} [1 - \sigma^{(n)}(s)] e^{\gamma(s-m_n)} \, \mathrm{d}W_s\right| > \epsilon\right).$$

Taking $\liminf_{n\to\infty} \sup_{x_0 \in B_{a_n}}$, the second term vanishes by (4.30) and the first term becomes $\nu(\dot{B}^\epsilon)$ by (4.26).

Now (4.29) follows from these limits:

$$\lim_{\epsilon\downarrow 0} \nu(B^\epsilon) = \lim_{\epsilon\downarrow 0} \nu(\dot{B}^\epsilon) = \nu(B). \tag{4.31}$$

To verify (4.31) let $\epsilon \downarrow 0$ and use that, obviously, $\nu(\partial B) = 0$. Then $\nu(B^\epsilon) \downarrow \nu(\mathrm{cl}(B)) = \nu(B)$ because $B^\epsilon \downarrow \mathrm{cl}(B)$, and $\nu(\dot{B}^\epsilon) \uparrow \nu(\dot{B}) = \nu(B)$ because $\dot{B}^\epsilon \uparrow \dot{B}$.

The proof of (4.29) and that of Theorem 4.1 are now complete. $\qquad\square$

4.7   On a possible proof of Conjecture 4.1

In this section we provide some discussion for the reader interested in a possible way of proving Conjecture 4.1.

The main difference relative to the attractive case is that, as we have mentioned earlier, in that case one does not need the spine change of measure as in the proof of Theorem 2.2. In the repulsive case however, one cannot bypass the spine change of measure. Essentially, an h-transform transforms the outward Ornstein-Uhlenbeck process into an inward Ornstein-Uhlenbeck process. Indeed, $\lambda_c = \gamma d$ for the outward O-U operator with parameter $\gamma < 0$ and one should use the corresponding positive harmonic function (ground state) $\phi_c(x) := \exp(\gamma|x|^2)$ for the h-transform. In the exponential branching clock setting (and with independent particles), this inward Ornstein-Uhlenbeck process becomes the 'spine.' A possible way of proving Conjecture 4.1 would be to try to adapt the spine change of measure to unit time branching and dependent particles.

4.8   The proof of Lemma 4.7 and that of (4.27)

4.8.1   *Proof of Lemma 4.7*

The proof of the first part is a bit tedious, the proof of the second part is very simple. We recall that $\{\mathcal{F}_t\}_{t\geq 0}$ denotes the canonical filtration for Y.

(a): Throughout the proof, we may (and will) assume that, the growth of the support of Y is bounded from above by the function a, because this happens with probability one. That is, we assume that

$$\exists n_0(\omega) \in \mathbb{N} \text{ such that } \forall n \geq n_0 \; \forall \xi, \widetilde{\xi} \in \Pi_n : \; |\xi_0|, |\widetilde{\xi}_0| \leq C_0\sqrt{n}. \quad (4.32)$$

(Recall that C_0 is not random.)

First assume $d = 1$.

Next, note that given $\mathcal{F}_n$ (or, what is the same[13], given Z_n), ξ_{m_n} and $\widetilde{\xi}_{m_n}$ have joint normal distribution. This is because by Remark 4.3, $(Z_t^1, Z_t^2, ..., Z_t^{n_t})$ given Z_n is a.s. joint normal for $t > n$, and $(\xi_{m_n}, \widetilde{\xi}_{m_n})$ is a projection of $(Z_t^1, Z_t^2, ..., Z_t^{n_t})$. Therefore, denoting $\widehat{x} := x - \xi_0$, $\widehat{y} := y - \widetilde{\xi}_0$, the joint (conditional) density of ξ_{m_n} and $\widetilde{\xi}_{m_n}$ (given $\mathcal{F}_n$) on $\mathbb{R}^2$ is of the form

$$f^{(n)}(x,y) = f(x,y)$$
$$= \frac{1}{2\pi\sigma_x\sigma_y\sqrt{1-\rho^2}} \exp\left(-\frac{1}{2(1-\rho^2)}\left[\frac{\widehat{x}^2}{\sigma_x^2} + \frac{\widehat{y}^2}{\sigma_y^2} - \frac{2\rho\widehat{x}\widehat{y}}{\sigma_x\sigma_y}\right]\right),$$

---

[13]Given $\mathcal{F}_n$, the distribution of $\left(\xi_{m_n}, \widetilde{\xi}_{m_n}\right)$ will not change by specifying Z_n, that is, specifying $\overline{Z}_n$.

where σ_x^2, σ_y^2 and $\rho = \rho_n$ denote the (conditional) variances of the marginals and the (conditional) correlation[14] between the marginals, respectively, given $\mathcal{F}_n$. Abbreviating $\kappa := \frac{1}{\sigma_x \sigma_y}$, one has

$$f(x,y) = \frac{1}{2\pi\sigma_x\sigma_y\sqrt{1-\rho^2}} \exp\left(-\frac{1}{2(1-\rho^2)}\left[\frac{\widehat{x}^2}{\sigma_x^2} + \frac{\widehat{y}^2}{\sigma_y^2}\right]\right) \exp\left(\frac{\rho}{1-\rho^2}\kappa\widehat{xy}\right).$$

Let $f_1^{(n)} = f_1$ and $f_2^{(n)} = f_2$ denote the (conditional) marginal densities of f, given $\mathcal{F}_n$. We now show that P-a.s., for all large enough n,

$$|f(x,y) - f_1(x)f_2(y)| \le K(B)n\rho, \text{ with some } K(B) > 0 \text{ on } B, \qquad (4.33)$$

and that P-a.s.,

$$\rho = \rho_n = E\left[(\xi_{m_n} - E(\xi_{m_n} \mid \mathcal{F}_n))(\widetilde{\xi}_{m_n} - E(\widetilde{\xi}_{m_n} \mid \mathcal{F}_n)) \mid \mathcal{F}_n\right] \le \frac{3}{\gamma}\cdot 2^{-n}, \ n \ge 1. \qquad (4.34)$$

Clearly, (4.33) and (4.34) imply the statement in (a):

$$\left|\int_{B\times B} f(x,y) - f_1(x)f_2(y)dxdy\right|$$

$$\le \int_{B\times B} |f(x,y) - f_1(x)f_2(y)|dxdy \le |B|^2 K(B)n\rho_n = |B|^2 K(B)\frac{3}{\gamma}\cdot n2^{-n}.$$

To see (4.33), write

$$f(x,y) - f_1(x)f_2(y)$$

$$= \left\{f(x,y) - \frac{1}{2\pi\sigma_x\sigma_y}\exp\left(-\frac{1}{2}\left[\frac{\widehat{x}^2}{\sigma_x^2} + \frac{\widehat{y}^2}{\sigma_y^2}\right]\right)\exp\left(\frac{\rho}{1-\rho^2}\kappa\widehat{xy}\right)\right\}$$

$$+ \left\{\frac{1}{2\pi\sigma_x\sigma_y}\exp\left(-\frac{1}{2}\left[\frac{\widehat{x}^2}{\sigma_x^2} + \frac{\widehat{y}^2}{\sigma_y^2}\right]\right)\exp\left(\frac{\rho}{1-\rho^2}\kappa\widehat{xy}\right) - f_1(x)f_2(y)\right\} =: I + II.$$

Now,

$$|I| = \frac{1}{2\pi\sigma_x\sigma_y}\exp\left(-\frac{1}{2}\left[\frac{\widehat{x}^2}{\sigma_x^2} + \frac{\widehat{y}^2}{\sigma_y^2}\right]\right)\exp\left(\frac{\rho}{1-\rho^2}\kappa\widehat{xy}\right)$$

$$\cdot\left|\left(\frac{1}{\sqrt{1-\rho^2}}e^{\frac{1}{2}\left[\frac{\widehat{x}^2}{\sigma_x^2} + \frac{\widehat{y}^2}{\sigma_y^2}\right] - \frac{1}{2(1-\rho^2)}\left[\frac{\widehat{x}^2}{\sigma_x^2} + \frac{\widehat{y}^2}{\sigma_y^2}\right]} - 1\right)\right|$$

$$\le \frac{1}{2\pi\sigma_x\sigma_y}\exp\left(\frac{\rho}{1-\rho^2}\kappa\widehat{xy}\right)$$

$$\cdot\left|\left(\frac{1}{\sqrt{1-\rho^2}}\exp\left\{\frac{1}{2}\left[\frac{\widehat{x}^2}{\sigma_x^2} + \frac{\widehat{y}^2}{\sigma_y^2}\right]\left(1 - \frac{1}{(1-\rho^2)}\right)\right\} - 1\right)\right|.$$

---

[14]Provided, of course, that $\rho_n \ne 1$, but we will see in (4.34) below that $\lim_{n\to\infty}\rho_n = 0$.

Since B is a fixed bounded measurable set, using (4.32) along with the approximations $1 - e^{-a} \approx a$ as $a \to 0$, and $1 - \sqrt{1 - \rho^2} \approx \rho^2/2$ as $\rho \to 0$, one can see that *if (4.34) holds*, then there exists a $K(B) > 0$ such that P-a.s.,

$$|I| \leq K(B)n\rho^2 \text{ for all large enough } n.$$

To see that the presence of the $\mathcal{F}_n$-dependent σ_x, σ_y do not change this fact, recall that ξ and $\tilde{\xi}$ are both (time inhomogeneous) Ornstein-Uhlenbeck processes (see Section 4.4.1), and so σ_x and σ_y are bounded between two positive (absolute) constants for $n \geq 1$. (Recall that the variance of an Ornstein-Uhlenbeck process is bounded between two positive constants, which depend on the parameters only, on the time interval (ϵ, ∞), for $\epsilon > 0$.)

A similar (but simpler) computation shows that *if (4.34) holds*, then there exists a $K(B) > 0$ (we can choose the two constants the same, so this one will be denoted by $K(B)$ too) such that P-a.s.,

$$|II| \leq K(B)n\rho, \ \forall x, y \in B \text{ for all large enough } n.$$

These estimates of I and II yield (4.33).

Thus, it remains to prove (4.34). Recall that we assume $d = 1$. Using similar notation as in Subsection 4.4.1, let $\widetilde{W}^{(i)}$ $(i = 1, 2)$ be Brownian motions, which, satisfy for $s \in [k, k+1)$, $0 \leq k < m_n$,

$$\sigma_{n+k}\widetilde{W}_s^{(1)} = \bigoplus_{i \in I_{n+k}} 2^{-n-k} W_s^{k,i} \oplus (1 - 2^{-n-k}) W_s^{k,1}, \qquad (4.35)$$

$$\sigma_{n+k}\widetilde{W}_s^{(2)} = \bigoplus_{i \in J_{n+k}} 2^{-n-k} W_s^{k,i} \oplus (1 - 2^{-n-k}) W_s^{k,2},$$

where the $W^{k,i}$ are 2^{n+k} independent standard Brownian motions, and $I_{n+k} := \{i : 2 \leq i \leq 2^{n+k}\}$, $J_{n+k} := \{i : 1 \leq i \leq 2^{n+k}, i \neq 2\}$. Recall that, by (4.11), given $\mathcal{F}_n$, Y and $\tilde{Y}$ are Ornstein-Uhlenbeck processes driven by $\widetilde{W}^{(1)}$ and $\widetilde{W}^{(2)}$, respectively, and $\widetilde{W}^{(1)}$ and $\widetilde{W}^{(2)}$ are independent of $\mathcal{F}_n$.

Notation 4.2. We are going to use the following (slight abuse of) notation. For $r > 0$, the expression $\sum_{j=0}^{r-1} \int_j^{j+1} f(s)\, dW_s$ will mean $\sum_{j=0}^{\lfloor r \rfloor - 1} \int_j^{j+1} f(s)\, dW_s + \int_{\lfloor r \rfloor}^r f(s)\, dW_s$, where W is Brownian motion.

Using this notation with $r = m_n$ and recalling that $\sigma^{(n)}(s) := \sigma_{n+l}$ for $s \in [l, l+1)$, one has

$$\xi_{m_n} - E(\xi_{m_n} \mid \mathcal{F}_n) = \int_0^{m_n} \sigma^{(n)}(s) e^{\gamma(s - m_n)}\, d\widetilde{W}_s^{(1)}$$

$$= \sum_{j=0}^{m_n - 1} \sigma_{n+j} \int_j^{j+1} e^{\gamma(s - m_n)}\, d\widetilde{W}_s^{(1)}$$

and

$$\widetilde{\xi}_{m_n} - E(\widetilde{\xi}_{m_n} \mid \mathcal{F}_n) = \int_0^{m_n} \sigma^{(n)}(s) e^{\gamma(s-m_n)} \, d\widetilde{W}_s^{(2)}$$

$$= \sum_{j=0}^{m_n-1} \sigma_{n+j} \int_j^{j+1} e^{\gamma(s-m_n)} \, d\widetilde{W}_s^{(2)},$$

where, of course, $E(\xi_{m_n} \mid \mathcal{F}_n) = e^{-\gamma m_n} \xi_0$ and $E(\widetilde{\xi}_{m_n} \mid \mathcal{F}_n) = e^{-\gamma m_n} \widetilde{\xi}_0$. Writing out $\sigma_{n+j} \, d\widetilde{W}_s^{(1)}$ and $\sigma_{n+j} \, d\widetilde{W}_s^{(2)}$ according to (4.35), one obtains, that given $\mathcal{F}_n$,

$$I := \xi_{m_n} - E(\xi_{m_n} \mid \mathcal{F}_n)$$

$$= \sum_{j=0}^{m_n-1} \left[\sum_{i \in I_{n+j}} 2^{-n-j} \int_j^{j+1} e^{\gamma(s-m_n)} dW_s^{j,i} + (1 - 2^{-n-j}) \int_j^{j+1} e^{\gamma(s-m_n)} dW_s^{j,1} \right],$$

$$II := \widetilde{\xi}_{m_n} - E(\widetilde{\xi}_{m_n} \mid \mathcal{F}_n)$$

$$= \sum_{j=0}^{m_n-1} \left[\sum_{i \in J_{n+j}} 2^{-n-j} \int_j^{j+1} e^{\gamma(s-m_n)} dW_s^{j,i} + (1 - 2^{-n-j}) \int_j^{j+1} e^{\gamma(s-m_n)} dW_s^{j,2} \right].$$

Because I and II are jointly independent of $\mathcal{F}_n$, one has

$$E(I \cdot II \mid \mathcal{F}_n) = E(I \cdot II).$$

Since the Brownian motions $W^{j,i}$ are independent for fixed j and different i's, and the Brownian increments are also independent for different j's, therefore one has $E(I \cdot II) = E \sum_{j=0}^{m_n-1}(III + IV)$, where

$$III := (2^{n+j} - 2) 2^{-2(n+j)} \left(\int_j^{j+1} e^{\gamma(s-m_n)} \, dB_s \right)^2 ;$$

$$IV := 2^{1-n-j}(1 - 2^{-n-j}) \left(\int_j^{j+1} e^{\gamma(s-m_n)} \, dB_s \right)^2 ,$$

and B is a generic Brownian motion. By Itô's isometry (Proposition 1.5),

$$E(I \cdot II)$$

$$= \sum_{j=0}^{m_n-1} \left[(2^{n+j} - 2) 2^{-2(n+j)} + 2^{1-n-j}(1 - 2^{-n-j}) \right] \int_j^{j+1} e^{2\gamma(s-m_n)} ds$$

$$= \frac{1}{2\gamma} \sum_{j=0}^{\lfloor m_n \rfloor - 1} \left[3 \cdot 2^{-(n+j)} - 4 \cdot 2^{-2(n+j)} \right] \left[e^{2\gamma(j+1-m_n)} - e^{2\gamma(j-m_n)} \right] + R_n$$

$$= \frac{1}{2\gamma} 2^{-n} \sum_{j=0}^{\lfloor m_n \rfloor - 1} \left[3 \cdot 2^{-j} - 4 \cdot 2^{(-n-2j)} \right] \left[e^{2\gamma(j+1-m_n)} - e^{2\gamma(j-m_n)} \right] + R_n,$$

where

$$R_n := \frac{1}{2\gamma} 2^{-n} \cdot \left[3 \cdot 2^{-\lfloor m_n \rfloor} - 4 \cdot 2^{(-n-2\lfloor m_n \rfloor)} \right] \left[1 - e^{2\gamma(\lfloor m_n \rfloor - m_n)} \right] < \frac{3}{2\gamma} 2^{-n}.$$

(Note that $3 \cdot 2^{-j} > 4 \cdot 2^{(-n-2j)}$ and $\gamma > 0$.) Hence

$$0 < E(I \cdot II)$$
$$< \frac{3}{2\gamma} 2^{-n} \sum_{j=0}^{\lfloor m_n \rfloor - 1} \left[e^{2\gamma(j+1-m_n)} - e^{2\gamma(j-m_n)} \right] + R_n < \frac{3}{2\gamma} 2^{-n} (2 - e^{-2\gamma m_n}),$$

and so (4.34) follows, finishing the proof of part (a) for $d = 1$.

Assume that $d \geq 2$. It is clear that (4.34) follows from the one-dimensional case. As far as (4.33) is concerned, the computation is essentially the same as in the one-dimensional case. Note, that although the formulæ are lengthier in higher dimension, the $2d$-dimensional covariance matrix is block-diagonal because of the independence of the d coordinates (Lemma 4.2), and this simplifies the computation significantly. We leave the simple details to the reader.

(b): Write

$$\mathrm{Var}\left(1_{\{\xi_{m_n} \in B\}} \mid \mathcal{F}_n \right) = P(\xi_{m_n} \in B \mid \mathcal{F}_n) - P^2(\xi_{m_n} \in B \mid \mathcal{F}_n),$$

and note that $P(\xi_{m_n} \in B \mid \xi_0 = x) = q_n(x, B, n+m_n)$, and ξ_0 is the location of the parent particle at time n. Hence, (4.29) together with (4.24) implies the limit in (b).   □

4.8.2   Proof of (4.27)

We will assume that $\nu(B) > 0$ (i.e. $C(B) = \nu(B) - (\nu(B))^2 > 0$), or equivalently, that B has positive Lebesgue measure. This does not cause any loss of generality, since otherwise the $\mathcal{Z}_i$'s vanish a.s. and (4.27) is trivially true.

Now let us estimate $E\left[\mathcal{Z}_i \mathcal{Z}_j \mid \mathcal{F}_n\right]$ and $E\left[\mathcal{Z}_i^2 \mid \mathcal{F}_n\right]$. The calculation is based on Lemma 4.7 as follows. First, by part (a) of Lemma 4.7, it holds P-a.s. that for all large enough n,

$$P(Y_{m_n}^{1,j} \in B, Y_{m_n}^{2,k} \in B \mid \mathcal{F}_n) - P(Y_{m_n}^{1,j} \in B \mid \mathcal{F}_n)\ P(Y_{m_n}^{2,k} \in B \mid \mathcal{F}_n)$$
$$\leq C(B, \gamma) \cdot n 2^{-n}.$$

Therefore, recalling that $\ell_n = 2^{\lfloor m_n \rfloor}$, one has that P-a.s., for all large enough n,

$$\ell_n^2 E[\mathcal{Z}_1 \mathcal{Z}_2 \mid \mathcal{F}_n]$$

$$= \sum_{j,k=1}^{\ell_n} \left\{ P(Y_{m_n}^{1,j} \in B, Y_{m_n}^{2,k} \in B \mid \mathcal{F}_n) - P(Y_{m_n}^{1,j} \in B \mid \mathcal{F}_n) P(Y_{m_n}^{2,k} \in B \mid \mathcal{F}_n) \right\}$$

$$\leq C(B, \gamma) \cdot n 2^{-n} \ell_n^2.$$

This estimate holds when $Y_{m_n}^{1,j}$ and $Y_{m_n}^{2,k}$ are replaced by any $Y_{m_n}^{p,j}$ and $Y_{m_n}^{r,k}$, where $p \neq r$ and $1 \leq p, r \leq 2^n$; consequently, if

$$I_n := \sum_{1 \leq i \neq j \leq 2^n} E\left[\mathcal{Z}_i \mathcal{Z}_j \mid \mathcal{F}_n\right]$$

(which is the left-hand side of the inequality in (4.27)) then one has that P-a.s., for all large enough n,

$$\ell_n^2 I_n \leq 2^n \cdot (2^n - 1) C(B, \gamma) \cdot n 2^{-n} \ell_n^2 < C(B, \gamma) \cdot n 2^n \ell_n^2.$$

Hence, to finish the proof, it is sufficient to show that[15]

$$\ell_n^2 J_n = \Theta\left(n 2^n \ell_n^2\right) \text{ a.s.,} \tag{4.36}$$

for

$$J_n = n \ell_n \sum_{i=1}^{2^n} E[\mathcal{Z}_i^2 \mid \mathcal{F}_n]$$

(which is the right-hand side of the inequality in (4.27) without the constant). To this end, we essentially repeat the argument in the proof of Claim 4.2. The only difference is that we now *also use the assumption* $C(B) > 0$, and obtain that

$$\ell_n^2 E[\mathcal{Z}_1^2 \mid \mathcal{F}_n] = \mathcal{O}(n 2^{-n} \ell_n^2) + \Theta(\ell_n),$$

as $n \to \infty$, a.s.

Just like in the proof of Claim 4.2, replacing 1 by i, the estimate holds uniformly for $1 \leq i \leq 2^n$, and so

$$\ell_n^2 \sum_{i=1}^{2^n} E[\mathcal{Z}_i^2 \mid \mathcal{F}_n] = \mathcal{O}(n \ell_n^2) + \Theta(2^n \ell_n) = \Theta(2^n \ell_n) \text{ a.s.,}$$

where in the last equality we used that $\ell_n = 2^{\lfloor m_n \rfloor}$ and $m_n = o(n)$. From here, (4.36) immediately follows:

$$\ell_n^2 J_n = n \ell_n^3 \sum_{i=1}^{2^n} E[\mathcal{Z}_i^2 \mid \mathcal{F}_n] = \Theta(n 2^n \ell_n^2) \text{ a.s.,}$$

and the proof of (4.27) is complete.   □

---

[15]What we mean here is that there exist $c, C > 0$ absolute constants such that for all $n \geq 1$, $c < J_n / 2^n < C$ a.s.

4.9 The center of mass for supercritical super-Brownian motion

In Lemma 4.1 we have shown that $\overline{Z}_t := \frac{1}{n_t} \sum_{i=1}^{n_t} Z_t^i$, the center of mass for Z satisfies $\lim_{t\to\infty} \overline{Z}_t = N$, where $N \sim \mathcal{N}(0, 2\mathbf{I}_d)$. In fact, the proof reveals that $\overline{Z}$ moves like a Brownian motion, which is nevertheless slowed down tending to a final limiting location (see Lemma 4.1 and its proof).

Since this is also true for $\gamma = 0$ (BBM with unit time branching and no self-interaction), our first natural question is whether we can prove a similar result for the supercritical super-Brownian motion.

Let X be the $(\frac{1}{2}\Delta, \beta, \alpha; \mathbb{R}^d)$-superdiffusion with $\alpha, \beta > 0$ (supercritical super-Brownian motion), and let P_μ denote the corresponding probability when the initial finite measure is μ. (We will use the abbreviation $P := P_{\delta_0}$.) Let us restrict Ω to the survival set

$$S := \{\omega \in \Omega \mid X_t(\omega) > 0, \ \forall t > 0\}.$$

Since $\beta > 0$, $P_\mu(S) > 0$ for all $\mu \neq \mathbf{0}$. (In fact, using the log-Laplace equation, it is easy to derive that $P(S) = 1 - \exp\left(-\frac{\beta}{\alpha}\|\mu\|\right)$.)

It turns out that on the survival set the center of mass for X stabilizes:

Theorem 4.2. *Let $\alpha, \beta > 0$ and let $\overline{X}$ denote the center of mass process for the $(\frac{1}{2}\Delta, \beta, \alpha; \mathbb{R}^d)$-superdiffusion X, that is let*

$$\overline{X} := \frac{\langle \mathrm{id}, X\rangle}{\|X\|},$$

where $\langle f, X\rangle := \int_{\mathbb{R}^d} f(x)\, X(\mathrm{d}x)$ and $\mathrm{id}(x) = x$. Then, on S, $\overline{X}$ is continuous and converges P-almost surely.

Remark 4.12. A heuristic argument for the convergence is as follows. Obviously, the center of mass is invariant under H-transforms whenever H is spatially (but not temporarily) constant. Let $H(t) := e^{-\beta t}$. Then X^H is a $(\frac{1}{2}\Delta, 0, e^{-\beta t}\alpha; \mathbb{R}^d)$-superdiffusion, that is, a critical super-Brownian motion with a clock that is slowing down. Therefore, heuristically it seems plausible that $\overline{X^H}$, the center of mass for the transformed process stabilizes, because after some large time T, if the process is still alive, it behaves more or less like the heat flow ($e^{-\beta t}\alpha$ is small), under which the center of mass does not move. ◇

4.9.1 *Proof of Theorem 4.2*

Since α, β are constant, the branching is independent of the motion, and therefore N defined by

$$N_t := e^{-\beta t} \|X_t\|$$

is a nonnegative martingale (positive on S) tending to a limit almost surely. It is straightforward to check that it is uniformly bounded in L^2 and is therefore uniformly integrable (UI). Write

$$\overline{X}_t = \frac{e^{-\beta t}\langle \mathrm{id}, X_t\rangle}{e^{-\beta t}\|X_t\|} = \frac{e^{-\beta t}\langle \mathrm{id}, X_t\rangle}{N_t}.$$

We now claim that $N_\infty > 0$ a.s. on S. Let $A := \{N_\infty = 0\}$. Clearly $S^c \subset A$, and so if we show that $P(A) = P(S^c)$, then we are done. As mentioned above, $P(S^c) = e^{-\beta/\alpha}$. On the other hand, a standard martingale argument (see the argument after formula (20) in [Engländer (2008)]) shows that $0 \leq u(x) := -\log P_{\delta_x}(A)$ must solve the equation

$$\frac{1}{2}\Delta u + \beta u - \alpha u^2 = 0, \tag{4.37}$$

but since $P_{\delta_x}(A) = P(A)$ constant, therefore $-\log P_{\delta_x}(A)$ solves $\beta u - \alpha u^2 = 0$. Since N is uniformly integrable, 'no mass is lost in the limit,' giving $P(A) < 1$. So $u > 0$, which in turn implies that $-\log P_{\delta_x}(A) = \beta/\alpha$.

Once we know that $N_\infty > 0$ a.s. on S, it is sufficient to focus on the term $e^{-\beta t}\langle \mathrm{id}, X_t\rangle$: we are going to show that it converges almost surely. Clearly, it is enough to prove this coordinate-wise.

Recall the 'transformation to critical superprocess': if X is an $(L, \beta, \alpha; \mathbb{R}^d)$-superdiffusion, and $H(x,t) := e^{-\lambda t}h(x)$, where h is a positive solution of $(L + \beta)h = \lambda h$, then X^H is a $(L + a\frac{\nabla h}{h} \cdot \nabla, 0, e^{-\lambda t}\alpha h; \mathbb{R}^d)$-superdiffusion.

In our case $\beta(\cdot) \equiv \beta$. So choosing $h(\cdot) \equiv 1$ and $\lambda = \beta$, we have $H(t) = e^{-\beta t}$ and X^H is a $(\frac{1}{2}\Delta, 0, e^{-\beta t}\alpha; \mathbb{R}^d)$-superdiffusion, that is, a critical super-Brownian motion with a clock that is slowing down. Since, as noted above, it is enough to prove the convergence coordinate-wise, we can assume that $d = 1$. One can write

$$e^{-\beta t}\langle \mathrm{id}, X_t\rangle = \langle \mathrm{id}, X_t^H\rangle.$$

Let $\{\mathcal{S}_s\}_{s\geq 0}$ be the 'expectation semigroup' for X, that is, the semigroup corresponding to the operator $\frac{1}{2}\Delta + \beta$. The expectation semigroup $\{\mathcal{S}_s^H\}_{s\geq 0}$ for X^H satisfies $T_s := \mathcal{S}_s^H = e^{-\beta s}\mathcal{S}_s$ and thus it corresponds to Brownian motion. In particular then

$$T_s[\mathrm{id}] = \mathrm{id}. \tag{4.38}$$

(One can pass from bounded continuous functions to $f := \mathrm{id}$ by defining $f_1 := f\mathbf{1}_{x>0}$ and $f_2 := f\mathbf{1}_{x\leq 0}$, then noting that by monotone convergence, $E_{\delta_x}\langle f_i, X_t^H\rangle = \mathbb{E}_x f_i(W_t) \in (-\infty, \infty)$, $i = 1, 2$, where W is a Brownian motion with expectation $\mathbb{E}$, and finally taking the sum of the two equations.) Therefore $M := \langle \mathrm{id}, X^H\rangle$ is a martingale:

$$E_{\delta_x}(M_t \mid \mathcal{F}_s) = E_{\delta_x}(\langle \mathrm{id}, X_t^H\rangle \mid \mathcal{F}_s) = E_{X_s}\langle \mathrm{id}, X_t^H\rangle$$

$$= \int_{\mathbb{R}} E_{\delta_y}\langle \mathrm{id}, X_t^H\rangle X_s^H(\mathrm{d}y) = \int_{\mathbb{R}} y\, X_s^H(\mathrm{d}y) = M_s.$$

We now show that M is UI and even uniformly bounded in L^2, verifying its a.s. convergence, and that of the center of mass. To achieve this, define g_n by $g_n(x) = |x| \cdot \mathbf{1}_{\{|x|<n\}}$. Then we have

$$E\langle \mathrm{id}, X_t^H\rangle^2 = E|\langle \mathrm{id}, X_t^H\rangle|^2 \leq E\langle |\mathrm{id}|, X_t^H\rangle^2,$$

and by the monotone convergence theorem we can continue with

$$= \lim_{n\to\infty} E\langle g_n, X_t^H\rangle^2.$$

Since g_n is compactly supported, there is no problem to use the moment formula and continue with

$$= \lim_{n\to\infty} \int_0^t \mathrm{d}s\, e^{-\beta s}\langle \delta_0, T_s[\alpha g_n^2]\rangle = \alpha \lim_{n\to\infty} \int_0^t \mathrm{d}s\, e^{-\beta s} T_s[g_n^2](0).$$

Recall that $\{T_s; s \geq 0\}$ is the Brownian semigroup, that is, $T_s[f](x) = \mathbb{E}_x f(W_s)$, where W is Brownian motion. Since $g_n(x) \leq |x|$, therefore we can trivially upper estimate the last expression by

$$\alpha \int_0^t \mathrm{d}s\, e^{-\beta s}\mathbb{E}_0(W_s^2) = \alpha \int_0^t \mathrm{d}s\, s e^{-\beta s} = \alpha \left(\frac{1 - e^{-\beta t}}{\beta^2} - \frac{t e^{-\beta t}}{\beta}\right) < \frac{\alpha}{\beta^2}.$$

Since this upper estimate is independent of t, we are done:

$$\sup_{t\geq 0} E\langle \mathrm{id}, X_t^H\rangle^2 \leq \frac{\alpha}{\beta^2}.$$

Finally, we show that $\overline{X}$ has continuous paths. To this end we first note that we can (and will) consider a version of X where all the paths are continuous in the weak topology of measures. We now need a simple lemma.

Lemma 4.8. *Let $\{\mu_t,\ t \geq 0\}$ be a family in $\mathcal{M}_f(\mathbb{R}^d)$ and assume that $t_0 > 0$ and $\mu_t \overset{w}{\Rightarrow} \mu_{t_0}$ as $t \to t_0$. Assume furthermore that*

$$C = C_{t_0,\epsilon} := \mathrm{cl}\left(\bigcup_{t=t_0-\epsilon}^{t_0+\epsilon} \mathrm{supp}(\mu_t)\right)$$

is compact with some $\epsilon > 0$. Let $f : \mathbb{R}^d \to \mathbb{R}^d$ be a continuous function. Then $\lim_{t\to t_0}\langle f, \mu_t\rangle = \langle f, \mu_{t_0}\rangle$.

Proof.   First, if $f = (f_1, ..., f_d)$ then all f_i are $\mathbb{R}^d \to \mathbb{R}$ continuous functions and $\lim_{t \to t_0} \langle f, \mu_t \rangle = \langle f, \mu_{t_0} \rangle$ simply means that $\lim_{t \to t_0} \langle f_i, \mu_t \rangle = \langle f_i, \mu_{t_0} \rangle$. Therefore, it is enough to prove the lemma for a $\mathbb{R}^d \to \mathbb{R}$ continuous function. Let k be so large that $C \subset I_k := [-k, k]^d$. Using a mollified version of $\mathbf{1}_{[-k,k]}$, it is trivial to construct a continuous function $\widehat{f} =: \mathbb{R}^d \to \mathbb{R}$ such that $\widehat{f} = f$ on I_k and $\widehat{f} = 0$ on $\mathbb{R}^d \setminus I_{2k}$. Then,

$$\lim_{t \to t_0} \langle f, \mu_t \rangle = \lim_{t \to t_0} \langle \widehat{f}, \mu_t \rangle = \langle \widehat{f}, \mu_{t_0} \rangle = \langle f, \mu_{t_0} \rangle,$$

since $\widehat{f}$ is a bounded continuous function.   □

Returning to the proof of the theorem, let us invoke the fact that for

$$C_s(\omega) := \mathrm{cl}\left(\bigcup_{z \le s} \mathrm{supp}(X_z(\omega)) \right),$$

we have $P(C_s \text{ is compact}) = 1$ for all fixed $s \ge 0$ (compact support property; see [Engländer and Pinsky (1999)]). By the monotonicity in s, there exists an $\Omega_1 \subset \Omega$ with $P(\Omega_1) = 1$ such that for $\omega \in \Omega_1$,

$$C_s(\omega) \text{ is compact } \forall s \ge 0.$$

Let $\omega \in \Omega_1$ and recall that we are working with a continuous path version of X. Then letting $f := \mathrm{id}$ and $\mu_t = X_t(\omega)$, Lemma 4.8 implies that for $t_0 > 0$, $\lim_{t \to t_0} \langle \mathrm{id}, X_t(\omega) \rangle = \langle \mathrm{id}, X_{t_0}(\omega) \rangle$. The right continuity at $t_0 = 0$ is similar.   □

4.10   Exercises

(1) Show that the coordinate processes of Z are independent one-dimensional interactive branching processes of the same type as Z.
(2) Derive equation (4.37).
(3) Write out the detailed proof of the following two statements:
   (a) That the growth rate of the support of Y satisfies (4.24) with a sufficiently large $C_0 = C_0(\gamma)$.
   (b) That (4.25) holds with a sufficiently large $C_1 = C_1(\gamma)$.
   Hint: Check how the expectation calculations for the non-interacting case carry through.
(4) **(Model with drift)** Consider the same attractive model as in this chapter, but now change the underlying motion process by adding a

drift term to Brownian motion. That is, let the motion process correspond to

$$L := \frac{1}{2}\Delta + \mathbf{b} \cdot \nabla \quad \text{on } \mathbb{R}^d.$$

Can you describe the behavior of the system when the drift term $\mathbf{b}$:
$\mathbb{R}^d \to \mathbb{R}^d$ is constant? Can you still treat some cases when $\mathbf{b}$ is spatially
dependent? (This latter part of the exercise is more like a research
project though.)

4.11   Notes

This chapter is based on [Engländer (2010)]. It turns out that S. Harris and the
author of this book had independently thought about this type of model. After
submitting the paper [Engländer (2010)], the author also became aware of an
independently discovered proof of the convergence of the center of mass for BBM
by O. Adelman with J. Berestycki and S. Harris.

The proof of Theorem 4.1 reveals that actually

$$2^{-t_n} Z_{t_n}(\mathrm{d}y) \stackrel{w}{\Rightarrow} \left(\frac{\gamma}{\pi}\right)^{d/2} \exp\left(-\gamma |y - x|^2\right) \mathrm{d}y$$

holds P^x-a.s. for any given sequence $\{t_n\}$ with $t_n \uparrow \infty$ as $n \to \infty$. This, of course,
is still weaker than P^x-a.s. convergence as $t \to \infty$, but one can probably argue,
using the method of Asmussen and Hering, as in the previous chapter, to upgrade
it to continuous time convergence. Nevertheless, since our model is defined with
unit time branching anyway, we felt satisfied with (4.14).

As a next step, it seems natural to replace the linearity of the interaction by
a more general rule. That is, to define and analyze the system where (4.2) is
replaced by

$$\mathrm{d}Z_t^i = \mathrm{d}W_t^{m,i} + 2^{-m} \sum_{1 \leq j \leq 2^m} g(|Z_t^j - Z_t^i|) \frac{Z_t^j - Z_t^i}{|Z_t^j - Z_t^i|} \, \mathrm{d}t; \ i = 1, 2, \ldots, 2^m,$$

where the function $g : \mathbb{R}_+ \to \mathbb{R}$ has some nice properties. (In this chapter we
treated the $g(x) = \gamma x$ case.) The analysis of this general model is still to be
achieved.

A further natural goal is to construct and investigate the properties of a
superprocess with representative particles that are attracted to or repulsed from
its center of mass.

There is one work in this direction we are aware of: motivated by the material
presented here, H. Gill [Gill (2010)] has constructed a *superprocess with attraction
to its center of mass*. More precisely, Gill constructs a supercritical interacting
measure-valued process with representative particles that are attracted to or repulsed from its center of mass using Perkins's *historical stochastic calculus*.

In the attractive case, Gill proves the equivalent of our Theorem 4.1 (see later): on S, the mass normalized process converges almost surely to the stationary distribution of the Ornstein-Uhlenbeck process centered at the limiting value of its center of mass; in the repulsive case, he obtains substantial results concerning the equivalent of our Conjecture 4.1 (see later), using [Engländer and Winter (2006)]. In addition, a version of Tribe's result on the 'last surviving particle' [Tribe (1992)] is presented in [Gill (2010)].

In [Balázs, Rácz and Tóth (2014)] a one-dimensional particle system is considered with interaction via the center of mass. There is a kind of attraction towards the center of mass in the following sense: each particle jumps to the right according to some common distribution F, but the rate at which the jump occurs is a monotone decreasing function of the signed distance between the particle and the mass center. Particles being far ahead slow down, while the laggards catch up.

Chapter 5

Branching in random environment: Trapping of the first/last particle

Recall the problem of a single Brownian particle trying to 'survive' in a Poissonian system of traps (obstacles) from Section 1.12. In many models (for example those in population biology or nuclear physics), the particles also have an additional feature: *branching*. It seems therefore quite natural to ask whether an asymptotics similar to (1.29) can be obtained for branching processes. Analogously to the single particle case, one may want to study, for example, the probability that no reaction has occurred up to time $t > 0$ between the family generated by a single particle and the traps.

Thus, in this chapter we will study a branching Brownian motion on $\mathbb{R}^d$ with branching rate $\beta > 0$, in a Poissonian field of traps.

5.1 The model

Our 'system' consists of the following two components.

BBM component: Let $Z = (Z_t)_{t \geq 0}$ be the d-dimensional dyadic (two offspring) branching Brownian motion with constant branching rate $\beta > 0$, starting with a single particle at the origin. Write P_0 to denote the law of Z, indicating that the initial particle starts at 0.

PPP component: Let ω be the Poisson point process on $\mathbb{R}^d$ with a spatially dependent locally finite intensity measure ν such that

$$\frac{\mathrm{d}\nu}{\mathrm{d}x} \sim \frac{\ell}{|x|^{d-1}}, \quad |x| \to \infty, \qquad \ell > 0, \tag{5.1}$$

i.e., the integral of $\mathrm{d}\nu/\mathrm{d}x$ over large spheres centered at the origin is asymptotically linear. Write $\mathbb{P}$ to denote the law of ω, and $\mathbb{E}$ to denote the corresponding expectation. Just like in Section 1.12, for $a > 0$, let

$$K = K_a(\omega) = \bigcup_{x_i \in \mathrm{supp}(\omega)} \overline{B}_a(x_i) \tag{5.2}$$

be the a-neighborhood of ω, which is to be thought of as a configuration of traps attached to ω (here $B_a(x_i)$ is the ball of radius a centered at x).

The reason we study this particular decay[1] is that this order of decay turns out to be the interesting one: it serves as a 'threshold' rate, where phenomena only depend on the 'fine tuning constant' ℓ. Taking a smaller or larger order of decay would result in features similar to those in the 'small ℓ' and the 'large ℓ' regimes, respectively. When the critical value of ℓ is determined, the threshold will thus divide a 'low intensity' regime from a 'high intensity' one.

5.2   A brief outline of what follows

For $A \subseteq \mathbb{R}^d$ Borel and $t \geq 0$, $Z_t(A)$ is the number of particles located in A at time t, and $|Z_t|$ is the total number of particles. For $t \geq 0$, let

$$R_t = \bigcup_{s \in [0,t]} \operatorname{supp}(Z_s) \tag{5.3}$$

denote the union of all the particle trajectories up to time t (= the range of Z up to time t). Let T be the first time that Z hits a trap, i.e.,

$$T := \inf\{t \geq 0 : Z_t(K) \geq 1\} = \inf\{t \geq 0 : R_t \cap K \neq \emptyset\}. \tag{5.4}$$

The event $\{T > t\}$ is thus the *survival of the branching particle system* up to time t of Z *among the Poissonian traps* (i.e., no particle hits a trap up to time t).

Suppose now that, instead of considering the trapping time in (5.4), we kill the process when *all* the particles are absorbed/killed by a trap (extinction). That is, if $Z^K = (Z_t^K)_{t \geq 0}$ denotes the BBM with killing at the boundary of the trap set K, then we define

$$\tilde{T} = \inf\{t \geq 0 : |Z_t^K| = 0\} \tag{5.5}$$

and we pick $\{\tilde{T} > t\}$ as the survival up to time t.

We are interested in the *annealed* probabilities of the events $\{T > t\}$ and $\{\tilde{T} > t\}$.

The first probability will be shown to decay like $\exp[-I(\ell, \beta, d)t + o(t)]$ as $t \to \infty$, where the rate constant $I(\ell, \beta, d)$ is determined in terms of a variational problem. It turns out that this rate constant exhibits a crossover at a critical value $\ell_{cr} = \ell_{cr}(\beta, d)$.

---

[1] Of course, this isn't really 'decay' in one dimension.

The second probability, on the other hand, will be shown to tend to a positive limit as $t \to \infty$.

Focussing on the first problem, the next natural question concerns the *optimal survival strategy*. That is, we are interested in the behavior of the system, conditioned on the unlikely event of survival.

Remark 5.1 (Terminology). As usual in the theory of large deviations, one often uses a heuristic language and talks about 'strategies' and 'costs.' An event of low (high) probability has a 'high cost' ('low cost'). If the event of survival follows from some other event (a strategy) for which it is easier to compute its probability, and that probability is relatively high (that is, the strategy has 'low cost'), then the strategy is considered good. There might be strategies which have the same cost on the scale one is working on; by the uniqueness of the optimal strategy one means a conditional limit theorem: conditioned on survival, the probability of the event (strategy) tends to one as time goes to infinity.

The term 'super-exponentially small' will often be abbreviated by 'SES.' Finally, we will frequently use the informal term 'system' to refer to the two components (BBM and PPP) together.

We will see that, conditional on survival until time t, the following properties hold with probability tending to one as $t \to \infty$. For $\ell < \ell_{cr}$, a ball of radius $\sqrt{2\beta}\, t$ around the origin is emptied, the BBM stays inside this ball and branches at rate β. For $\ell > \ell_{cr}$, on the other hand, the 'system' exhibits the following behavior.

- $d = 1$: suppresses the branching until time t, empties a ball of radius $o(t)$ around the origin (i.e., a ball whose radius is larger than the trap radius but smaller than order t), and stays inside this ball;
- $d \geq 2$: suppresses the branching until time $\eta^* t$, empties a ball of radius $\sqrt{2\beta}\,(1-\eta^*)t$ around a point at distance $c^* t$ from the origin, and during the remaining time $(1 - \eta^*)t$ branches at the original rate β. Here, $0 < \eta^* < 1$ and $c^* > 0$ are the minimizers of the variational problem for $I(\ell, \beta, d)$.

In the latter case, we will show that *one* optimal survival strategy is the following: the 'system' completely suppresses the branching until time $\eta^* t$, i.e., only the initial particle is alive at time $\eta^* t$, within a small empty tube moves the initial particle to a point at distance $c^* t$ from the origin, empties a ball of radius $\sqrt{2\beta}\,(1-\eta^*)t$ around that point, stays inside this ball during the remaining time $(1 - \eta^*)t$ and branches at rate β.

This does not rule out the existence of other survival strategies with the same exponential cost, which cannot be distinguished without a higher-order analysis. (Note, for example, that not branching at all or producing, say, $t^n, n \geq 1$ particles by time t, have the same cost on a logarithmic scale.) However, we will prove the uniqueness of some parts of this strategy later — see Remark 5.4.

Last but not least, we will see a surprising feature of the crossover at the critical value: η^* and c^* tend to a strictly positive limit as $\ell \downarrow \ell_{cr}$, i.e., the crossover at ℓ_{cr} is *discontinuous*. Moreover, $c^* > \sqrt{2\beta}\,(1 - \eta^*)$ for all $\ell > \ell_{cr}$, i.e., the empty ball does not contain the origin.

5.3   The annealed probability of $\{T > t\}$

To formulate the main results, we need some more notation. For $r, b \geq 0$, define

$$f_d(r, b) = \int_{B_r(0)} \frac{dx}{|x + be|^{d-1}}, \qquad (5.6)$$

where $e = (1, 0, \ldots, 0)$. For $\eta \in [0, 1]$ and $c \in [0, \infty)$, let

$$k_{\beta,d}(\eta, c) = \lim_{t \to \infty} \frac{1}{\ell t}\, \nu\left(B_{\sqrt{2\beta}\,(1-\eta)t}(cte)\right)$$

$$= f_d\left(\sqrt{2\beta}\,(1 - \eta), c\right) \qquad (5.7)$$

(recall (5.1)). Let

$$\ell_{cr}^* = \ell_{cr}^*(\beta, d) = \frac{1}{s_d}\sqrt{\frac{\beta}{2}} \qquad (5.8)$$

with s_d the surface of the d-dimensional unit ball:

$$s_1 = 2, s_2 = 2\pi, s_3 = 4\pi, \ldots, s_d = \frac{2\pi^{d/2}}{\Gamma(\frac{d}{2})}.$$

Define

$$\ell_{cr} = \ell_{cr}(\beta, d) = \begin{cases} \ell_{cr}^* & \text{if } d = 1, \\ \alpha_d \ell_{cr}^* & \text{if } d \geq 2, \end{cases} \qquad (5.9)$$

with

$$\alpha_d = \frac{-1 + \sqrt{1 + 4M_d^2}}{2M_d^2} \in (0, 1), \qquad (5.10)$$

where

$$M_d = \frac{1}{2s_d} \max_{R \in (0,\infty)} [f_d(R,0) - f_d(R,1)]. \tag{5.11}$$

The next theorem expresses the probability of survival in terms of a variational problem.[2] To this end, we now define the following, three-variable function of the parameters:

$$I(\ell, \beta, d) := \min_{\eta \in [0,1],\, c \in [0,\infty)} \left\{ \beta\eta + \frac{c^2}{2\eta} + \ell k_{\beta,d}(\eta, c) \right\}. \tag{5.12}$$

For $\eta = 0$ put $c = 0$ and $k_{\beta,d}(0,0) = f_d(\sqrt{2\beta}, 0) = s_d\sqrt{2\beta}$ and delete the $\frac{c^2}{2\eta}$ term. (Formally, we can define the term after the minimum $+\infty$ for $\eta = 0$ and $c > 0$.)

Theorem 5.1 (Variational formula). *Given d, β and a, the ofllowing holds for any $\ell > 0$:*

$$\lim_{t \to \infty} \frac{1}{t} \log(\mathbb{E} \times P_0)(T > t) = -I(\ell, \beta, d). \tag{5.13}$$

Remark 5.2 (Interpretation of Theorem 5.1). Fix β, d and η, c.

- The probability to completely suppress the branching[3] until time ηt is

$$\exp[-\beta\eta t]. \tag{5.14}$$

- If $c, \eta > 0$, then the likelihood for the initial particle to move to a site at distance ct from the origin during time ηt is (recall Lemma 1.3)

$$\exp\left[-\frac{c^2}{2\eta}t + o(t)\right]. \tag{5.15}$$

- Under (5.1), the probability to empty a $\sqrt{2\beta}\,(1 - \eta)t$-ball around a site at distance ct from the origin is (see (5.7))

$$\exp[-\ell k_{\beta,d}(\eta, c)t + o(t)]. \tag{5.16}$$

The probability to empty a 'small tube' in $\mathbb{R}^d$, connecting the origin with this site is $\exp[o(t)]$; for the initial particle to remain inside this tube up to time ηt is also $\exp[o(t)]$. (See Section 5.4.1.)

- The probability for the offspring of the initial particle present at time ηt to remain inside the $\sqrt{2\beta}\,(1 - \eta)t$-ball during the remaining time $(1 - \eta)t$ is $\exp[o(t)]$ as well. (See Section 1.14.7.)

The total cost of these three large deviation events gives rise to the sum under the minimum in (5.12); the 'minimal cost' is therefore determined by the minimizers of (5.12).

---

[2]Variational problems are ubiquitous in 'large deviations' literature.

[3]That is, the initial particle has never produced any offspring.

5.4   Proof of Theorem 5.1

5.4.1   *Proof of the lower bound*

Fix β, d and η, c. Recall from Remark 5.2 the type of strategy the lower bound is based on. Considering the 'grand total' of all the costs and minimizing it over the parameters η and c will yield the required lower estimate.

The only cost in the strategy that is not completely obvious to see, is the one related to the 'small tube' when $d \geq 2$. We explain this part below.

Let $d \geq 2$. Evidently, we may assume, that the specific site has first coordinate ct and all other coordinates zero. Pick $t \mapsto r(t)$ such that $\lim_{t \to \infty} r(t) = \infty$ but $r(t) = o(t)$ as $t \to \infty$. Pick also $k > c$ and define the two-sided cylinder ('small tube')[4] as

$$T_t = \left\{ x = (x_1, \ldots, x_d) \in \mathbb{R}^d : \ |x_1| \leq kt, \ \sqrt{x_2^2 + \cdots + x_d^2} \leq r(t) \right\}.$$

Claim 5.1. *The probability to empty T_t is $\exp[o(t)]$ as $t \to \infty$.*

Proof.   Recall (5.1). Abbreviate $r := \sqrt{x_2^2 + \cdots + x_d^2}$ and $y := x_1$ and use polar coordinates in $(d-1)$-dimension (resulting in a factor r^{d-2}). Then the claim easily follows from the estimate

$$\left| \int_p^q \int_u^v \frac{r^{d-2}}{\left(\sqrt{r^2 + y^2} \right)^{d-1}} \, dr dy \right| \leq \int_p^q \int_u^v \left| \frac{1}{\sqrt{r^2 + y^2}} \left(\frac{r}{\sqrt{r^2 + y^2}} \right)^{d-2} \right| dr dy$$

$$\leq \int_p^q \int_u^v \frac{1}{\sqrt{r^2 + y^2}} \, dr dy = \int_p^q \log \left(2 \left(r + \sqrt{r^2 + y^2} \right) \right) \Big|_{r=u}^v \, dy,$$

for $p \leq q$, $u \leq v$; $p, q, u, v \in \mathbb{R}$, along with the formula

$$\int \log \left(\left(c + \sqrt{c^2 + y^2} \right) \right) dy$$

$$= -y + y \log(c + \sqrt{c^2 + y^2}) + c \log \left[2(x + \sqrt{c^2 + y^2}) \right], \ c > 0,$$

plus some careful, but completely elementary calculation.   □

Moreover, if P denotes the d-dimensional Wiener measure, W^1 denotes the first coordinate of the d-dimensional Brownian motion and

$$A_t := \{ct \leq W_{\eta t}^1 \leq ct + r(t)\},$$
$$B_t := \{|W_s^1| \leq kt \ \forall 0 \leq s \leq \eta t\}, \tag{5.17}$$

---

[4]Recall that the tube is not needed for $d = 1$.

then

$$P(A_t \cap B_t) = \exp\left[-\frac{c^2}{2\eta}t + o(t)\right], \tag{5.18}$$

because

$$\exp\left[-\frac{c^2}{2\eta}t + o(t)\right] = P(A_t) \geq P(A_t \cap B_t) \geq P(A_t) - P(B_t^c)$$

$$= \exp\left[-\frac{c^2}{2\eta}t + o(t)\right] - \exp\left[-\frac{k^2}{2\eta}t + o(t)\right] = \exp\left[-\frac{c^2}{2\eta}t + o(t)\right].$$

Recalling that W^1 is the first coordinate process, decompose the Brownian motion W into an independent sum $W = W^1 \oplus W^{d-1}$, and let

$$C_t = \{|W_s^{d-1}| \leq r(t) \ \forall 0 \leq s \leq \eta t\}. \tag{5.19}$$

Since $r(t) \to \infty$, we have $P(C_t) = \exp[o(t)]$. This, along with (5.18), and the independence of W^1 and W^{d-1}, implies

$$P(A_t \cap B_t \cap C_t) = \exp\left[-\frac{c^2}{2\eta}t + o(t)\right]. \tag{5.20}$$

That is, *emptying T_t, confining the Brownian particle to T_t up to time ηt, and moving it to distance $ct + o(t)$ from the origin at time ηt, has total cost* $\exp[-\frac{c^2}{2\eta}t + o(t)]$. (As a matter of fact, the first two of these do not contribute on our logarithmic scale.) The fact that the tube T_t intersects the ball to be emptied does not affect the argument.

5.4.2   *Proof of the upper bound*

Fix β, d and $\epsilon > 0$ small. Recall that $|Z_t|$ is the number of particles at time t. For $t > 1$, define

$$\eta_t = \sup\left\{\eta \in [0,1]: |Z_{\eta t}| \leq \lfloor t^{d+\epsilon} \rfloor\right\}. \tag{5.21}$$

Then, for all $n \in \mathbb{N}$,

$$(\mathbb{E} \times P_0)(T > t)$$

$$= \sum_{i=0}^{n-1} (\mathbb{E} \times P_0)\left(\{T > t\} \cap \left\{\frac{i}{n} \leq \eta_t < \frac{i+1}{n}\right\}\right)$$

$$+ \exp[-\beta t + o(t)]$$

$$\leq \sum_{i=0}^{n-1} \exp\left[-\beta\frac{i}{n}t + o(t)\right] (\mathbb{E} \times P_t^{(i,n)})(T > t)$$

$$+ \exp[-\beta t + o(t)], \tag{5.22}$$

where we used (1.45) and the conditional probabilities

$$P_t^{(i,n)}(\cdot) := P_0 \left(\ \cdot \ \Big| \ \frac{i}{n} \le \eta_t < \frac{i+1}{n} \right), \qquad i = 0, 1, \ldots, n-1. \quad (5.23)$$

Note that $\eta_t < \frac{i+1}{n}$ implies $\left| Z_{\frac{i+1}{n}t} \right| > \lfloor t^{d+\epsilon} \rfloor$. Let $A_t^{(i,n)}$, $i = 0, 1, \ldots, n-1$, denote the event that, among the $\left| Z_{\frac{i+1}{n}t} \right|$ particles alive at time $\frac{i+1}{n}t$, there are $\le \lfloor t^{d+\epsilon} \rfloor$ particles such that the ball with radius

$$\rho_t^{(i,n)} = (1 - \epsilon)\sqrt{2\beta} \left(1 - \frac{i+1}{n} \right) t \quad (5.24)$$

around the particle is non-empty (i.e. receives at least one point from ω). It is plain that

$$(\mathbb{E} \times P_t^{(i,n)})(T > t) \le (\mathbb{E} \times P_t^{(i,n)})(A_t^{(i,n)})$$
$$+ (\mathbb{E} \times P_t^{(i,n)})(T > t \mid [A_t^{(i,n)}]^c). \quad (5.25)$$

Consider now the BBM's emanating from the 'parent particles' alive at time $\frac{i+1}{n}t$. The distributions of each of these BBM's are clearly radially symmetric with respect to their starting points. Using this fact, along with their independence and Proposition 1.16, we now show that the second term on the right-hand side of (5.25) is bounded above by

$$\left[1 - \frac{C_1}{[\rho_t^{(i,n)}]^{d-1}} \right]^{\lfloor t^{d+\epsilon} \rfloor} \le \exp\left[-C_2(\epsilon)t^{1+\epsilon} \right] \quad (5.26)$$

uniformly in all parameters, which is super-exponentially small (SES).

Indeed, on the event $[A_t^{(i,n)}]^c$, there are more than $\lfloor t^{d+\epsilon} \rfloor$ balls containing traps, and in the remaining time $(1 - \frac{i+1}{n})t$, the BBM emanating from the center of each ball leaves this ball with a probability tending to 1 as $t \to \infty$ (by Proposition 1.16 and (5.24)). Moreover, by radial symmetry, the trap inside the ball has a probability $C_1/[\rho_t^{(i,n)}]^{d-1}$ to be hit by the BBM when exiting.

To estimate the first term on the right-hand side of (5.25), the trick is to 'randomly' pick $\lfloor t^{d+\epsilon} \rfloor + 1$ particles from the $\left| Z_{\frac{i+1}{n}t} \right|$ particles alive at time $\frac{i+1}{n}t$. By 'randomly' we mean to do this independently of their spatial position and according to some probability distribution $\mathbb{Q}$. A 'concrete' way to realize $\mathbb{Q}$ is to mark a random ancestral line by tossing a coin at each branching time and choosing the 'nicer' or the 'uglier' offspring according to the outcome. This way we choose a 'random' particle from the offspring. Repeat this procedure independently so many times until it produces $\lfloor t^{d+\epsilon} \rfloor + 1$ different particles.[5] Since the particles are chosen independently from the motion process, each of them is at a 'random location' whose spatial distribution is identical to that of $W(\frac{i+1}{n}t)$, where $(W = W(s))_{s \geq 0}$ denotes standard Brownian motion.

Recall that $A_t^{(i,n)}$ is the event that, among the $\left| Z_{\frac{i+1}{n}t} \right|$ particles alive at time $\frac{i+1}{n}t$, there are no more than $\lfloor t^{d+\epsilon} \rfloor$ particles such that the ball with radius $\rho_t^{(i,n)}$ around the particle is non-empty (has a trap). Hence, by the 'pigeon-hole principle,' on the event $A_t^{(i,n)}$, at least one of the $\lfloor t^{d+\epsilon} \rfloor + 1$ particles picked at random must have an empty ball around it.

In order to bound $(\mathbb{E} \times P_t^{(i,n)})(T > t)$, consider $C_{i,n,t}$, the collection of the centers of the empty balls at time $\frac{i+1}{n}t$ and let

$$x_0^{(i,n,t)} := \mathrm{argmin}_{x \in C_{i,n,t}} |x|$$

(i.e. the one closest to the origin). We will achieve our estimate by 'slicing the probability space' according to the location of $x_0^{(i,n,t)}$.

We have that for any $0 \leq c < \infty$ and $\delta > 0$,

$$(\mathbb{E} \times P_t^{(i,n)}) \left(\left\{ ct \leq |x_0^{(i,n,t)}| \leq (c+\delta)t \right\} \cap A_t^{(i,n)} \right)$$

$$\leq (\mathbb{E} \times P_t^{(i,n)} \times \mathbb{Q}) \left(\mathrm{Event} \right)$$

$$\leq (\lfloor t^{d+\epsilon} \rfloor + 1) \exp \left[-\frac{c^2}{2(i+1)/n} t + o(t) \right]$$

$$\times \exp \left[-\ell f_d \left((1-\epsilon)\sqrt{2\beta} \left(1 - \frac{i+1}{n} \right), c \right) t + O(\delta)t + o(t) \right] \quad (5.27)$$

where

$$\mathrm{Event} := \Big(\exists \, \mathrm{a\ random\ point\ at\ distance\ between}\ ct\ \mathrm{and}\ (c+\delta)t,$$

$$\mathrm{and\ the\ ball\ with\ radius}\ \rho_t^{(i,n)}\ \mathrm{around\ this\ random\ point\ is\ empty} \Big).$$

Armed with the bounds (5.25) and (5.27), we are finally in the position to

---

[5]The procedure is similar to the one when a random line of descent is chosen in the Spine Construction in Chapter 2.

estimate $(\mathbb{E} \times P_t^{(i,n)})(T > t)$. Indeed, one has

$$(\mathbb{E} \times P_t^{(i,n)})(T > t)$$

$$\leq \sum_{j=0}^{n-1} (\mathbb{E} \times P_t^{(i,n)}) \left(\left\{ \frac{j}{n} \sqrt{2\beta}\, t \leq |x_0^{(i,n,t)}| < \frac{j+1}{n} \sqrt{2\beta}\, t \right\} \cap A_t^{(i,n)} \right)$$

$$+ \exp[-\beta t + o(t)] + \text{SES}$$

$$\leq \sum_{j=0}^{n-1} (\lfloor t^{d+\epsilon} \rfloor + 1) \exp\left[\frac{-\beta j^2/n^2}{(i+1)/n} t + o(t) \right]$$

$$\times \exp\left[-\ell f_d \left((1-\epsilon)\sqrt{2\beta} \left(1 - \frac{i+1}{n} \right), \frac{j}{n} \right) t \right.$$

$$\left. + O(1/n)t + o(t) \right]$$

$$+ \exp[-\beta t + o(t)] + \text{SES}. \tag{5.28}$$

Here, the SES term comes from the second term on the right-hand side of (5.25), while the restriction on the sum is taken care of by the middle term. Indeed, $|x_0^{(i,n,t)}| \geq \sqrt{2\beta}\, t$ means that all the centers of the empty balls are at distance $\geq \sqrt{2\beta}\, t$. The probability of this event is bounded above by the probability that a single Brownian particle is at distance $\geq \sqrt{2\beta}\, t$ at time t (by our way of constructing $\mathbb{Q}$), which is $\exp[-\beta t + o(t)]$, by Lemma 1.3.

To finish the argument, substitute (5.28) into (5.22), optimize over $i, j \in \{0, 1, \ldots, n-1\}$, let $n \to \infty$ followed by $\epsilon \downarrow 0$, and obtain

$$\limsup_{t \to \infty} \frac{1}{t} \log(\mathbb{E} \times P_0)(T > t) \leq -I(\ell, \beta, d). \tag{5.29}$$

(In (5.28), put $\eta = i/n$ and $c = j/n$ before letting $n \to \infty$, and use the continuity of the functional the minimum of which is taken.)   □

5.5 Crossover at the critical value

The second major statement of this chapter concerns the existence of a *critical value* for the 'fine-tuning constant' $\ell > 0$, separating two, qualitatively different regimes. The reader should note the difference between the $d = 1$ and the $d \geq 2$ settings.

Theorem 5.2 (Crossover). *Fix β, a. Then the following holds.*

(i) *For $d \geq 1$ and all $\ell \neq \ell_{cr}$, the variational problem in (5.12) has a unique pair of minimizers, denoted by $\eta^* = \eta^*(\ell, \beta, d)$ and $c^* = c^*(\ell, \beta, d)$.*

(ii) *For $d = 1$,*

$$\ell \leq \ell_{cr} : I(\ell, \beta, d) = \beta \frac{\ell}{\ell_{cr}^*},$$

$$\ell > \ell_{cr} : I(\ell, \beta, d) = \beta, \tag{5.30}$$

and

$$\ell < \ell_{cr} : \eta^* = 0, \, c^* = 0,$$

$$\ell > \ell_{cr} : \eta^* = 1, \, c^* = 0. \tag{5.31}$$

(iii) *For $d \geq 2$,*

$$\ell \leq \ell_{cr} : I(\ell, \beta, d) = \beta \frac{\ell}{\ell_{cr}^*},$$

$$\ell > \ell_{cr} : I(\ell, \beta, d) < \beta \left(1 \wedge \frac{\ell}{\ell_{cr}^*} \right), \tag{5.32}$$

and

$$\ell < \ell_{cr} : \eta^* = 0, \, c^* = 0,$$

$$\ell > \ell_{cr} : 0 < \eta^* < 1, \, c^* > 0. \tag{5.33}$$

(iv) *For $d \geq 2$, the function $\ell \mapsto I(\ell, \beta, d)$ is continuous and strictly increasing, with*

$$\lim_{\ell \to \infty} I(\ell, \beta, d) = \beta.$$

(See Fig. 5.1.)

(v) *For $d \geq 2$, the functions $\ell \mapsto \eta^*(\ell, \beta, d)$ and $\ell \mapsto c^*(\ell, \beta, d)$ are both discontinuous at ℓ_{cr} and continuous on (ℓ_{cr}, ∞), and their asymptotic behavior is given by*

$$\lim_{\ell \to \infty} \frac{1 - \eta^*(\ell, \beta, d)}{c^*(\ell, \beta, d)} = 1, \qquad \lim_{\ell \to \infty} c^*(\ell, \beta, d) = 0. \tag{5.34}$$

Moreover, $c^* > \sqrt{2\beta} \, (1 - \eta^*)$ for all $\ell > \ell_{cr}$.

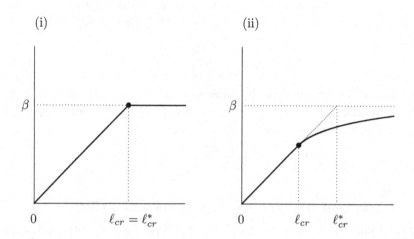

Fig. 5.1 The function $\ell \mapsto I(\ell, \beta, d)$ for: (i) $d = 1$; (ii) $d \geq 2$.

Remark 5.3 (Interpretation of Theorem 5.2). We see that (5.12) exhibits a crossover at the critical value $\ell_{cr} = \ell_{cr}(\beta, d)$ defined in (5.9), separating a low intensity from a high intensity regime. In the low intensity regime the minimizers are trivial (extreme), while in the high intensity regime they are only trivial for $d = 1$. There are two peculiar facts that should be pointed out:

(1) for $d \geq 2$ the minimizers are discontinuous at ℓ_{cr},
(2) in the high intensity regime the empty ball inside which the BBM branches freely does not contain the origin.

Consequently, at the crossover, the center of the empty ball is 'jumping away' from the origin, whereas the radius is 'jumping down.'

5.6   Proof of Theorem 5.2

5.6.1   *Proof of Theorem 5.2(i)*

For $\eta \in [0, 1]$, $c \in [0, \infty)$, define

$$F_d(\eta, c) := \beta\eta + \frac{c^2}{2\eta} + \ell f_d\left(\sqrt{2\beta}\,(1 - \eta), c\right), \qquad (5.35)$$

with the understanding that $F_d(0,0) := \ell s_d \sqrt{2\beta}$ and $F_d(0,c) := +\infty$ for $c > 0$. (Note that $F_d(1,c) = \beta + c^2/2$ for $c \geq 0$.) Then (5.12) reads (insert (5.7))

$$I(\ell, \beta, \eta) = \min_{\eta \in [0,1],\, c \in [0,\infty)} F_d(\eta, c). \tag{5.36}$$

To see the *existence* of the minimizers η^*, c^* of (5.36), note that F_d diverges uniformly in η as $c \to \infty$, since $F_d(\eta, c) \geq c^2/2$, and that F_d is lower semicontinuous, since

$$\lim_{(\eta,c)\to(0,0)} \beta\eta + \ell f_d\left(\sqrt{2\beta}\,(1-\eta), c\right) = \ell s_d \sqrt{2\beta}$$

and $\frac{c^2}{2\eta} \geq 0$.

Our next task is to verify the *uniqueness* of (η^*, c^*) when $\ell \neq \ell_{cr}$.

$\underline{d = 1}$: Since $f_1(r, b) = 2r$, we have $f_1(\sqrt{2\beta}\,(1 - \eta), c) = 2\sqrt{2\beta}\,(1 - \eta)$, which does not depend on c. Hence the minimum over c in (5.36) is taken at $c^* = 0$, so that (5.36) reduces to

$$I(\ell, \beta, 1) = \min_{\eta \in [0,1]} \left\{\beta\eta + \ell 2\sqrt{2\beta}\,(1 - \eta)\right\}. \tag{5.37}$$

The function under the minimum in (5.37) is linear in η, and changes its slope from positive to negative as ℓ moves upwards through the critical value ℓ_{cr} given by $\beta = \ell_{cr} 2\sqrt{2\beta}$. This identifies ℓ_{cr} as in (5.9). The minimizer of (5.37) changes from $\eta^* = 0$ to $\eta^* = 1$, proving (5.31), while $I(\ell, \beta, 1)$ changes from $\ell 2\sqrt{2\beta}$ to β, proving (5.30).

$\underline{d \geq 2}$: We have

$$F_d(\eta, c) - F_d(0, 0) = \beta\eta + \frac{c^2}{2\eta} + \ell A_{d,\beta}(\eta, c) \tag{5.38}$$

where

$$A_{\beta,d}(\eta, c) := f_d(\sqrt{2\beta}\,(1 - \eta), c) - f_d(\sqrt{2\beta}, 0) \leq 0 \tag{5.39}$$

with equality if and only if $(\eta, c) = (0, 0)$.[6] Suppose that $(\eta, c) = (0, 0)$ is a minimizer when $\ell = \ell_0$. Then the right-hand side of (5.38) is nonnegative for all (η, c) when $\ell = \ell_0$. Consequently, for all $\ell < \ell_0$ the right-hand side of (5.38) is zero when $(\eta, c) = (0, 0)$ and strictly positive otherwise. Therefore we conclude that there must exist an $\ell_{cr} \in [0, \infty]$ such that

(i) $(\eta, c) = (0, 0)$ is the *unique* minimizer when $\ell < \ell_{cr}$.

(ii) $(\eta, c) = (0, 0)$ is not a minimizer when $\ell > \ell_{cr}$.

---

[6]The latter statement is easily deduced from (5.6).

In Section 5.6.2 we will identify ℓ_{cr} as given by (5.9)–(5.11), and this will show that actually $\ell_{cr} \in (0, \infty)$. It remains to prove that when $\ell > \ell_{cr}$ the minimizers are unique, which is done in Steps I–III below.

I. Minimizers in the interior: First, we can rule out the combination $\eta^* = 0, c^* > 0$, as $F(0, c) = \infty$ for all $c > 0$ (see the remark below (5.12)). The same holds for the case $\eta^* > 0, c^* = 0$, because $F(\eta, 0)$ takes its minimum either at $\eta = 0$ or $\eta = 1$, and so $\eta^* > 0$ would imply $\eta^* = 1$; however, $\eta^* = 1$ can be excluded via the following lemma.

Lemma 5.1. *For every $\ell_0 > 0$ there exists a $\delta_0 = \delta_0(\ell_0) > 0$ such that $\eta^* \leq 1 - \delta_0$ for all $\ell \leq \ell_0$.*

Proof.   Since $F_d(1, c) = \beta + \frac{c^2}{2}$, a minimizer $\eta^* = 1$ would necessarily come with a minimizer $c^* = 0$, yielding the minimal value β. However, one can do better: note from (5.6) that $f_d(r, b) \sim v_d r^d / b^{d-1}$ as $r \downarrow 0$ and $r/b \downarrow 0$, with v_d the volume of the d-dimensional unit ball. Pick $\eta = 1 - \delta$ and $c = \delta^{3/4}$. Then, for $\delta \downarrow 0$,

$$F_d(1 - \delta, \delta^{3/4}) = \beta(1 - \delta) + \frac{\delta^{3/2}}{2(1 - \delta)} + \ell f_d(\sqrt{2\beta}\,\delta, \delta^{3/4})$$

$$= \beta(1 - \delta) + \left[\frac{1}{2}\delta^{3/2} + \ell v_d(2\beta)^{d/2}\delta^{(d+3)/4}\right](1 + o(1))$$

$$= \beta(1 - \delta) + o(\delta). \tag{5.40}$$

For δ small enough, the right-hand side is strictly decreasing in δ, showing that the minimum cannot occur at $\delta = 0$. In fact, the above expansion shows that $\delta \geq \delta_0(\ell_0)$ for any $\ell \leq \ell_0$. $\square$

Our conclusion from the above is that for $\ell > \ell_{cr}$ it is sufficient to consider $0 < \eta < 1$ and $c > 0$. Hence, we continue with checking the stationary points for F_d.

II. Stationary points: For $R \geq 0$, let

$$f_d(R) = \int_{B_R(0)} \frac{dx}{|x + e|^{d-1}}. \tag{5.41}$$

Then we may write (5.35) as

$$F_d(\eta, c) = \beta\eta + \frac{c^2}{2\eta} + \ell c f_d\left(\frac{\sqrt{2\beta}\,(1 - \eta)}{c}\right). \tag{5.42}$$

The stationary points are the solutions of the equations

$$0 = \beta - \frac{c^2}{2\eta^2} - \ell\sqrt{2\beta} f_d'\left(\frac{\sqrt{2\beta}(1-\eta)}{c}\right),$$

$$0 = \frac{c}{\eta} + \ell\left[f_d\left(\frac{\sqrt{2\beta}(1-\eta)}{c}\right)\right.$$
$$\left. - \frac{\sqrt{2\beta}(1-\eta)}{c} f_d'\left(\frac{\sqrt{2\beta}(1-\eta)}{c}\right)\right]. \tag{5.43}$$

Eliminating f_d', we obtain

$$\ell f_d\left(\frac{\sqrt{2\beta}(1-\eta^*)}{c^*}\right) = -\frac{c^*}{\eta^*} + \frac{1-\eta^*}{c^*}\left[\beta - \frac{c^{*2}}{2\eta^{*2}}\right] \tag{5.44}$$

and hence

$$F_d(\eta^*, c^*) = \beta - \frac{c^{*2}}{2\eta^{*2}}. \tag{5.45}$$

Putting

$$u = \frac{\sqrt{2\beta}(1-\eta)}{c}, \qquad v = \frac{c}{\sqrt{2\beta}\,\eta}, \tag{5.46}$$

we may rewrite (5.43) as

$$0 = \beta - \beta v^2 - \ell\sqrt{2\beta} f_d'(u),$$

$$0 = \sqrt{2\beta}\,v + \ell[f_d(u) - u f_d'(u)], \tag{5.47}$$

and (5.45) as

$$F_d(u^*, v^*) = \beta(1 - v^{*2}). \tag{5.48}$$

III. Uniqueness: Suppose that (u_1, v_1) and (u_2, v_2) give the same minimum. Then, by (5.48), we have $v_1 = v_2$. Suppose that $u_1 \neq u_2$. Then from the first line of (5.47) it follows that $f_d'(u_1) = f_d'(u_2)$. Using this in the second line of (5.47), we get

$$f_d(u_1) - u_1 f_d'(u_1) = f_d(u_2) - u_2 f_d'(u_1), \tag{5.49}$$

or

$$\frac{f_d(u_1) - f_d(u_2)}{u_1 - u_2} = f_d'(u_1) = f_d'(u_2). \tag{5.50}$$

This in turn implies that there must exist a third value u_3, strictly between u_1 and u_2, such that

$$f_d'(u_3) = f_d'(u_1) = f_d'(u_2). \tag{5.51}$$

Uniqueness now follows from the following property of f_d, implying that f'_d does not attain the same value at three different points. (The singularity of f'_d at 1 does not affect the above argument.)

Lemma 5.2. *The function* $g_d := f'_d$ *is strictly increasing on* $(0,1)$, *infinity at* 1, *and strictly decreasing on* $(1,\infty)$. *Furthermore,* $\lim_{R \to \infty} g_d(R) = s_d$. *($s_d$ is the surface of the unit ball.)*

Proof.   If ω is the angle between the vectors x and e in $\mathbb{R}^d$, then using polar coordinates, we can write (5.41) as

$$f_d(R) = C(d) \int_0^R dr \, r^{d-1} \int_0^\pi d\omega \, \left(1 + r^2 - 2r\cos\omega\right)^{-\frac{d-1}{2}}, \qquad (5.52)$$

where the other angle variables besides ω (if $d \geq 3$) just contribute to $C(d)$, as the integrand in (5.41) only depends on r and ω. It is easy to see that $C(d) = s_d/\pi$. (Change the integrand in the definition of f_d to one and use that $s_d = v_d \cdot d$, where v_d is the volume of the unit ball, to check this.)
    Hence

$$g_d(R) = (s_d/\pi)R^{d-1} \int_0^\pi d\omega \, \left(1 + R^2 - 2R\cos\omega\right)^{-\frac{d-1}{2}}. \qquad (5.53)$$

Set $S = 1/R$ to write

$$g_d(1/S) = s_d/\pi \int_0^\pi d\omega \, \left(1 + S^2 - 2S\cos\omega\right)^{-\frac{d-1}{2}}, \qquad (5.54)$$

and note that we have to verify our statements about g_d for $S \mapsto g_d(1/S)$, swapping the words 'increasing' and 'decreasing' and changing $R \to \infty$ to $S \to 0$.
    For $S > 1$, the integrand is strictly decreasing in S for all $\omega \in [0,\pi]$; thus $S \mapsto g_d(1/S)$ is strictly increasing on $(1,\infty)$. At $S = 1$, the integral diverges. Last, for $0 < S < 1$, (5.54) yields

$$g_d(1/S) = s_d F(\nu, \nu; 1, S^2) = s_d \sum_{k=0}^\infty S^{2k} \left(\frac{\prod_{l=0}^{k-1}(\nu + l)}{k!}\right)^2, \qquad (5.55)$$

where F is the *hypergeometric function* and $\nu := \frac{d-1}{2}$ (see [Gradhsteyn and Ryzhik (1980)], formulae (3.665.2) and (9.100)). The above summands for each $k \geq 0$ are strictly increasing functions of S; thus $S \mapsto g_d(1/S)$ is strictly decreasing on $(0,1)$.
    The last statement follows from the fact that

$$1 - S < \left(1 + R^2 - 2R\cos\omega\right)^{\frac{1}{2}} < 1 + S$$

for small positive S's. $\square$

5.6.2 *Proof of Theorem 5.2(ii)–(iii)*

Part (ii) is an immediate consequence of the calculation for $d = 1$ in Section 5.6.1. Part (iii) partly follows from the calculation for $d \geq 2$ in Section 5.6.1. The remaining items are proved here.

Recall that s_d is the surface of the unit ball in $\mathbb{R}^d$.

First note that by eliminating $f'_d(u)$ from the condition for stationarity (5.47), a simple computation leads to the equation

$$\zeta(u) = \varpi_v(u), \tag{5.56}$$

where

$$\zeta(u) := \frac{\ell}{\sqrt{2\beta}} f_d(u) = \frac{\ell}{\ell^*_{cr}} \frac{1}{2s_d} f_d(u),$$

(recall (5.8)) and

$$\varpi_v(u) := -v + \frac{1}{2} u(1 - v^2).$$

From (5.36) and (5.48) we see that if the optimum is attained in the interior $(c > 0, \, 0 < \eta < 1, \, u \in (0, \infty))$, then

$$I(\ell, \beta, d) = \beta(1 - v^{*2}) \tag{5.57}$$

with v^* being the maximal value of v for which (5.56) is soluble for at least one $u \in (0, \infty)$.

Let us now analyze the function ζ a bit. Note that $f_d(u) \sim s_d u$ as $u \to \infty$ by (5.41); this, along with Lemma 5.2 implies that the function ζ

(1) is positive and strictly increasing,
(2) is strictly convex on $(0, 1)$,
(3) has an infinite slope at 1,
(4) is strictly concave on $(1, \infty)$,
(5) has limiting derivative

$$\lim_{u \to \infty} \zeta'(u) = \frac{1}{2} \frac{\ell}{\ell^*_{cr}}.$$

(See Fig. 5.2 for illustration.)

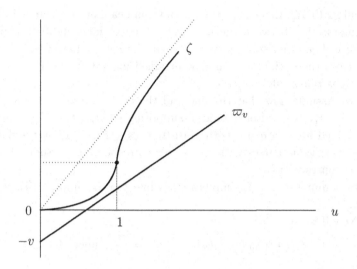

Fig. 5.2   Qualitative plot of ζ and ϖ_v. The dotted line is $u \mapsto \frac{1}{2}\frac{\ell}{\ell_{cr}^*}u$.

We now claim that the above analysis leads to the representation of ℓ_{cr} as

$$\ell_{cr} = \sup\left\{\ell > 0\colon \ \zeta(u) > -\sqrt{1 - \frac{\ell}{\ell_{cr}^*}} + \frac{1}{2}\frac{\ell}{\ell_{cr}^*}u \ \ \forall u \in (0, \infty)\right\}. \quad (5.58)$$

To see this, consider the graph (a straight line) of ϖ_v when $v = v_\ell$ is chosen so that $1 - v^2 = \ell/\ell_{cr}^*$, that is,

$$v_\ell = \sqrt{1 - \frac{\ell}{\ell_{cr}^*}},$$

in which case its slope is exactly $\ell/(2\ell_{cr}^*)$. Call this graph 'the line.'

Now, recall from the discussion after (5.39), that if $\ell < \ell_{cr}$, then the unique optimum is attained when $\eta = c = 0$, hence when $u = \infty$, while for $\ell > \ell_{cr}$, it is attained when $\eta \in (0, 1)$ and $c > 0$, in which case $0 < u, v < \infty$. Recall also, that $F_d(0, 0) = \ell s_d \sqrt{2\beta}$, by definition.

(i) Assume that the line and the graph of ζ (we will call it 'the curve') are disjoint. Then the line, nevertheless, will intersect the curve when we start decreasing v, since then the line has a higher 'y-intercept' and also a slope larger than $\ell/(2\ell_{cr}^*)$, while $\ell/(2\ell_{cr}^*)$ is the limiting slope for the curve.

Recalling (5.57), this means that an optimum cannot be attained by having an intersection between the line and the curve for some $0 < u, v < \infty$, because then other stationary points with larger v-values exist.[7] Hence, in this case, the optimum cannot be attained for any $0 < u, v < \infty$. Thus, $\ell > \ell_{cr}$ is ruled out.

(ii) Assume now that the line and the curve are not disjoint. In this case, $\ell < \ell_{cr}$ is ruled out. Indeed, supposing $\ell < \ell_{cr}$, the *unique* optimum is attained for $c = \eta = 0$ with $F_d(0,0) = \ell s_d \sqrt{2\beta}$. But, by the definition of v_ℓ, this value agrees with $\beta(1 - v_\ell)$, corresponding to some stationary point; contradiction.

In summary: $\ell < \ell_{cr}$ implies (i), while $\ell > \ell_{cr}$ implies (ii), verifying (5.58).

Now if we set

$$\hat{f}_d(u) = s_d u - f_d(u), \qquad M_d = \frac{1}{2s_d} \max_{u \in (0,\infty)} \hat{f}_d(u), \qquad (5.59)$$

then (5.58) reads

$$\ell_{cr} = \sup\left\{ \ell > 0 : \sqrt{1 - \frac{\ell}{\ell_{cr}^*}} > \frac{\ell}{\ell_{cr}^*} M_d \right\}, \qquad (5.60)$$

yielding

$$\sqrt{1 - \frac{\ell_{cr}}{\ell_{cr}^*}} = \frac{\ell_{cr}}{\ell_{cr}^*} M_d, \qquad (5.61)$$

leading to a quadratic equation for $\alpha_d = \frac{\ell_{cr}}{\ell_{cr}^*}$. This verifies (5.9)–(5.11) and Theorem 5.2(ii)–(iii).   □

5.6.3   *Proof of Theorem 5.2(iv)–(v)*

The properties below are easily deduced from Fig. 5.2. (Note that ζ/ℓ does not depend on ℓ.)

(a) $\ell \mapsto u^*(\ell, \beta, d)$ and $\ell \mapsto v^*(\ell, \beta, d)$ are continuous on (ℓ_{cr}, ∞);
(b) $\ell \mapsto v^*(\ell, \beta, d)$ is strictly decreasing on (ℓ_{cr}, ∞);
(c)

$$\lim_{\ell \downarrow \ell_{cr}} u^*(\ell, \beta, d) = u_d, \qquad \lim_{\ell \downarrow \ell_{cr}} v^*(\ell, \beta, d) = \sqrt{1 - \alpha_d}, \qquad (5.62)$$

where u_d is the unique maximizer of the variational problem in (5.59) and α_d is given by (5.10);

---

[7]The supremum that can be achieved by increasing v, is not a maximum, since when $1 - v^2 = \ell/\ell_{cr}^*$, the line and the curve 'meet at infinity,' corresponding to, $u = \infty$.

(d) $\lim_{\ell \to \infty} u^*(\ell, \beta, d) = \lim_{\ell \to \infty} v^*(\ell, \beta, d) = 0$;

(e) $u^*(\ell, \beta, d) \in (0, 1)$ for all $\ell \in (\ell_{cr}, \infty)$.

(iv) Items (a) and (b) in combination with (5.36) and (5.48) imply that $\ell \mapsto I(\ell, \beta, d)$ is continuous and strictly increasing on (ℓ_{cr}, ∞). To see that it is continuous at ℓ_{cr}, use item (c) to get $\lim_{\ell \downarrow \ell_{cr}} I(\ell, \beta, d) = \beta \alpha_d = \beta \frac{\ell_{cr}}{\ell_{cr}^*}$ (recall (5.9)), which coincides with the limit from below. Item (d) in combination with (5.36) yields $\lim_{\ell \to \infty} I(\ell, \beta, d) = \beta$.

(v) Since (recall (5.46))

$$\eta^* = \frac{1}{1 + u^* v^*}, \qquad c^* = \sqrt{2\beta} \, \frac{v^*}{1 + u^* v^*}, \tag{5.63}$$

item (a) implies that $\ell \mapsto \eta^*(\ell, \beta, d)$ and $\ell \mapsto c^*(\ell, \beta, d)$ are continuous on (ℓ_{cr}, ∞). Clearly, (5.62) and (5.63) imply that η^* and c^* tend to a strictly positive limit as $\ell \downarrow \ell_{cr}$, which shows a discontinuity from their value zero for $\ell < \ell_{cr}$. Item (d) shows that $(1 - \eta^*)/c^*$ and c^* tend to zero as $\ell \to \infty$. Finally, from item (e) we obtain that $c^* > \sqrt{2\beta} \, (1 - \eta^*)$ (recall (5.46)), completing the proof. $\qquad\qquad\square$

5.7   Optimal annealed survival strategy

The last major step is to identify the optimal annealed survival strategy. Here is the result:

Theorem 5.3 (Optimal survival strategy). *Fix β, a. For $r, b > 0$ and $t \geq 0$, define*

$$C(t; r, b) = \{\exists x_0 \in \mathbb{R}^d \colon |x_0| = b, B_{rt}(x_0 t) \cap K = \emptyset\}, \tag{5.64}$$

that is, the event that there is a clearing at distance bt with radius rt. Then the following holds.

(i) For $d = 1$, $\ell < \ell_{cr}$ or $d \geq 2$, any ℓ, and $0 < \epsilon < 1 - \eta^*$,

$$\lim_{t \to \infty} (\mathbb{E} \times P_0) \Big(C \Big(t; \sqrt{2\beta} \, (1 - \eta^* - \epsilon), c^* \Big) \mid T > t \Big) = 1,$$

$$\lim_{t \to \infty} (\mathbb{E} \times P_0) \Big(|Z_t| \geq \lfloor e^{\beta(1 - \eta^* - \epsilon)t} \rfloor \mid T > t \Big) = 1. \tag{5.65}$$

(ii) For $d \geq 1$, $\ell < \ell_{cr}$ and $\epsilon > 0$,

$$\lim_{t \to \infty} (\mathbb{E} \times P_0) \Big(B_{(1+\epsilon)\sqrt{2\beta} \, t}(0) \cap K \neq \emptyset \mid T > t \Big) = 1,$$

$$\lim_{t \to \infty} (\mathbb{E} \times P_0) \Big(R_t \subseteq B_{(1+\epsilon)\sqrt{2\beta} \, t}(0) \mid T > t \Big) = 1,$$

$$\lim_{t \to \infty} (\mathbb{E} \times P_0) \Big(R_t \not\subseteq B_{(1-\epsilon)\sqrt{2\beta} \, t}(0) \mid T > t \Big) = 1. \tag{5.66}$$

(iii) For $d \geq 1$, $\ell > \ell_{cr}$ and $0 < \epsilon < \eta^*$,

$$\lim_{t\to\infty} (\mathbb{E} \times P_0)\Big(|Z((\eta^* - \epsilon)t)| \leq \lfloor t^{d+\epsilon}\rfloor \mid T > t\Big) = 1. \qquad (5.67)$$

(iv) For $d = 1$, $\ell > \ell_{cr}$ and $\epsilon > 0$,

$$\lim_{t\to\infty} (\mathbb{E} \times P_0)\Big(B_{\epsilon t}(0) \cap K \neq \emptyset \mid T > t\Big) = 1,$$

$$\lim_{t\to\infty} (\mathbb{E} \times P_0)\Big(R_t \subseteq B_{\epsilon t}(0) \mid T > t\Big) = 1. \qquad (5.68)$$

Remark 5.4 (Interpretation of Theorem 5.3). What we see here is that in the low intensity regime $\ell < \ell_{cr}$, the system empties a ball of radius $\sqrt{2\beta}\, t$, and until time t stays inside this ball and branches at rate β, whereas in the high intensity regime $\ell > \ell_{cr}$,

- $\underline{d = 1}$: The system empties an $o(t)$-ball (i.e., a ball with radius $> a$ but $\ll t$), and until time t suppresses the branching (i.e., produces a polynomial number of particles) and stays inside this ball.
- $\underline{d \geq 2}$: The system empties a ball of radius $\sqrt{2\beta}\,(1 - \eta^*)t$ around a point at distance $c^* t$ from the origin, suppresses the branching until time $\eta^* t$, and during the remaining time $(1 - \eta^*)t$ branches at rate β.

The reason Theorem 5.3 says nothing about some further properties,[8] is that those are too delicate to be distinguished on a logarithmic scale.

5.8 Proof of Theorem 5.3

The proofs of the various statements in Theorem 5.3 all rely on the following simple consequence of Theorem 5.1. Let $\{\mathcal{E}_t\}_{t\geq 0}$ be a family of events satisfying

$$\limsup_{t\to\infty} \frac{1}{t} \log(\mathbb{E} \times P_0)(\{T > t\} \cap \mathcal{E}_t^c) < -I(\ell, \beta, d). \qquad (5.69)$$

Then

$$\lim_{t\to\infty} (\mathbb{E} \times P_0)(\mathcal{E}_t \mid T > t) = 1. \qquad (5.70)$$

Since all the statements in Theorem 5.3 have the form of (5.70), they may be demonstrated by showing the corresponding inequality of type (5.69). The proofs below are based on Section 5.4.2. We use the notation of that section freely.

---

[8]For example, concerning the existence or exact shape of a 'small tube' in which the BBM reaches the clearing.

5.8.1　*Proof of Theorem 5.3(iii)*

Let $\ell > \ell_{cr}$ and $0 < \epsilon < \eta^*$. Abbreviate

$$K_t = \left\{ |Z((\eta^* - \epsilon)t)| \le \lfloor t^{d+\epsilon} \rfloor \right\}. \tag{5.71}$$

Since $K_t^c = \{\eta_t < \eta^* - \epsilon\}$ (recall (5.21)), we have similarly as in (5.22) that

$$(\mathbb{E} \times P_0)\left(\{T > t\} \cap K_t^c\right)$$

$$\le \sum_{i=0}^{\lceil n(\eta^* - \epsilon) \rceil - 1} (\mathbb{E} \times P_0)\left(\{T > t\} \cap \left\{\frac{i}{n} \le \eta_t < \frac{i+1}{n}\right\}\right)$$

$$\le \sum_{i=0}^{\lceil n(\eta^* - \epsilon) \rceil - 1} \exp\left[-\beta \frac{i}{n} t + o(t)\right] (\mathbb{E} \times P_t^{(i,n)})(T > t). \tag{5.72}$$

To continue the estimate, substitute (5.28) into (5.72) and optimize over $j \in \{0, 1, \ldots, n-1\}$, but with the constraint

$$i \in \{0, 1, \ldots, \lceil n(\eta^* - \epsilon) \rceil - 1\}. \tag{5.73}$$

By Theorem 2(i), the variational problem defining $I(\ell, \beta, d)$ has a unique pair of minimizers. However, under the optimization, the parameter $\eta = i/n$ is bounded away from η^* because of (5.73). Consequently,

$$\limsup_{t \to \infty} \frac{1}{t} \log(\mathbb{E} \times P_0)\left(\{T > t\} \cap K_t^c\right) < -I(\ell, \beta, d). \tag{5.74}$$

5.8.2　*Proof of Theorem 5.3(i)*

The proof of the first limit in (5.65) is very similar to that of part (iii). Let $\epsilon' > 0$ and $\delta > 0$ be so small that

$$\sqrt{2\beta}\,\epsilon > \sqrt{2\beta}\,\epsilon' + \delta. \tag{5.75}$$

Then, obviously, if

$$C_t := \{\exists x_0 \in \mathbb{R}^d : |\,|x_0| - c^*\,| < \delta, \; B_{\sqrt{2\beta}\,(1 - \eta^* - \epsilon')t}(x_0 t) \cap K = \emptyset\} \tag{5.76}$$

then $C_t \subseteq C(t; \sqrt{2\beta}\,(1 - \eta^* - \epsilon), c^*)$, so it suffices to prove the claim for C_t. Consider the optimization procedure in the proof in Section 5.4.2, but now for the probability

$$(\mathbb{E} \times P_0)\left(\{T > t\} \cap C_t^c\right). \tag{5.77}$$

Similarly to the proof of part (iii), the vector parameter $(\eta, c) = (i/n, j/n)$ is again bounded away from its optimal value. The difference is that, instead of (5.73), now $(i/n, j/n)$ is bounded away from the set

$$(\eta^* - \epsilon', \eta^* + \epsilon') \times (c^* - \delta, c^* + \delta). \tag{5.78}$$

Again, it follows from the uniqueness of the minimizers that

$$\limsup_{t\to\infty} \frac{1}{t} \log(\mathbb{E} \times P_0)\left(\{T > t\} \cap C_t^c\right) < -I(\ell, \beta, d). \qquad (5.79)$$

To prove the second limit in (5.65), abbreviate

$$\overline{K}_t := \left\{ |Z_t| \geq \lfloor e^{\beta(1-\eta^*-\epsilon)t} \rfloor \right\}. \qquad (5.80)$$

First note that, by (1.45), for any $\epsilon > 1/m$ and $k \geq 1 - \eta^* - 1/m$,

$$\sup_{x\in\mathbb{R}^d} P_x\left(|Z(kt)| < \lfloor e^{\beta(1-\eta^*-\epsilon)t} \rfloor\right)$$

$$\leq e^{-\beta(\epsilon-1/m)t}[1 + o(1)]. \qquad (5.81)$$

The probability $(\mathbb{E} \times P_0)(T > t)$ was already estimated through (5.22) and (5.28). To estimate $(\mathbb{E} \times P_0)(\{T > t\} \cap \overline{K}_t^c)$, use the analogue of (5.22), but modify the estimate in (5.28) as follows. First, observe that for

$$\frac{i+1}{n} \leq \eta^* + \frac{1}{m}$$

we can use the Markov property at time $\frac{i+1}{n}$ together with (5.81), to obtain an estimate that is actually stronger than the one in (5.28):

$$(\mathbb{E} \times P_t^{(i,n)})(\{T > t\} \cap \overline{K}_t^c)$$

$$\leq \exp\left[-\beta\left(\epsilon - \frac{1}{m}\right)t + o(t)\right]$$

$$\times \sum_{j=0}^{n-1}(\lfloor t^{d+\epsilon}\rfloor + 1)\exp\left[\frac{-\beta j^2/n^2}{(i+1)/n}t + o(t)\right]$$

$$\times \exp\left[-\ell f_d\left((1-\epsilon)\sqrt{2\beta}\left(1 - \frac{i+1}{n}\right), \frac{j}{n}\right)t\right.$$

$$\left. + O(1/n)t + o(t)\right]$$

$$+ \exp[-\beta t + o(t)] + \text{SES}. \qquad (5.82)$$

Compare now (5.28) with (5.82). The presence of the extra factor $\exp\left[-\beta\left(\epsilon - \frac{1}{m}\right)t + o(t)\right]$ in (5.82) means that when the parameter $\eta = i/n$ is close to its optimal (for (5.28)) value η^*, the optimum obtained from (5.82) is strictly smaller than the one obtained from (5.28). Since, on the other hand, η^* is the unique minimizer for (5.28), this is already enough to conclude that

$$\limsup_{t\to\infty} \frac{1}{t} \log (\mathbb{E} \times P_0)\left(\{T > t\} \cap \overline{K}_t^c\right) < -I(\ell, \beta, d). \qquad (5.83)$$

5.8.3 Proof of Theorem 5.3(ii)

To prove the second limit in (5.66), abbreviate

$$D_t := \{R_t \subseteq B_{(1+\epsilon)\sqrt{2\beta}\,t}(0)\} \tag{5.84}$$

and note that, since

$$\frac{1}{2}\left((1+\epsilon)\sqrt{2\beta}\right)^2 > \beta, \tag{5.85}$$

the same argument as in the proof of (1.60) gives us that

$$\limsup_{t\to\infty} \frac{1}{t} \log P_0(D_t^c) \le -(1+\epsilon)^2\beta + \beta = -(2+\epsilon)\epsilon\beta. \tag{5.86}$$

Pick $\epsilon' > 0$ such that $\epsilon'\beta\frac{\ell}{\ell_{cr}^*} = (2+\epsilon)\epsilon\beta$. Then (5.86) says that

$$\limsup_{t\to\infty} \frac{1}{t} \log P_0(D_t^c) \le -\epsilon'\beta\frac{\ell}{\ell_{cr}^*}. \tag{5.87}$$

Using the first limit in (5.65) with $\epsilon = \epsilon'/2$, we find that (recall $\eta^* = c^* = 0$ and (5.8))

$$
\begin{aligned}
&\limsup_{t\to\infty} \frac{1}{t} \log \left(\mathbb{E}\times P_0\right)\left(\{T>t\}\cap D_t^c\right)\\
&\le \limsup_{t\to\infty} \frac{1}{t} \log \left(\mathbb{E}\times P_0\right)\left(C\left(t;\sqrt{2\beta}\left(1-\epsilon'/2\right),0\right)\cap D_t^c\right)\\
&= \limsup_{t\to\infty} \frac{1}{t}\left[\log \mathbb{P}\left(C\left(t;\sqrt{2\beta}\left(1-\epsilon'/2\right),0\right)\right) + \log P_0\left(D_t^c\right)\right]\\
&\le -\left(1-\epsilon'/2+\epsilon'\right)\beta\frac{\ell}{\ell_{cr}^*}\\
&= -\left(1+\epsilon'/2\right)I(\ell,\beta,d)\\
&< -I(\ell,\beta,d),
\end{aligned}
\tag{5.88}
$$

where the second inequality uses (5.1) and (5.87), and the second equality uses the first line of (5.32).

To prove the third limit in (5.66), let $0 < \epsilon' < \epsilon$ and introduce the shorthands

$$
\begin{aligned}
A_t^1 &:= \{B_{(1-\epsilon)\sqrt{2\beta}\,t}(0)\cap K = \emptyset\},\\
A_t^2 &:= \{B_{(1-\epsilon')\sqrt{2\beta}\,t}(0)\cap K = \emptyset\},\\
D_t^1 &:= \{R_t \not\subseteq B_{(1-\epsilon)\sqrt{2\beta}\,t}(0)\}.
\end{aligned}
\tag{5.89}
$$

Estimate

$$
\begin{aligned}
(\mathbb{E}\times P_0)(\{T>t\}\cap [D_t^1]^c) &\le (\mathbb{E}\times P_0)(\{T>t\}\cap [D_t^1]^c\cap A_t^2)\\
&+(\mathbb{E}\times P_0)\left(\{T>t\}\cap [A_t^2]^c\right).
\end{aligned}
\tag{5.90}
$$

From (5.79) we have that

$$\limsup_{t\to\infty} \frac{1}{t} \log(\mathbb{E} \times P_0)\left(\{T > t\} \cap [A_t^2]^c\right) < -I(\ell, \beta, d). \qquad (5.91)$$

Clearly,

$$(\mathbb{E} \times P_0)([D_t^1]^c \cap A_t^1) = P_0([D_t^1]^c)\mathbb{P}(A_t^1),$$
$$(\mathbb{E} \times P_0)([D_t^1]^c \cap A_t^2) = P_0([D_t^1]^c)\mathbb{P}(A_t^2), \qquad (5.92)$$

and

$$\lim_{t\to\infty} \frac{1}{t} \log \mathbb{P}(A_t^2) < \lim_{t\to\infty} \frac{1}{t} \log \mathbb{P}(A_t^1). \qquad (5.93)$$

Hence

$$\limsup_{t\to\infty} \frac{1}{t} \log(\mathbb{E} \times P_0)([D_t^1]^c \cap A_t^2)$$

$$< \limsup_{t\to\infty} \frac{1}{t} \log(\mathbb{E} \times P_0)([D_t^1]^c \cap A_t^1) \le -I(\ell, \beta, d), \quad (5.94)$$

where the last inequality follows from Theorem 5.1 and the fact that $\{[D_t^1]^c \cap A_t^1\} \subseteq \{T > t\}$. By (5.90)–(5.91), and (5.94), we obtain that

$$\limsup_{t\to\infty} \frac{1}{t} \log(\mathbb{E} \times P_0)(\{T > t\} \cap [D_t^1]^c) < -I(\ell, \beta, d). \qquad (5.95)$$

The proof of the first limit in (5.66) is a slight adaptation of the previous argument. Let $0 < \epsilon' < \epsilon$. Let D_t be as in (5.84) but replace ϵ by ϵ', and introduce

$$A_t^1 := \{B_{(1+\epsilon)\sqrt{2\beta}\,t}(0) \cap K \ne \emptyset\},$$
$$A_t^2 := \{B_{(1+\epsilon')\sqrt{2\beta}\,t}(0) \cap K \ne \emptyset\}. \qquad (5.96)$$

Estimate

$$(\mathbb{E} \times P_0)(\{T > t\} \cap [A_t^1]^c) \le (\mathbb{E} \times P_0)(\{T > t\} \cap D_t \cap [A_t^1]^c)$$
$$+(\mathbb{E} \times P_0)\left(\{T > t\} \cap [D_t]^c\right). \qquad (5.97)$$

Now the statement follows from (5.97) and (5.88) along with the estimate

$$\limsup_{t\to\infty} \frac{1}{t} \log(\mathbb{E} \times P_0)(D_t \cap [A_t^1]^c)$$

$$< \limsup_{t\to\infty} \frac{1}{t} \log(\mathbb{E} \times P_0)(D_t \cap [A_t^2]^c)$$

$$\le -I(\ell, \beta, d). \qquad (5.98)$$

5.8.4 *Proof of Theorem 5.3(iv)*

An event (depending on t) $\mathcal{E}_t$ will be called *negligible*, if, as $t \to \infty$,

$$\log(\mathbb{E} \times P)(\mathcal{E}_t) = o[\log(\mathbb{E} \times P)(T > t)].$$

We will consider the two statements in the reversed order. For the second statement in (5.68), first note that, by Theorem 5.2(ii), we have $\eta^* = 1$. Now recall the definition of K_t from (5.71). The estimate in (5.74) with $\eta^* = 1$ says that the event $\{T > t\} \cap K_t^c$ is negligible, i.e., considering survival, we may also assume that there are polynomially many particles only at time $t(1 - \epsilon)$ $(0 < \epsilon < 1)$.

The strategy of the rest of the proof is to show two facts:

(a) *no particle* has left the $\epsilon t/2$-ball around the origin up to time $t(1 - \epsilon)$ (let F_t denote this event);

(b) each BBM emanating from one of the 'parent' particles at time $t(1-\epsilon)$ is to be contained in an $\epsilon t/2$-ball around the position of the parent particle (let G_t denote this event).

For (a), note that trivially, $K_t \cap F_t^c$ has an exponentially small probability (because the polynomial factor does not affect the exponential estimate), but we must in fact show that $\{T > t\} \cap K_t \cap F_t^c$ is negligible. We now sketch how to modify (5.88) to prove this and leave the obvious details to the reader. To estimate $\{T > t\} \cap K_t \cap F_t^c$, replace $(1 + \epsilon)\sqrt{2\beta}$ by $\epsilon/2$ and, instead of the first limit in (5.65) (regarding the existence of the empty ball), use Theorem 5.3(iii) along with the fact that the branching is independent of the motion.

For (b), we must show that $\{T > t\} \cap K_t \cap G_t^c$ is negligible. The proof is similar to the one in the previous paragraph: (5.88) should be appropriately modified. The difference is that now we must use the Markov property at time $t(1 - \epsilon)$ and deal with several particles at that time. However, this is no problem because on the event K_t we have polynomially many particles only. (The use of Theorem 5.3(iii) is just like in the previous paragraph.)

The first statement in (5.68) follows after replacing $(1 + \epsilon)\sqrt{2\beta}$ and $(1 + \epsilon')\sqrt{2\beta}$ by ϵ resp. ϵ' in (5.96)–(5.98), and using the second statement in (5.68) instead of (5.88).

5.9   Non-extinction

What if 'survival' means *non-extinction?* What we mean here is that we only require that there is at least one particle not absorbed/killed by traps. In this case it is appropriate to think of the model as one where the motion component (Brownian motion) is being replaced by a new one (Brownian motion killed at the boundary of the random set $K_a(\omega)$); in this sense, $\{\tilde{T} > t\}$ is indeed the non-extinction of this modified BBM.

Although the model itself makes sense, the 'tail asymptotics' does not:

Theorem 5.4 (No tail). *Fix d, β, a. For any locally finite intensity measure ν,*

$$\lim_{t \to \infty} (\mathbb{E} \times P_0)(\tilde{T} > t) > 0. \qquad (5.99)$$

Let λ_c^R denote the generalized principal eigenvalue (which is just the classical Dirichlet eigenvalue) of $\Delta/2$ on the ball $B_R(0)$.

Remark 5.5. Heuristically, (5.99) follows from the fact that the system may survive by emptying a ball with a finite radius $R > R_0$, where R_0 is chosen such that the branching rate β balances against the 'killing rate' $\lambda_c^{R_0}$, that is, $-\lambda_c^{R_0} = \beta$. The rigorous proof is below. ◇

Proof.   Let $-\lambda_c^{R_0} = \beta$. Since

$$\lambda_c(\Delta/2 + \beta; B_R(0)) > 0$$

for any $R > R_0$, thus by Example 3.1, the Strong Law (Theorem 2.2) holds. In particular, the probability (denoted by p_R) that at least one particle has not left $B_R(0)$ ever, is positive. Consequently, a lower bound for the left-hand side of (5.99) is $\sup_{R>R_0} \{p_R \exp[-\nu(B_R(0))]\}$. □

5.10   Notes

This chapter follows very closely the article [Engländer and den Hollander (2003)], which was motivated by [Engländer (2000)]. In the latter it had been shown that if $d \geq 2$ and $d\nu/dx \equiv \ell$, then the annealed survival probability decays like $\exp[-\beta t + o(t)]$. Intuitively, this means that the system suppresses the branching until time t in order to avoid the traps. The corresponding asymptotics for $d = 1$ was left open in [Engländer (2000)] and found in [Engländer and den Hollander (2003)].

For the case when the dyadic branching law is replaced by a general one, with probability generating function G, the following is shown in the upcoming paper

[Öz, Çağlar and Engländer (2014)]. Let m be the expectation corresponding to G, assume that it is finite, and let $m^* := m - 1$. Let p_0 be the probability of producing zero offspring, $p_0 = G'(0)$. If $p_0 > 0$, the extinction of the process is possible. Let τ denote the extinction time for the process, and $\mathcal{E}$ the event of extinction. Define

$$I(\ell, f, \beta, d) = \min_{\eta \in [0,1], c \in [0,\sqrt{2\beta}]} \left\{ \beta \alpha \eta + \frac{c^2}{2\eta} + \ell f_d(\sqrt{2\beta m^*}(1 - \eta), c) \right\},$$

where f_d is as in this chapter, $\alpha := 1 - G'(q)$, and q is the probability[9] of $\mathcal{E}$. (For $\eta = 0$ put $c = 0$ and $f_d(\sqrt{2\beta m^*}, 0) = s_d \sqrt{2\beta m^*}$.) If T is the first time the trap configuration is hit, then:

(1) If $p_0 = 0$, then

$$\lim_{t \to \infty} \frac{1}{t} \log(\mathbb{E} \times P)(T > t) = -I(l, f, \beta, d).$$

(2) If $p_0 > 0$ and $m^* > 0$, then

$$\lim_{t \to \infty} \frac{1}{t} \log(\mathbb{E} \times P)(T > t \mid \mathcal{E}^c) = -I(l, f, \beta, d).$$

(3) If $p_0 > 0$ and $m^* \leq 0$, then

$$\lim_{t \to \infty} (\mathbb{E} \times P)(T > t) = (\mathbb{E} \times P)(T > \tau) > 0. \tag{5.100}$$

The proof utilizes a 'prolific backbone' decomposition. Again, solving the variational problem, leads to a 'cutoff' value for ℓ, separating two regimes. In particular, when the branching is strictly dyadic, one gets the results of this chapter.

For critical branching, a somewhat similar setting is considered in [Le Gall and Véber (2012)]. The authors consider a d-dimensional critical branching Brownian motion in a random Poissonian environment. Inside the obstacles, 'soft killing' is taking place: each particle is killed at rate $\epsilon > 0$, where ϵ is small. The basic question is: what is $p_\epsilon(R) := P$(the BBM ever visits the complement of the R-ball) if R is large? It turns out that the answer depends on how ϵR^2 behaves:

(1) If ϵ is small compared to R^{-2} then we are back in the no obstacle regime and $p_\epsilon(R) \approx C/R^2$.

(2) If R^{-2} is small compared to ϵ then the above probability decays exponentially in $R\sqrt{\epsilon}$.

(3) If the two are on the same order, then one gets back the no obstacle case, but with C replaced by a different fine tuning constant, given by the value at the origin of the solution of a certain semilinear PDE with singular boundary condition. The first order term in that PDE depends on the limit of ϵR^2.

---

[9]Of course, $q = 0$ if and only if $p_0 = 0$, in which case $\alpha = 1$.

Some suggested open problems, related to the results of this chapter, are as follows.

(a) Concerning $\{T > t\}$:

- For $d = 1$ and $\ell > \ell_{cr}$, what is the radius of the $o(t)$-ball that is emptied and how many particles are there inside this ball at time t?

- For $d \geq 2$ and $\ell > \ell_{cr}$, what is the shape of the "small tube" in which the system moves its particles away from the origin while suppressing the branching? How many particles are alive at time $\eta^* t$?

- What can be said about the optimal survival strategy at $\ell = \ell_{cr}$?

- Instead of letting the trap density decay to zero at infinity, another way to make survival easier is by providing the Brownian motion with an *inward drift*, while keeping the trap density field constant. Suppose that $d\nu/dx \equiv \ell$ and that the inward drift *radially increases* like $\sim \kappa |x|^{d-1}$, $|x| \to \infty$, $\kappa > 0$. Is there again a crossover in ℓ at some critical value $\ell_{cr} = \ell_{cr}(\kappa, \beta, d)$? And what is the optimal survival strategy?

(b) Concerning $\{\widetilde{T} > t\}$:

- What is the limit in (5.99), say, when $d\nu/dx$ is spherically symmetric?

- Assume that the Brownian motion has an outward drift. For what values of the drift does the survival probability decay to zero?

Finally, an open problem, related to the general offspring distribution: What is the value of the right-hand side of (5.100) for $G(s) = (1/2) + (1/2)s^2$? And for general G?

Chapter 6

Branching in random environment: Mild obstacles

In Section 1.12, we reviewed the problem of the survival asymptotics for a single Brownian particle among Poissonian obstacles — an important topic, thoroughly studied in the last few decades. As mentioned in the notes to the previous chapter, more recently[1] a model of a spatial *branching* process in a random environment has been introduced; 'hard' obstacles were considered, and instantaneous killing of the branching process once any particle hits the trap configuration K. In the previous chapter we have seen how the model can result in some surprising phenomena when the trap intensity varies in space, and we also discussed some questions when killing is defined as the absorption of the last particle.

The difference between those models and the one we are going to consider now is that this time, instead of killing particles, we choose a 'milder' mechanism as follows. We will again study a spatial branching model, where the underlying motion is d-dimensional ($d \geq 1$) Brownian motion and the branching is dyadic. Instead of killing at the boundary of a Poissonian trap system, however, now it is the *branching rate* which is affected by a Poissonian collection of reproduction suppressing sets, which we will dub *mild*[2] *obstacles.*

The main result of this chapter will be a Quenched Law of Large Numbers for the population for all $d \geq 1$. In addition, we will show that the branching Brownian motion with mild obstacles *spreads less quickly* in space than ordinary BBM. When the underlying motion is a generic diffusion process, we will obtain a *dichotomy* for the quenched local growth that is independent of the Poissonian intensity. Lastly, we will also discuss gen-

---

[1]In [Engländer (2000)].

[2]The adjective 'mild' was chosen to differentiate from 'soft.' The reader should recall that the latter means that particles are killed at a certain rate inside the obstacles.

eral offspring distributions (beyond the dyadic one considered in the main theorems) as well as mild obstacle models for superprocesses.

6.1   Model

Our purpose is to study a spatial branching model with the property that the branching rate is decreased in a certain random region. Similarly to the previous chapter, we will use a Poissonian model for the random environment.

Let ω be a Poisson point process (PPP) on $\mathbb{R}^d$ with constant intensity $\nu > 0$ and let $\mathbf{P}$ denote the corresponding law. Furthermore, let $a > 0$ and $0 < \beta_1 < \beta_2$ be fixed. We define the *branching Brownian motion (BBM) with a mild Poissonian obstacle*, or the '$(\nu, \beta_1, \beta_2, a)$-BBM' as follows. Let $K = K_\omega$ be as in (5.2), but now consider K as a *mild obstacle configuration* attached to ω. This means that given ω, we define P^ω as the law of the strictly dyadic (precisely two offspring) BBM on $\mathbb{R}^d$, $d \geq 1$ with spatially dependent branching rate

$$\beta(x, \omega) := \beta_1 1_{K_\omega}(x) + \beta_2 1_{K_\omega^c}(x).$$

The informal definition is that as long as a particle is in K^c, it obeys the branching rule with rate β_2, while in K its reproduction is suppressed and it branches with the lower rate β_1. (We assume that the process starts with a single particle at the origin.) We will call the process under P^ω a *BBM with mild Poissonian obstacles* and denote it by Z. As before, the total mass process will be denoted by $|Z|$; W will denote d-dimensional Brownian motion with probabilities $\{\mathbb{P}_x,\ x \in \mathbb{R}^d\}$.

Invoking here the paragraph after Lemma 1.5, we record below the comparison between an ordinary 'free' BBM and the one with mild obstacles, which is useful to keep in mind.

Remark 6.1 (Comparison). (i) Let $\widehat{P}$ correspond to ordinary BBM with branching rate β_2 everywhere. Then for all $t \geq 0$, all $B \subseteq \mathbb{R}^d$ Borel, and all $k \in \mathbb{N}$,

$$P^\omega(Z_t(B) < k) \geq \widehat{P}(Z_t(B) < k)$$

holds $\mathbf{P}$-a.s., that is, the 'free' BBM is 'everywhere stochastically larger' than the BBM with mild obstacles, $\mathbf{P}$-a.s.

(ii) More generally, let P correspond to the $(L, V; \mathbb{R}^d)$-branching diffusion and let Q correspond to the $(L, W; \mathbb{R}^d)$-branching diffusion. If $V \leq W$, then for all $t \geq 0$, all $B \subseteq \mathbb{R}^d$ Borel, and all $k \in \mathbb{N}$,

$$P(Z_t(B) < k) \geq Q(Z_t(B) < k).$$

that is, the second branching diffusion is 'everywhere stochastically larger' than the first one. ◇

6.2 Connections to other problems

Let us see some further problems motivating the study of our particular setting.

- **Law of Large Numbers:** Even though, in Chapter 2, we have proven a Strong Law of Large Numbers for branching diffusions, we did this under certain assumptions on the corresponding operator. As already explained in the Notes after Chapter 2, to prove the (Strong) LLN[3] for a generic, locally surviving branching diffusion is a highly nontrivial open problem. In our situation the scaling is not purely exponential. In general, the analysis is difficult in such a case[4] and it is interesting that in our setup *randomization* will help in proving a kind of LLN — see more in Section 6.4.1.

- **Wave-fronts in random medium; KPP-equation:** The spatial spread of our process is related to a work of Lee-Torcaso [Lee and Torcaso (1998)] on wave-front propagation for a random KPP[5] equation and to earlier work of Freidlin [Freidlin (1985)] on KPP equation with random coefficients — see Section 6.6 of this chapter, and also [Xin (2000)] for a survey.

- **Catalytic spatial branching.** An alternative view on our setting is as follows. Arguably, the model can be viewed as a *catalytic* BBM as well — the catalytic set is then K^c (in the sense that branching is 'intensified' there). Catalytic spatial branching (mostly for superprocesses though) has been the subject of vigorous research in the last few decades, initiated by D. Dawson, K. Fleischmann and others — see the survey papers [Klenke (2000)] and [Dawson and Fleischmann (2002)] and

---

[3]That is, that the process behaves asymptotically as its expectation.

[4]See for example [Engländer and Winter (2006); Evans and Steinsaltz (2006); Fleischmann, Mueller and Vogt (2007)] and references therein.

[5]'KPP' is an abbreviation for Kolmogorov-Petrovskii-Piscounov; also 'FKPP' is used, to give credit to R. A. Fisher's contribution. The equation is of the form $\dot{u} = \beta u(1-u) + u_{xx}$, where $\beta > 0$ is either a constant or a function. The higher dimensional analogue is $\dot{u} = \beta u(1-u) + \Delta u$.

references therein. In those models the individual branching rates of particles migrating in space depend on the amount of contact between the particle ('reactant') and a certain random medium called the catalyst. The random medium is usually assumed to be a 'thin' random set (that could even be just one point, like a 'point source,' for example) or another superprocess. In some work, 'mutually' or even 'cyclically' catalytic[6] branching is considered [Dawson and Fleischmann (2002)]. Our model is simpler than most catalytic models as our catalytic/blocking areas are fixed, whereas in several catalytic models they are moving. On the other hand, while for catalytic settings studied so far results were mostly only qualitative, we are aiming to get quite sharp *quantitative* result.

Notwithstanding the paucity of results for the *discrete setting*, one example is given in the notes at the end of this chapter. The discrete setting has the advantage that when $\mathbb{R}^d$ is replaced by $\mathbb{Z}^d$, the difference between the sets K and K^c is no longer relevant. Indeed, the equivalent of a Poisson trap configuration is an i.i.d. trap configuration on the lattice, and then its complement is also i.i.d. (with a different parameter). So, in the discrete case 'Poissonian mild obstacles' give the same type of model as 'Poissonian catalysts' would. This convenient 'self-duality' is lost in the continuous setting as the 'Swiss cheese' K^c is not the same type of geometric object as K.

- **Population models:** Mild obstacles appear to be relevant as a model in biology (see the notes at the end of the chapter for more elaboration).

Returning to our mathematical model, consider the following natural questions (both in the annealed and the quenched sense):

(1) What can one say about the growth of the total population size?
(2) What are the large deviations? (For instance, what is the probability of producing an atypically small population.)
(3) What can we say about the *local* population growth?

As far as (1) is concerned, recall that the total population of an ordinary (free) BBM grows a.s. and in expectation as $e^{\beta_2 t}$. Indeed, for ordinary BBM, the spatial component plays no role, and hence the total mass is just a β_2-rate pure birth process (Yule's process) X. By the Kesten-Stigum Theorem (Theorem 1.15), the limit $N := \lim_{t\to\infty} e^{-\beta_2 t} X_t$ exists a.s. and in mean, and $P(0 < N < \infty) = 1$.

---

[6]That is, 'A' catalyses 'B,' which catalyses 'C,' and finally 'C' catalyses 'A.'

In our model of BBM with the reproduction blocking mechanism, how much will the suppressed branching in K slow the global reproduction down? Will it actually change the exponent β_2? We will see that although the global reproduction does slow down, the slowdown is captured by a sub-exponential factor, being different for the quenched and the annealed case.

Consider now (2). Some further motivation may be bolstered by the following argument. Let us assume for simplicity that $\beta_1 = 0$ and ask the simplest question: what is the probability that there is no branching at all up to time $t > 0$? In order to avoid branching the first particle has to 'resist' the branching rate β_2 inside K^c. Therefore this question is fairly similar to the survival asymptotics for a single Brownian motion among 'soft obstacles' — but of course in order to prevent branching the particle seeks for large islands covered by K rather than the usual 'clearings'. In other words, the particle now prefers to avoid the K^c instead of K. Hence, (2) above is a possible generalization of this (modified) soft obstacle problem for a single particle, and the presence of branching creates new type of challenges.

As far as (3) is concerned, we will see how it is related to the local extinction/local exponential growth dichotomy for branching diffusions discussed in Subsection 1.15.5.

The result on the (quenched) growth of the total population size will be utilized when investigating the spatial spread of the process.

6.3 Some preliminary claims

Let us first see some results that are straightforward to derive from others in the literature.

6.3.1 *Expected global growth and dichotomy for local growth*

Concerning the expected global growth rate we have the following result.

Claim 6.1 (Expected global growth rate). *Let $|Z_t|$ denote the size of the total population at $t \geq 0$. Then, on a set of full $\mathbf{P}$-measure, as $t \to \infty$,*

$$E^\omega |Z_t| = \exp\left[\beta_2 t - k(d, \nu)\frac{t}{(\log t)^{2/d}}(1 + o(1))\right], \qquad (6.1)$$

(quenched asymptotics), and

$$(\mathbf{E} \otimes E^\omega)\,|Z_t| = \exp\left[\beta_2 t - c(d,\nu)t^{d/(d+2)}(1 + o(1))\right], \qquad (6.2)$$

(annealed asymptotics), where $k(d,\nu)$ and $c(d,\nu)$ are as in (1.33) and (1.30), respectively.

Notice that

(1) β_1 does not appear in the formulas,
(2) the higher the dimension, the smaller the expected population size.

Proof. Since $\beta := \beta_1 1_K + \beta_2 1_{K^c} = \beta_2 - (\beta_2 - \beta_1)1_K$, the Many-to-One Formula (1.37), applied ω-wise, yields that

$$E^\omega|Z_t| = e^{\beta_2 t}\mathbb{E}\exp\left[-\int_0^t (\beta_2 - \beta_1)1_K(W_s)\,\mathrm{d}s\right].$$

The expectation on the right-hand side is precisely the survival probability among soft obstacles of 'height' $\beta_2 - \beta_1$ (that is $V := (\beta_2 - \beta_1)\mathbf{1}_{B(0,1)}$), except that we do not sum the shape functions on the overlapping balls. This, however, does not make any difference as far as the asymptotic behavior is concerned (see [Sznitman (1998)], Remark 4.2.2). The statements thus follow from (1.29) and (1.32) along with Subsection 1.12.2. □

To understand the difference between the annealed and the quenched case, invoke Proposition 1.11. In the annealed case large clearings (far away) are automatically (that is, **P**-a.s.) present. Hence, similarly to the single Brownian particle problem, the difference between the two asymptotics is due to the fact that even though there is an appropriate clearing far away **P**-a.s., there is one around the origin with a small (but not too small) probability. Still, the two cases will have a similar element when, in Theorem 6.1 we drop the expectations and investigate the process itself, and show that inside such a clearing a large population is going to flourish (see the proof of Theorem 6.1 for more on this). This also leads to the intuitive explanation for the *decrease of the population size as the dimension increases*. The radial drift in dimension d is $(d-1)/2$. The more transient the motion (the larger d), the harder for the particles to stay in the appropriate clearings.

Remark 6.2 (Main contribution in the annealed case). Let us pretend for a moment that we are talking about an ordinary BBM with rate β_2. As explained, at time t the process has roughly $e^{\beta_2 t}$ particles if t is large (the population size divided by $e^{\beta_2 t}$ has a limit as $t \to \infty$, a.s. and in mean). Let us say that all the trajectories of the particles up to t form a 'blue' tree.

The BBM with mild obstacles (forming, say, a 'red' tree) can be coupled with this. Namely, inside the obstacles the rate is β_1, and so if we consider, independent, additional branches at rate $\beta_2 - \beta_1$ inside K, we get the blue tree. In other words, the red tree is part of the blue tree.

For t fixed, take a ball $B = B(0, R(t))$ and suppose that it is a clearing ($B \subset K^c$). The particles that have been confined to B up to time t are not affected by the blocking effect of K: the tree formed by these particles belongs entirely to the red one.

Let us consider now the annealed setting. If the probability of staying inside the ball for a single particle is p_t, then the expected size of the red tree (and thus, also that of the blue tree) can be estimated from below by $|Z_t^*| p_t \sim e^{\beta_2 t} p_t$, where Z^* is the ordinary ('blue') BBM with rate β_2.

Optimize $R(t)$ with respect to the cost of having such a clearing and the probability of confining a single Brownian motion to it. (This is a simple computation, and is the same as in the classical single particle problem for the annealed, soft obstacle case.) Hence, one gets the expectation in the theorem as a lower estimate, which is the same as $e^{\beta_2 t}$ times the classical Donsker-Varadhan asymptotics.

Now, from the classical annealed soft obstacle problem, it is known that the optimal strategy for a single particle to survive (which has the same probability as staying 'red' in our setting) is precisely to empty a ball of optimal radius and remain inside it up to time t. Similarly here, the main contribution to the expectation in (6.2) is coming from the expectation on the event of having a clearing with optimal radius $R(t)$ and confining *all* the particles to the ball up to t. Indeed, if the **P**-probability of having the optimal clearing is π_t, then, on this event, the conditional expectation of the size of the red tree is at least $p_t e^{\beta_2 t}$. That is, the contribution to the expected size is already maximal (roughly $e^{\beta_2 t} \pi_t p_t$) on this event. Of course, this does not rule out another, equal contribution using another 'strategy' for producing many particles.

In fact, one suspects that roughly $p_t e^{\beta_2 t}$ particles will *typically* stay inside the optimal clearing up to time t (i.e. with probability tending to one as $t \to \infty$). The intuitive reasoning is as follows. If we had *independent* particles instead of BBM, then, by a 'Law of Large Numbers type argument' (using Chebyshev inequality and the fact that $\lim_{t \to \infty} p_t e^{\beta_2} = \infty$), roughly $p_t e^{\beta_2 t}$ particles out of the total $e^{\beta_2 t}$ would stay in the $R(t)$-ball up to time t with probability tending to 1 as $t \uparrow \infty$. One suspects then that the branching system behaves similarly too, because the particles are 'weakly correlated' only. This kind of argument (in the quenched case though) will be made precise in the proof of our main theorem by estimating certain covariances.   ◇

The following claim concerns the quenched local population size. It identifies the rate on a logarithmic scale.

Claim 6.2 (Quenched local exponential growth). *The following holds on a set of full* **P***-measure: For any* $\epsilon > 0$ *and any bounded open set* $\emptyset \neq B \subset \mathbb{R}^d$,

$$P^\omega \left(\limsup_{t\uparrow\infty} e^{-(\beta_2-\epsilon)t} Z_t(B) = \infty \right) > 0,$$

but

$$P^\omega \left(\limsup_{t\uparrow\infty} e^{-\beta_2 t} Z_t(B) < \infty \right) = 1.$$

Proof.   Since in this case $L = \Delta/2$ and since $\lambda_c(\Delta/2, \mathbb{R}^d) = 0$, the statement is a particular case of Claim 6.3 (see the next claim below).   □

Problem 6.1. Can we obtain results on a finer scale? The quenched asymptotics of the expected logarithmic global growth rate suggests that perhaps the rate is $\beta_2 - c(d,\nu)(\log t)^{-2/d}$ at time t, in some sense. This problem will be addressed in Section 6.4.1. We will prove an appropriate formulation of the statement when the limit is meant *in probability*.

We now show how Claim 6.2 can be generalized for the case when the underlying motion is a diffusion. Let **P** be as before but replace the Brownian motion by an L-diffusion on $\mathbb{R}^d$, where L is a second order elliptic operator, satisfying Assumption 1.1. The branching L-diffusion with the Poissonian obstacles can be defined analogously to the case of BBM. Let $\lambda_c(L) \leq 0$ denote the generalized principal eigenvalue for L on $\mathbb{R}^d$.

The following assertion shows that the local behavior of the process exhibits a *dichotomy*. The crossover is given in terms of the local branching rate β_2 and $\lambda_c(L)$: local extinction occurs when the branching rate inside the 'free region' K^c is not sufficiently large to compensate the transience of the underlying L-diffusion; if it is strong enough, then local mass grows exponentially. Note an interesting feature of the result: *neither β_1 nor the intensity ν of the obstacles plays any role.*

Given the environment ω, denote by P^ω the law of the branching L-diffusion.

Claim 6.3 (Quenched exponential growth vs. local extinction). *There are two possibilities, for $\nu > 0$ and arbitrary $\beta_1 \in (0, \beta_2)$.*

(i) $\underline{\beta_2 > -\lambda_c(L)}$: *On a set of full* **P**-*measure, for any $\epsilon > 0$ and any bounded open set $\emptyset \neq B \subset \mathbb{R}^d$,*

$$P^\omega \left(\limsup_{t \uparrow \infty} e^{(-\beta_2 - \lambda_c(L) + \epsilon)t} Z_t(B) = \infty \right) > 0,$$

but

$$P^\omega \left(\limsup_{t \uparrow \infty} e^{(-\beta_2 - \lambda_c(L))t} Z_t(B) < \infty \right) = 1.$$

(ii) $\underline{\beta_2 \leq -\lambda_c(L)}$: *On a set of full* **P**-*measure, local extinction holds.*

Proof. Recall Lemma 2.1 about the local extinction/local exponential growth dichotomy for branching diffusions.

In order to be able to use that lemma, we compare the rate β with another, *smooth* (i.e. C^γ) function V. Recalling that $K = K_a(\omega) := \bigcup_{x_i \in \mathrm{supp}(\omega)} \overline{B}(x_i, a)$, let us enlarge the obstacles by factor two:

$$K^* = K_a^*(\omega) := \bigcup_{x_i \in \mathrm{supp}(\omega)} \overline{B}(x_i, 2a).$$

Then $(K^*)^c \subset K^c$. Recall that $\beta(x) := \beta_1 1_K(x) + \beta_2 1_{K^c}(x) \leq \beta_2$ and let $V \in C^\gamma$ ($\gamma \in (0, 1]$) with

$$\beta_2 1_{(K^*)^c} \leq V \leq \beta. \tag{6.3}$$

Consider the operator $L + V$ on $\mathbb{R}^d$ and let $\lambda_c = \lambda_c(\omega)$ denote its generalized principal eigenvalue. Since $V \in C^\gamma$, we are in the setting of Lemma 2.1. Since $V \leq \beta_2$, one has $\lambda_c \leq \lambda_c(L) + \beta_2$ for every ω.

On the other hand, one gets a lower estimate on λ_c as follows. Fix $R > 0$. Since $\beta_2 1_{(K^*)^c} \leq V$, by the homogeneity of the Poisson point process, for almost every environment the set $\{x \in \mathbb{R}^d \mid V(x) = \beta_2\}$ contains a clearing of radius R. Hence, by comparison, $\lambda_c \geq \lambda^{(R)}$, where $\lambda^{(R)}$ is the principal eigenvalue of $L + \beta_2$ on a ball of radius R. Since R can be chosen arbitrarily large and since $\lim_{R \uparrow \infty} \lambda^{(R)} = \lambda_c(L) + \beta_2$, we conclude that $\lambda_c \geq \lambda_c(L) + \beta_2$ for almost every environment.

From the lower and upper estimates, we obtain that

$$\lambda_c = \lambda_c(L) + \beta_2 \quad \text{for a.e. } \omega. \tag{6.4}$$

Consider now the branching processes with the same motion component L but with rate V, respectively constant rate β_2. The statements (i) and (ii) are true for these two processes by (6.4) and Lemma 1.11. As far as the original process (with rate β) is concerned, (i) and (ii) now follow by Remark 6.1.   □

Remark 6.3. The existence of a *continuous* function satisfying (6.3) would of course immediately follow from Uryson's Lemma.[7] In fact it is easy to see the existence of such functions which are even C^∞ by writing $\beta = \beta_2 - (\beta_2 - \beta_1)1_K$ and considering the function $V := \beta_2 - (\beta_2 - \beta_1)f$, where $f \geq 1_{K^*}$ and f is a C^∞-function obtained as a sum of compactly supported C^∞-functions f_n, $n \geq 1$, with disjoint support, where supp(f_n) is in the ϵ_n-neighborhood of the nth connected component of 1_{K^*}, with appropriately small $0 < \epsilon_n$'s.   ◇

6.4   Law of large numbers and spatial spread

6.4.1   *Quenched asymptotics of global growth; LLN*

We are now going to investigate the behavior of the (quenched) global growth rate.

As already mentioned in the introduction of this chapter, it is a notoriously difficult problem to prove the law of large numbers for general, locally surviving spatial branching systems, and, in particular, the not purely exponential case is harder.

To further elucidate this point, let us consider a generic $(L, \beta; D)$-branching diffusion, $D \subset \mathbb{R}^d$, and let $0 \not\equiv f$ be a nonnegative compactly supported bounded measurable function on D. If λ_c, the generalized principal eigenvalue of $L + \beta$ on D is positive and $T = \{T_t\}_{t \geq 0}$ denotes the semigroup corresponding to $L + \beta$ on D, then $(T_t f)(\cdot)$ grows (point-wise) as $\lambda_c t$ on a logarithmic scale. However, in general, the scaling is not precisely exponential due to the presence of a sub-exponential term.

Recall, that the operator $L + \beta - \lambda_c$ is called product-critical (or product-L^1-critical), if $\langle \phi, \tilde{\phi} \rangle < \infty$ (ϕ and $\tilde{\phi}$ are the ground state and the adjoint ground state, respectively), and in this case we pick ϕ and $\tilde{\phi}$ with the normalization $\langle \phi, \tilde{\phi} \rangle = 1$. This is equivalent to the positive recurrence (ergodicity) of the diffusion corresponding to the h-transformed operator $(L + \beta - \lambda_c)^\phi$. Since the density for this diffusion process has a limit as

---

[7]See e.g. Lemma 4.4 in [Kelley (1955)].

$t \to \infty$, $(T_t f)(\cdot)$ scales precisely with $e^{\lambda_c t}$.

If product-criticality fails, however, then the h-transform does not produce a positive recurrent diffusion (it is either null recurrent or transient), and the corresponding density tends to zero as $t \to \infty$. Consequently, $(T_t f)(\cdot)$ does not scale precisely with $e^{\lambda_c t}$, but rather has a sub-exponential factor. This latter scenario holds in the case of the operator $\frac{1}{2}\Delta + \beta$: $(T_t f)(\cdot)$ does not scale precisely exponentially **P**-a.s. (We have $\lambda_c = \beta$, and the sub-exponential factor is $t^{-d/2}$.)

Replacing f by the function $g \equiv 1$, it is still true that the growth is not precisely exponential – this is readily seen in Claim 6.1 and its proof.

Since the process in expectation is related to the semigroup T, therefore the lack of purely exponential scaling indicates the same type of behavior for the expectation of the process (locally or globally). As we have already mentioned, it is the randomization of the branching rate β that helps in this not purely exponential scaling.[8] Indeed, β has some 'nice' properties for almost every environment ω, i.e. the 'irregular' branching rates 'sit in the **P**-zero set'.

Define now the *average growth rate* by

$$r_t = r_t(\omega) := \frac{\log |Z_t(\omega)|}{t}.$$

Next, replace $|Z_t(\omega)|$ by its expectation $\overline{Z}_t := E^\omega |Z_t(\omega)|$ and define

$$\widehat{r}_t = \widehat{r}_t(\omega) := \frac{\log \overline{Z}_t}{t}.$$

Recall from Claim 6.1 that

$$\lim_{t \to \infty} (\log t)^{2/d} (\widehat{r}_t - \beta_2) = -c(d, \nu) \tag{6.5}$$

holds on a set of full **P**-measure. We are going to show that an analogous statement holds for r_t itself.

Theorem 6.1 (Quenched LLN for global mass). *On a set of full* **P***-measure,*

$$\lim_{t \to \infty} (\log t)^{2/d} (r_t - \beta_2) = -c(d, \nu) \tag{6.6}$$

holds in P^ω*-probability.*

One interprets Theorem 6.1 as a kind of quenched LLN. Loosely speaking,

$$r_t \approx \beta_2 - c(d, \nu)(\log t)^{-2/d} \approx \widehat{r}_t, \qquad \text{as } t \to \infty,$$

on a set of full **P**-measure.

The lengthy proof will be carried out in the next section.

---

[8]See the proof of Theorem 6.1 below.

6.5 Proof of Theorem 6.1

We give an upper and a lower estimate separately.

6.5.1 *Upper estimate*

Let $\epsilon > 0$. Using the Markov inequality along with the expectation formula (6.1), we have that on a set of full **P**-measure:

$$P^\omega \left[(\log t)^{2/d}(r_t - \beta_2) + c(d, \nu) > \epsilon \right]$$

$$= P^\omega \left\{ |Z_t| > \exp \left[t \left(\beta_2 - c(d, \nu)(\log t)^{-2/d} + \epsilon(\log t)^{-2/d} \right) \right] \right\}$$

$$\leq E^\omega |Z_t| \cdot \left(\exp \left[t \left(\beta_2 - c(d, \nu)(\log t)^{-2/d} + \epsilon(\log t)^{-2/d} \right) \right] \right)^{-1}$$

$$= \exp \left[-\epsilon t (\log t)^{-2/d} + o \left(t(\log t)^{-2/d} \right) \right] \to 0, \qquad \text{as } t \to \infty.$$

6.5.2 *Lower estimate*

We give a 'bootstrap argument': we start with a trivial and very crude lower estimate, and then we upgrade it to a refined one. For better readability, we broke the relatively long proof into three steps.[9]

6.5.2.1 *Step I: Rough exponential estimate*

Let $0 < \delta < \beta_1$. Then on a set of full **P**-measure

$$\lim_{t \to \infty} P^\omega(|Z_t| \geq e^{\delta t}) = 1. \tag{6.7}$$

This follows from Remark 6.1. (Compare the process with the one where $\beta \equiv \beta_1$.)

In fact, for recurrent dimensions ($d \leq 2$), δ can be taken anything in $(0, \beta_2)$ (see Theorem 6.3), but the proof of this statement requires more work, and we do not need it in our bootstrap argument.

6.5.2.2 *Step II: Time scales*

Let $\epsilon > 0$. We have to show that on a set of full **P**-measure,

$$\lim_{t \to \infty} P^\omega \left[(\log t)^{2/d}(r_t - \beta_2) + c(d, \nu) < -\epsilon \right] = 0. \tag{6.8}$$

---

[9]The third step is much longer than the first two though.

To achieve this, we will define a particular function p_t (the definition is given in (6.15)) satisfying that as $t \to \infty$,

$$p_t = \exp\left[-c(d,\nu)\frac{t}{(\log t)^{2/d}} + o\left(\frac{t}{(\log t)^{2/d}}\right)\right]. \tag{6.9}$$

Using this function we are going to show a statement implying (6.8), namely, that for all $\epsilon > 0$ there is a set of full **P**-measure, where

$$\lim_{t\to\infty} P^\omega\left[\log|Z_t| < \beta_2 t + \log p_t - \epsilon t(\log t)^{-2/d}\right] = 0. \tag{6.10}$$

Let us first give an outline of the *strategy of our proof*. A key step will be introducing three different time scales, $\ell(t)$, $m(t)$ and t where $\ell(t) = o(m(t))$ and $m(t) = o(t)$ as $t \to \infty$. For the first, shortest time interval, we will use that there are 'many' particles produced and they are not moving 'too far away', for the second (of length $m(t) - \ell(t)$) we will use that one particle moves into a clearing of a certain size at a certain distance, and in the third one (of length $t - m(t)$) we will use that there is a branching tree emanating from that particle so that a certain proportion of particles of that tree stay in the clearing with probability tending to one.

To carry out this program, we will utilize Proposition 1.11 concerning the size of clearings, and we will also need two functions $\mathbb{R}_+ \to \mathbb{R}_+$, ℓ and m, satisfying the following, as $t \to \infty$:

(i) $\ell(t) \to \infty$,
(ii) $\log t / \log \ell(t) \to 1$,
(iii) $\ell(t) = o(m(t))$,
(iv) $m(t) = o(\ell^2(t))$,
(v) $m(t) = o(t(\log t)^{-2/d})$.

Note that (i)–(v) are in fact not independent, because (iv) follows from (ii) and (v). We now pick ℓ and m satisfying (i)–(v) as follows. Let $\ell(t)$ and $m(t)$ be arbitrarily defined for $t \in [0, e]$, and

$$\ell(t) := t^{1-1/(\log\log t)}, \quad m(t) := t^{1-1/(2\log\log t)}, \quad \text{for } t \geq t_0 > e.$$

6.5.2.3 *Step III: Completing the refined lower estimate*

Fix $\delta \in (0, \beta_1)$ and define

$$I(t) := \lfloor \exp(\delta\ell(t)) \rfloor.$$

Let A_t denote the following event:

$$A_t := \{|Z_{\ell(t)}| \geq I(t)\}.$$

By (6.7) we know that on a set of full **P**-measure,

$$\lim_{t \to \infty} P^\omega(A_t) = 1. \tag{6.11}$$

By (6.11), for t fixed we can work on $A_t \subset \Omega$ and consider $I(t)$ particles at time $\ell(t)$.

As a next step, we need some control on their spatial position. To achieve this, use Remark 6.1 to compare BBM's with and without obstacles.

Denote $\mathcal{Z}$ the BBM *without* obstacles (and hence with rate β_2) starting at the origin with a single particle. Let $R(t) = \cup_{s \in [0,t]} \text{supp}(\mathcal{Z}_s)$ denote the range of $\mathcal{Z}$ up to time t. Let

$$M(t) = \inf\{r > 0 : R(t) \subseteq B(0, r)\} \text{ for } d \geq 1, \tag{6.12}$$

be the radius of the minimal ball containing $R(t)$. Then, by Proposition 1.16, $M(t)/t$ converges to $\sqrt{2\beta_2}$ in probability as $t \to \infty$.

Return now to the BBM *with* obstacles, and to the set of $I(t)$ particles at time $\ell(t)$. Even though they are at different locations, (6.12) together with Remark 6.1 imply that for any $\epsilon' > 0$, with P^ω-probability tending to one, they are all inside the $(\sqrt{2\beta_2} + \epsilon')\ell(t)$-ball.

Invoking (1.27) from Proposition 1.11, we know that with **P**-probability one, there is a clearing $B = B(x_0, \hat{\rho}(t))$ such that $|x_0| \leq \ell(t)$, for all large enough $t > 0$, where

$$\hat{\rho}(t) := \rho(\ell(t)) = R_0[\log \ell(t)]^{1/d} - [\log \log \ell(t)]^2, \text{ for } t \geq t_0 > e^e.$$

Note that $t \mapsto \hat{\rho}(t)$ is monotone increasing for large t.

In the sequel we will assume the 'worst case', when $|x_0| = \ell(t)$. Indeed, it is easy to see that $|x_0| < \ell(t)$ would help in all the arguments below. (Of course, x_0 depends on t, but this dependence is suppressed in our notation.) By the previous paragraph, with P^ω-probability tending to one, the distance of x_0 from *each* of the $I(t)$ particles is at most

$$(1 + \sqrt{2\beta_2} + \epsilon')\ell(t).$$

Now, any such particle moves to $B(x_0, 1)$ in another $m(t) - \ell(t)$ time with probability q_t, where (using (iii) and (iv) along with the Gaussian density)

$$q_t = \exp\left(-\frac{[(1 + \sqrt{2\beta_2} + \epsilon')\ell(t)]^2}{2[m(t) - \ell(t)]} + o\left(\frac{[(1 + \sqrt{2\beta_2} + \epsilon')\ell(t)]^2}{2[m(t) - \ell(t)]}\right)\right) \to 0,$$

as $t \to \infty$. Let the particle positions at time $\ell(t)$ be $z_1, z_2, ..., z_{I(t)}$ and consider the independent system of Brownian particles

$$\{W_{z_i}; \ i = 1, 2, ..., I(t)\},$$

where $W_{z_i}(0) = z_i$; $i = 1, 2, ..., I(t)$. In other words, $\{W_{z_i}; i = 1, 2, ..., I(t)\}$ just describes the evolution of the $I(t)$ particles picked at time $\ell(t)$ without respect to their possible further descendants and (using the Markov property) by resetting the clock at time $\ell(t)$.

Let C_t denote the following event:

$$C_t := \{\exists i \in \{1, 2, ..., I(t)\}, \exists 0 \le s \le m(t) - \ell(t) \text{ such that } W_{z_i}(s) \in B(x_0, 1)\}.$$

By the independence of the particles,

$$\limsup_{t \to \infty} P^\omega (C_t^c \mid A_t) = \limsup_{t \to \infty} (1 - q_t)^{I(t)} = \limsup_{t \to \infty} \left[(1 - q_t)^{1/q_t} \right]^{q_t I(t)}. \tag{6.13}$$

Since (iii) implies that $\frac{\ell^2(t)}{m(t)} = o(\ell(t))$ as $t \to \infty$ and since (i) is assumed, one has

$$q_t e^{\delta \ell(t)} = \exp\left(-\frac{[\ell(t) + (\sqrt{2\beta_2} + \epsilon')\ell(t)]^2}{2[m(t) - \ell(t)]} + \delta\ell(t) + o(\ell(t)) \right) \to \infty \quad \text{as } t \to \infty.$$

In view of this, (6.13) implies that $\lim_{t \to \infty} P^\omega (C_t^c \mid A_t) = 0$. Using this along with (6.11), it follows that on a set of full $\mathbf{P}$-measure,

$$\lim_{t \to \infty} P^\omega (C_t) = 1. \tag{6.14}$$

Once we know (6.14), we proceed as follows. Recall that $B = B(x_0, \hat{\rho}(t))$ and that $\{\mathbb{P}_x; x \in \mathbb{R}^d\}$ denote the probabilities corresponding to a single generic Brownian particle W (being different from the W_{z_i} above). Let $\sigma_B^{x_0}$ denote the first exit time from B:

$$\sigma_B^{x_0} = \sigma_{B(x_0, \hat{\rho}(t))}^{x_0} := \inf\{s \ge 0 \mid W_s \notin B\}.$$

Abbreviate $t^* := t - m(t)$ and define

$$p_t := \sup_{x \in B(x_0, 1)} \mathbb{P}_x(\sigma_B^{x_0} \ge t^*) = \sup_{x \in B(0,1)} \mathbb{P}_x(\sigma_B \ge t^*), \tag{6.15}$$

where $\sigma_B := \sigma_B^0$. Recall that the radius of B is

$$\hat{\rho}(t) = R_0 [\log \ell(t)]^{1/d} - o\left([\log \ell(t)]^{1/d}\right)$$

$$= \sqrt{\frac{\lambda_d}{c(d,\nu)}} [\log \ell(t)]^{1/d} - o\left([\log \ell(t)]^{1/d}\right), \tag{6.16}$$

and recall the definition of λ_d from Claim 6.1. Then, as $t \to \infty$,

$$p_t = \exp\left[-\frac{\lambda_d \cdot t^*}{\hat{\rho}^2(t)} + o\left(\frac{\lambda_d \cdot t^*}{\hat{\rho}^2(t)}\right) \right]$$

$$= \exp\left[-c(d,\nu)\frac{t^*}{[\log \ell(t)]^{2/d}} + o\left(\frac{t^*}{[\log \ell(t)]^{2/d}}\right) \right]. \tag{6.17}$$

Using (ii) and (v), it follows that in fact

$$p_t = \exp\left[-c(d,\nu)\frac{t}{(\log t)^{2/d}} + o\left(\frac{t}{(\log t)^{2/d}}\right)\right]. \tag{6.18}$$

A little later we will also need the following notation:

$$p_s^t := \sup_{x \in B(x_0,1)} \mathbb{P}_x(\sigma_B^{x_0} \geq s) = \sup_{x \in B(0,1)} \mathbb{P}_x(\sigma_B \geq s). \tag{6.19}$$

With this notation,

$$p_t = p_{t^*}^t.$$

By slightly changing the notation, let Z^x denote the BBM starting with a single particle at $x \in B$; and let $Z^{x,B}$ denote the BBM starting with a single particle at $x \in B$ and with absorption at ∂B (and still branching at the boundary at rate β_2).

Since branching does not depend on motion, $|Z^{x,B}|$ is a non-spatial Yule's process (and of course it does not depend on x) and thus, by Theorem 1.15, for all $x \in B$,

$$\exists N := \lim_{t \to \infty} e^{-\beta_2 t}|Z_t^{x,B}| > 0 \tag{6.20}$$

almost surely.

Note that some particles of Z^x may re-enter B after exiting, whereas for $Z^{x,B}$ that may not happen. Hence, by a simple coupling argument, for all $t \geq 0$, the random variable $|Z_t^x(B)|$ is stochastically larger than $|Z_t^{x,B}(B)|$.

Recall that our goal is to show (6.10), and recall also (6.15) and (6.18). In fact, we will prove the following, somewhat stronger version of (6.10): we will show that if the function $\gamma : [0,\infty) \to [0,\infty)$ satisfies $\lim_{t\to\infty} \gamma_t = 0$, then on a set of full **P**-measure,

$$\lim_{t \to \infty} P^\omega\left(|Z_t| < \gamma_t \cdot e^{\beta_2 t^*}p_t\right) = 0. \tag{6.21}$$

Recalling $t^* = t - m(t)$, and setting

$$\gamma_t := \exp\left(m(t) - \epsilon\frac{t}{(\log t)^{2/d}}\right), \text{ for } t \geq t_0 > e,$$

a simple computation shows that (6.21) yields (6.10). Note that this particular γ satisfies $\lim_{t\to\infty} \gamma_t = 0$ because of the condition (v) on the function m.

By the comparison between $|Z_t^x(B)|$ and $|Z_t^{x,B}(B)|$ (discussed in the paragraph after (6.20)) along with (6.14) and the Markov property applied at time $m(t)$, we have that

$$\lim_{t \to \infty} P\left(|Z_t| < \gamma_t \cdot e^{\beta_2 t^*}p_t\right) \leq \lim_{t \to \infty} \sup_{x \in B} P\left(|Z_{t^*}^{x,B}(B)| < \gamma_t \cdot e^{\beta_2 t^*}p_t\right).$$

Consider now the $J(x, t) := |Z_{t^*}^{x,B}|$ Brownian paths starting at $x \in B$, which are correlated through common ancestry, and let us denote them by $W_1, ..., W_{J(x,t)}$. Let

$$n_t^x := \sum_{i=1}^{J(x,t)} 1_{A_i},$$

where

$$A_i := \{W_i(s) \in B, \ \forall \ 0 \le s \le t\}.$$

Then we have to show that

$$L := \lim_{t \to \infty} \sup_{x \in B} P\left(n_t^x < \gamma_t \cdot e^{\beta_2 t^*} p_t\right) = 0. \tag{6.22}$$

Clearly, for all $x \in B$,

$$L = \lim_{t \to \infty} \sup_{x \in B} P\left(\frac{n_t^x}{N e^{\beta_2 t^*}} < \frac{\gamma_t p_t}{N}\right)$$

$$\le \lim_{t \to \infty} \sup_{x \in B} \left[P\left(\frac{n_t^x}{N e^{\beta_2 t^*}} < \frac{1}{2} p_t\right) + P\left(N \le 2\gamma_t\right)\right]. \tag{6.23}$$

Using the fact that $\lim_{t \to \infty} \gamma_t = 0$ and that N is almost surely positive,

$$\lim_{t \to \infty} P\left(N \le 2\gamma_t\right) = 0;$$

hence it is enough to show that

$$\lim_{t \to \infty} \sup_{x \in B} P\left(\frac{n_t^x}{N e^{\beta_2 t^*}} < \frac{1}{2} p_t\right) = 0. \tag{6.24}$$

The strategy for the rest of the proof is conditioning on the value of the positive random variable N and then using Chebyshev's inequality, for which we will have to carry out some variance calculations. Since the particles are correlated through common ancestry, we will have to handle the distribution of the splitting time of the most common ancestor of two generic particles. Doing so, we will prove a lemma, while some further computations will be deferred to an appendix.

Let R denote the law of N and define the conditional laws

$$P^y(\cdot) := P(\cdot \mid N = y), \ y > 0.$$

Then

$$P\left(\frac{n_t^x}{N e^{\beta_2 t^*}} < \frac{1}{2} p_t\right) = \int_0^\infty R(dy) \, P^y\left(\frac{n_t^x}{y e^{\beta_2 t^*}} < \frac{1}{2} p_t\right).$$

Define the conditional probabilities

$$\widetilde{P}^y(\cdot) := P^y\left(\cdot \mid |Z_{t,x}^B| \ge \mu_t\right) = P\left(\cdot \mid N = y, |Z_{t,x}^B| \ge \mu_t\right), \ y > 0,$$

where $\mu_t = \mu_{t,y} := \lfloor \frac{3y}{4}e^{\beta_2 t^*} \rfloor$. Recall that (6.15) defines p_t by taking supremum over x and that $|Z_t^{x,B}|$ in fact does not depend on x. One has

$$P\left(\frac{n_t^x}{Ne^{\beta_2 t^*}} < \frac{1}{2}p_t\right) \tag{6.25}$$

$$\le \int_0^\infty R(dy)\left[\widetilde{P}^y\left(\frac{n_t^x}{ye^{\beta_2 t^*}} < \frac{1}{2}p_t\right) + P^y\left(e^{-\beta_2 t^*}|Z_t^{x,B}| < \frac{3}{4}y\right)\right].$$

As far as the second term of the integrand in (6.25) is concerned, the limit in (6.20) implies that

$$\lim_{t \to \infty}\int_{\mathbb{R}} R(dy)\, P^y\left(e^{-\beta_2 t^*}|Z_t^{x,B}| < \frac{3}{4}y\right)$$

$$= \lim_{t \to \infty} P\left(e^{-\beta_2 t^*}|Z_t^{x,B}| < \frac{3}{4}N\right) = 0.$$

Let us now concentrate on the first term of the integrand in (6.25). In fact, it is enough to prove that for each fixed $K > 0$,

$$\lim_{t \to \infty}\int_{1/K}^\infty R(dy)\, \widetilde{P}^y\left(\frac{n_t^x}{ye^{\beta_2 t^*}} < \frac{1}{2}p_t\right) = 0. \tag{6.26}$$

Indeed, once we know (6.26), we can write

$$\lim_{t \to \infty}\int_0^\infty R(dy)\, \widetilde{P}^y\left(\frac{n_t^x}{ye^{\beta_2 t^*}} < \frac{1}{2}p_t\right)$$

$$\le \lim_{t \to \infty}\int_{1/K}^\infty R(dy)\, \widetilde{P}^y\left(\frac{n_t^x}{ye^{\beta_2 t^*}} < \frac{1}{2}p_t\right) + R\left(\left[0, \frac{1}{K}\right]\right)$$

$$= R\left(\left[0, \frac{1}{K}\right]\right). \tag{6.27}$$

Since this is true for all $K > 0$, thus letting $K \uparrow \infty$,

$$\lim_{t \to \infty}\int_0^\infty R(dy)\, \widetilde{P}^y\left(\frac{n_t^x}{ye^{\beta_2 t^*}} < \frac{1}{2}p_t\right) = 0.$$

Returning to (6.26), let us pick randomly μ_t particles out of the $J(x,t)$ — this is almost surely possible under $\widetilde{P}^y$. (Again, 'randomly' means that the way we pick the particles is independent of their genealogy and their spatial position.) Let us denote the collection of these μ_t particles by M_t, and define

$$\widehat{n}_t^x := \sum_{i \in M_t} 1_{A_i}.$$

One then has

$$\widetilde{P}^y \left(\frac{n_t^x}{y e^{\beta_2 t^*}} < \frac{1}{2} p_t \right) \le \widetilde{P}^y \left(\frac{\widehat{n}_t^x}{y e^{\beta_2 t^*}} < \frac{1}{2} p_t \right). \qquad (6.28)$$

We are going to use Chebyshev's inequality and therefore we now calculate the variance. Recall that $p_t = \sup_{x \in B(0,1)} \mathbb{P}_x(\sigma_B \ge t^*)$. Using that for $x \in B(0,1)$,

$$\mathbb{P}_x(\sigma_B \ge t) - [\mathbb{P}_x(\sigma_B \ge t)]^2 \le \mathbb{P}_x(\sigma_B \ge t) \le \mathbb{P}_x(\sigma_B \ge t^*) \le p_t,$$

one has

$$\widetilde{\mathrm{Var}}^y (\widehat{n}_t^x) \le \mu_t p_t + \mu_t(\mu_t - 1) \frac{\sum_{(i,j) \in K(t,x)} \widetilde{\mathrm{cov}}^y (1_{A_i}, 1_{A_j})}{\mu_t(\mu_t - 1)},$$

where $K(t,x) := \{(i,j) \; : \; i \ne j, 1 \le i, j \le \mu_t\}$. Now observe that

$$\frac{\sum_{i,j \in K(t,x)} \widetilde{\mathrm{cov}}^y (1_{A_i}, 1_{A_j})}{\mu_t(\mu_t - 1)} = \mathbf{E} \, \widetilde{\mathrm{cov}}^y (1_{A_i}, 1_{A_j}) = (\mathbf{E} \otimes \widetilde{P}^y)(A_i \cap A_j) - p_t^2,$$

where under $\mathbf{P}$ the pair (i,j) is chosen randomly and uniformly over the $\mu_t(\mu_t - 1)$ possible pairs.

Let $Q^{t,y}$ and $Q^{(t)}$ denote the distribution of the *splitting time* of the most recent common ancestor of the ith and the jth particle under $\widetilde{P}^y$ and under $\widetilde{P}$, respectively. By the strong Markov property applied at this splitting time, one has

$$(\mathbf{E} \otimes \widetilde{P}^y)(A_i \cap A_j) = p_t \int_{s=0}^t \int_B p_{t-s,x}^t \widetilde{p}^{(t)}(0, s, \mathrm{d}x) \, Q^{t,y}(\mathrm{d}s),$$

where

$$\widetilde{p}^{(t)}(0, t, \mathrm{d}x) := \mathbb{P}_0(W_t \in \mathrm{d}x \mid W_z \in B, \; z \le t).$$

By the Markov property applied at time s,

$$p_s^t \int_B p_{t-s,x}^t \widetilde{p}^{(t)}(0, s, \mathrm{d}x) = p_t,$$

and thus

$$(\mathbf{E} \otimes \widetilde{P}^y)(A_i \cap A_j) = p_t \int_{s=0}^t \frac{p_t}{p_s^t} Q^{t,y}(\mathrm{d}s).$$

Hence

$$\widetilde{\mathrm{Var}}^y (\widehat{n}_t^x) \le \mu_t(p_t - p_t^2) + \mu_t(\mu_t - 1) p_t^2 \cdot (I_t - 1), \qquad (6.29)$$

where

$$I_t := \int_{s=0}^\infty [p_s^t]^{-1} Q^{t,y}(\mathrm{d}s).$$

Note that this estimate is uniform in x (see the definition of p_t in (6.15)). Define also

$$J_t := \int_{s=0}^{\infty} [p_s^t]^{-1} Q^{(t)}(\mathrm{d}s).$$

Lemma 6.1.

$$\lim_{t \to \infty} J_t = 1. \tag{6.30}$$

The proof of this lemma is deferred to the end of this section.

Once we know (6.30), we proceed as follows. Using Chebyshev's inequality, one has

$$\widetilde{P}^y \left(\frac{\widehat{n}_t^x}{ye^{\beta_2 t^*}} < \frac{1}{2} p_t \right) \le \widetilde{P}^y \left(|\widehat{n}_t^x - E^y \widehat{n}_t^x| > \frac{1}{4} p_t y e^{\beta_2 t^*} \right) \le 16 \frac{\widetilde{\mathrm{Var}}^y (\widehat{n}_t^x)}{p_t^2 y^2 e^{2\beta_2 t^*}}.$$

By (6.29), we can continue the estimate by

$$\le 16 \left[\frac{\mu_t p_t}{p_t^2 y^2 e^{2\beta_2 t^*}} + \frac{1}{2} \mu_t(\mu_t - 1) \cdot y^{-2} e^{-2\beta_2 t^*} \cdot (I_t - 1) \right].$$

Writing out μ_t, integrating against $R(\mathrm{d}y)$, and using that the lower limit in the integral is $1/K$, one obtains the upper estimate

$$\int_{1/K}^{\infty} R(\mathrm{d}y) \, \widetilde{P}^y \left(\frac{\widehat{n}_t^x}{ye^{\beta_2 t^*}} < \frac{1}{2} p_t \right) \tag{6.31}$$

$$\le 12 K p_t^{-1} e^{-\beta_2 t^*} + \int_{1/K}^{\infty} R(\mathrm{d}y) \frac{1}{2} \mu_t(\mu_t - 1) \cdot y^{-2} e^{-2\beta_2 t^*} \cdot (I_t - 1).$$

(Recall that I_t in fact depends on y.) Since $\lim_{t \to \infty} p_t e^{\beta_2 t^*} = \infty$, thus the first term on the right-hand side of (6.31) tends to zero as $t \to \infty$. Recall now that $\mu_t := \lfloor \frac{3ye^{\beta_2 t^*}}{4} \rfloor$. As far as the second term of (6.31) is concerned, it is easy to see that it also tends to zero as $t \to \infty$, provided

$$\lim_{t \to \infty} \int_0^{\infty} R(\mathrm{d}y)(I_t - 1) = 0.$$

But $\int_0^{\infty} R(\mathrm{d}y)(I_t - 1) = J_t - 1$ and so we are finished by recalling (6.30). Hence (6.26) follows. This completes the proof of the lower estimate in Theorem 6.1.   ∎

6.5.2.4 *Proof of Lemma 6.1*

Since $J_t \geq 1$, thus it is enough to prove that

$$\limsup_{t \to \infty} J_t \leq 1.$$

For $r > 0$ we denote by $\lambda_r^* := \lambda_c(\frac{1}{2}\Delta, B(0,r))$ the principal eigenvalue of $\frac{1}{2}\Delta$ on $B(0,r)$. Since λ_r^* tends to zero as $r \uparrow \infty$ we can pick an $R > 0$ such that $-\lambda_R^* < \beta_2$. Let us fix this R for the rest of the proof.

Let us also fix $t > 0$ for a moment. From the probabilistic representation of the principal eigenvalue (Proposition 1.6) we conclude the following: for $\hat{\epsilon} > 0$ fixed there exists a $T(\hat{\epsilon})$ such that for $s \geq T(\hat{\epsilon})$,

$$\log p_s^t \geq (\lambda_{\hat{\rho}(t)} - \hat{\epsilon})s.$$

Hence, for $\hat{\epsilon} > 0$ small enough ($\hat{\epsilon} < -\lambda_R^*$) and for all t satisfying $\lambda_{\hat{\rho}(t)} \geq \lambda_R^* + \hat{\epsilon}$ (recall that $\lim_{t \to \infty} \hat{\rho}(t) = \infty$) and $s \geq T(\hat{\epsilon}, t)$,

$$\log p_s^t \geq \lambda_R^* \cdot s. \tag{6.32}$$

Note that $T(\hat{\epsilon}, t)$ can be chosen uniformly in t because[10] $\lim_{t \to \infty} \hat{\rho}(t) = \infty$, and so we will simply write $T(\hat{\epsilon})$. Furthermore, clearly, $T(\hat{\epsilon})$ can be chosen in such a way that

$$\lim_{\hat{\epsilon} \downarrow 0} T(\hat{\epsilon}) = \infty. \tag{6.33}$$

Depending on $\hat{\epsilon}$ let us break the integral into two parts:

$$J_t = \int_{s=0}^{T(\hat{\epsilon})} [p_s^t]^{-1} Q^{(t)}(ds) + \int_{s=T(\hat{\epsilon})}^{t} [p_s^t]^{-1} Q^{(t)}(ds) =: J_t^{(1)} + J_t^{(2)}.$$

We are going to control the two terms separately.

Controlling $J_t^{(1)}$: We show that

$$\text{There exists the limit } \lim_{t \to \infty} J_t^{(1)} \leq 1. \tag{6.34}$$

First, it is easy to check that for all $t > 0$, $Q^{(t)}(ds)$ is absolutely continuous, i.e. $Q^{(t)}(ds) = g^{(t)}(s)\,ds$ with some $g^{(t)} \geq 0$. So

$$J_t^{(1)} = \int_{s=0}^{T(\hat{\epsilon})} [p_s^t]^{-1} Q^{(t)}(ds) = \int_{s=0}^{T(\hat{\epsilon})} [p_s^t]^{-1} g^{(t)}(s)\,ds.$$

---

[10]Recall that we picked a version of ℓ such that $\hat{\rho}(t) = \rho(\ell(t))$ is monotone increasing for large t's.

Evidently, one has $[p_s^t]^{-1} \downarrow$ as $t \to \infty$. Also, since $Q^{(t)}([a,b])$ is monotone non-increasing in t for $0 \le a \le b$, therefore $g^{(t)}(\cdot)$ is also monotone non-increasing in t. Hence, by monotone convergence,

$$\lim_{t \to \infty} J_t^{(1)} = \int_{s=0}^{T(\hat{\epsilon})} g(s)\mathrm{d}s = \lim_{t \to \infty} \int_{s=0}^{T(\hat{\epsilon})} g^{(t)}(s)\mathrm{d}s \le \lim_{t \to \infty} \int_{s=0}^{t} g^{(t)}(s)\mathrm{d}s = 1,$$

where $g := \lim_{t \to \infty} g^{(t)}$.

Controlling $J_t^{(2)}$: Recall that

$$\log p_s^t \ge \lambda_R^* \cdot s, \ \forall s \ge T(\hat{\epsilon}). \tag{6.35}$$

Thus,

$$J_t^{(2)} \le \int_{T(\hat{\epsilon})}^{t} \exp(-\lambda_R^* \cdot s) \, Q^{(t)}(\mathrm{d}s).$$

We will show that

$$\lim_{\hat{\epsilon} \downarrow 0} \lim_{t \to \infty} \int_{T(\hat{\epsilon})}^{t} \exp(-\lambda_R^* \cdot s) \, Q^{(t)}(\mathrm{d}s)$$

$$= \lim_{\hat{\epsilon} \downarrow 0} \lim_{t \to \infty} \int_{T(\hat{\epsilon})}^{t} e^{-(\lambda_R^* + \beta_2)s} \, e^{\beta_2 s} Q^{(t)}(\mathrm{d}s) = 0. \tag{6.36}$$

Recall that $0 < \beta_2 + \lambda_R^*$. In order to verify (6.36), we will show that given $t_0 > 0$ there exists some $0 < K = K(t_0)$ with the property that

$$g^{(t)}(s) \le Kse^{-\beta_2 s}, \ \text{for} \ t > t_0, \ s \in [t_0, t]. \tag{6.37}$$

Indeed, it will then follow that

$$\lim_{\hat{\epsilon} \downarrow 0} \lim_{t \to \infty} \int_{T(\hat{\epsilon})}^{t} \exp(-\lambda_R^* \cdot s) \, Q^{(t)}(\mathrm{d}s)$$

$$= \lim_{T(\hat{\epsilon}) \to \infty} \lim_{t \to \infty} \int_{T(\hat{\epsilon})}^{t} \exp(-\lambda_R^* \cdot s) \, g^{(t)}(s)(\mathrm{d}s)$$

$$\le K \lim_{T(\hat{\epsilon}) \to \infty} \int_{T(\hat{\epsilon})}^{\infty} s \, e^{-(\lambda_R^* + \beta_2)s} \, (\mathrm{d}s) = 0.$$

Recall that $Q^{(t)}$ corresponds to the conditional law $P(\cdot \mid |Z_{t,x}^B| \ge \mu_t)$. We now claim that we can work with $P(\cdot \mid |Z_{t,x}^B| \ge 2)$ instead of $P(\cdot \mid |Z_{t,x}^B| \ge \mu_t)$. This is because if $Q_0^{(t)}$ corresponds to $P(\cdot \mid |Z_{t,x}^B| \ge 2)$, then an easy computation reveals that for any $\epsilon > 0$ there exists a $\hat{t}_0 = \hat{t}_0(\epsilon)$ such that for all $t \ge \hat{t}_0$ and for all $0 \le a < b$,

$$\left| Q^{(t)}([a,b]) - Q_0^{(t)}([a,b]) \right| \le 2(1 + \epsilon)Q_0^{(t)}([a,b]);$$

thus, if

$$Q_0^{(t)}(\mathrm{d}s) \le Lse^{-\beta_2 s}\mathrm{d}s \text{ on } [t_0, t] \text{ for } t > t_0 \tag{6.38}$$

holds with some $L > 0$, then also

$$Q^{(t)}(\mathrm{d}s) = g^{(t)}(s)\,\mathrm{d}s \le Kse^{-\beta_2 s}\mathrm{d}s, \quad t > t_0 \vee \hat{t}_0, \ s \in [t_0, t] \tag{6.39}$$

holds with $K := L + 2(1 + \epsilon)$.

The proof of the bound (6.38) is relegated to Section 6.9.

It is now easy to finish the proof of (6.30). To make the dependence on $\hat{\epsilon}$ clear, let us write $J_t^{(i)} = J_t^{(i)}(\hat{\epsilon})$, $i = 1, 2$. Then by (6.34), one has that for *all* $\hat{\epsilon} > 0$,

$$\limsup_{t\to\infty} J_t \le 1 + \limsup_{t\to\infty} J_t^{(2)}(\hat{\epsilon}).$$

Hence, (6.36) yields

$$\limsup_{t\to\infty} J_t \le 1 + \lim_{\hat{\epsilon}\downarrow 0}\limsup_{t\to\infty} J_t^{(2)}(\hat{\epsilon}) \le 1,$$

finishing the proof of the lemma. ∎

6.6 The spatial spread of the process

6.6.1 *The results of Bramson, Lee-Torcasso and Freidlin*

A natural question[11] concerns the spread of the system: *how much is the speed (spatial spread) of the free BBM reduced* due to the presence of the mild obstacles? Note that we are not talking about the bulk of the population (or the 'shape') but rather about *individual* particles traveling to very large distances from the origin (cf. Problem 6.3 later).

Recall[12] from Proposition 1.16, that ordinary 'free' branching Brownian motion with constant branching rate $\beta_2 > 0$ has radial speed $\sqrt{2\beta_2}$. Let N_t denote the population size at $t \ge 0$ and let ξ_k ($1 \le k \le N_t$) denote the position of the kth particle (with arbitrary labeling) in the population. Furthermore, let $m(t)$ denote a number for which $u(t, m(t)) = \frac{1}{2}$, where

$$u(x, t) := P\left[\max_{1 \le k \le N_t} \|\xi_k(t)\| \le x\right].$$

---

[11]The question was asked by L. Mytnik.

[12]See also [Kyprianou (2005)] for a review and a strong version of this statement.

In his classic paper, Bramson [Bramson (1978b)] considered the one dimensional case and proved[13] that as $t \to \infty$,

$$m(t) = t\sqrt{2\beta_2} - \frac{3}{2\sqrt{2\beta_2}} \log t + \mathcal{O}(1). \qquad (6.40)$$

Since the one-dimensional projection of a d-dimensional branching Brownian motion is a one-dimensional branching Brownian motion, we can utilize Bramson's result for the higher dimensional cases too. Namely, it is clear, that in high dimension the spread is *at least* as quick as in (6.40). In [Bramson (1978b)] the asymptotics (6.40) is derived for the case $\beta_2 = 1$; the general result can be obtained similarly. See also p. 438 in [Freidlin (1985)]. (It is also interesting to take a look at [Bramson (1978a)].) Studying the function u has significance in analysis too as u solves

$$\frac{\partial u}{\partial t} = \frac{1}{2}u_{xx} + \beta_2(u^2 - u), \qquad (6.41)$$

with initial condition

$$\lim_{t \downarrow 0} u(\cdot, t) = \mathbf{1}_{[0,\infty)}(\cdot). \qquad (6.42)$$

In this section we show that the branching Brownian motion with mild obstacles *spreads less quickly* than ordinary branching Brownian motion by giving an upper estimate on its speed.

A related result was obtained earlier by Lee-Torcaso [Lee and Torcaso (1998)], but, unlike in (6.40), only up to the linear term and moreover, for random walks instead of Brownian motions. The approach in [Lee and Torcaso (1998)] was to consider the problem as the description of wave-front propagation for a random KPP equation. They extended a result of Freidlin and Gärtner for KPP wave fronts to the case $d \geq 2$ for i.i.d. random media. In [Lee and Torcaso (1998)] the wave front propagation speed is attained for the discrete-space (lattice) KPP using a large deviation approach. Note that the 'speed' is only defined in a logarithmic sense. More precisely, let u denote the solution of the discrete-space KPP equation with an initial condition that vanishes everywhere except the origin. The authors define a bivariate function F on $\mathbb{R} \times \mathbb{R}^d \setminus \{\mathbf{0}\}$ and show that it satisfies

$$\lim_{t \to \infty} \frac{1}{t} \log u(t, tv\mathbf{e}) = -[F(v, \mathbf{e}) \vee 0],$$

for all $v > 0$ and $\mathbf{e} \in \mathbb{R}^d \setminus \{\mathbf{0}\}$. It turns out that there is a unique solution $v = v_{\mathbf{e}}$ to $F(v, \mathbf{e}) = 0$, and $v_{\mathbf{e}}$ defines the 'wave speed'. In particular, the speed is non-random.

---

[13]Recently a much shorter proof has been found by M. I. Roberts, by using spine methods (see [Roberts (2013)]).

Unfortunately, it does not seem to be easy to evaluate the variational formula for v_e given in [Lee and Torcaso (1998)], even in very simple cases.

It should be pointed out that the problem is greatly simplified for $d = 1$ and it had already been investigated by Freidlin in his classic text [Freidlin (1985)] (the *KPP equation with random coefficients* is treated in section VII.7.7). Again, it does not seem clear whether one can easily extract an explicit result for the speed of a branching RW with i.i.d branching coefficients which can only take two values, $0 < \beta_1 < \beta_2$ (bistable nonlinear term).

The description of wavefronts in random medium for $d > 1$ is still an open and very interesting problem. The above work of Torcaso and Lee concerning processes on $\mathbb{Z}^d$ is the only relevant article we are aware of. To the best of our knowledge, the problem is open; it is of special interest for a bistable nonlinear term.

Before turning to the upper estimate, we discuss the lower estimate.

6.6.2 *On the lower estimate for the radial speed*

We are going to show that, if in our model Brownian motion is replaced by Brownian motion with constant drift γ in a given direction, then any fixed non-empty ball is recharged infinitely often with positive probability, as long as the drift satisfies $|\gamma| < \sqrt{2\beta_2}$.

For simplicity, assume that $d = 1$ (the general case is similar). Fix the environment ω. Recall Doob's h-transform from Chapter 1:

$$L^h(\cdot) := \frac{1}{h} L(h \cdot).$$

Applying an h-transform with $h(x) := \exp(-\gamma x)$, a straightforward computation shows that the operator

$$L := \frac{1}{2} \frac{\mathrm{d}^2}{\mathrm{d}x^2} + \gamma \frac{\mathrm{d}}{\mathrm{d}x} + \beta$$

transforms into

$$L^h = \frac{1}{2} \frac{\mathrm{d}^2}{\mathrm{d}x^2} - \frac{\gamma^2}{2} + \beta.$$

Then, similarly to the proof of Claim 6.3, one can show that the generalized principal eigenvalue for this latter operator is $-\frac{\gamma^2}{2} + \beta_2$ for almost every environment. Since the generalized principal eigenvalue is invariant under h-transforms, it follows that $-\frac{\gamma^2}{2} + \beta_2 > 0$ is the generalized principal eigenvalue of L. Hence, by Claim 6.3, any fixed nonempty interval is recharged infinitely often with positive probability.

Turning back to our original setting, the application of the spine-technology seems also promising. In our case it is probably not very difficult to show the existence of a 'spine' particle (under a martingale-change of measure) that has drift γ as long as $|\gamma| < \sqrt{2\beta_2}$.

6.6.3   *An upper estimate on the radial speed*

Our main result in this section is an upper estimate on the speed of the process. We give an upper estimate in which the order of the correction term is *larger* than the $\mathcal{O}(\log t)$ term appearing in Bramson's result, namely it is $\mathcal{O}\left(\frac{t}{(\log t)^{2/d}}\right)$. (All orders are meant for $t \to \infty$.) We show that, loosely speaking, at time t the spread of the process is not more than

$$t\sqrt{2\beta_2} - c(d,\nu)\sqrt{\frac{\beta_2}{2}} \cdot \frac{t}{(\log t)^{2/d}}.$$

(Again, β_1 plays no role as long as $\beta_1 \in (0, \beta_2)$.) The precise statement is as follows.

Theorem 6.2. *Define the functions f and n on $[0, \infty)$ by*

$$f(t) := k(d,\nu)\frac{t}{(\log t)^{2/d}} \quad and \quad n(t) := t\sqrt{2\beta_2} \cdot \sqrt{1 - \frac{f(t)}{\beta_2 t}},$$

where we recall that $k(d,\nu) := \lambda_d \left(\frac{\nu\omega_d}{d}\right)^{2/d}$ and ω_d is the volume of the d-dimensional unit ball, while $-\lambda_d$ is the principal Dirichlet eigenvalue of $\frac{1}{2}\Delta$ on it. Then, as $t \to \infty$,

$$n(t) = t\sqrt{2\beta_2} - k(d,\nu)\sqrt{\frac{\beta_2}{2}} \cdot \frac{t}{(\log t)^{2/d}} + \mathcal{O}\left(\frac{t}{(\log t)^{4/d}}\right). \quad (6.43)$$

Furthermore, if

$$A_t := \{no\ particle\ has\ left\ the\ n(t)\text{-}ball\ up\ to\ t\}$$

$$= \left\{\bigcup_{0 \le s \le t} \operatorname{supp}(Z_s) \subseteq B(0, n(t))\right\},$$

then

$$\mathbf{P}\left[\liminf_{t \to \infty} P^\omega(A_t) > 0\right] = 1. \quad (6.44)$$

Proof.   First, equation (6.43) follows from the Taylor expansion $\sqrt{1-x} = 1 - \frac{x}{2} + \mathcal{O}(x^2)$, $x \approx 0$.

For (6.44), recall that $f(t) := c(d,\nu)\frac{t}{(\log t)^{2/d}}$ and the result on the quenched global population: there are roughly $\exp[\beta_2 t - f(t)]$ particles by time t. More precisely, on a set of full **P**-measure,

$$\lim_{t\to\infty} (\log t)^{2/d}(r_t - \beta_2) = -c(d,\nu) \quad \text{in } P^\omega\text{-probability.} \tag{6.45}$$

In particular, for all $\epsilon > 0$, as $t \to \infty$,

$$P^\omega\left(|Z_t| > e^{\beta_2 t - f(t) + \epsilon}\right) = o(\log t^{-2/d}). \tag{6.46}$$

The rest is a straightforward computation. We apply Corollary 1.1 with

$$g(t) := \lfloor e^{\beta_2 t - f(t) + \epsilon}\rfloor.$$

Denote $C_t := \{|Z_t| \le g(t)\}$. Using that $\lim_{t\to\infty} P^\omega(C_t) = 1$ and that

$$n^2(t) = 2t^2\left[\beta_2 - c(d,\nu)(\log t)^{-2/d}\right] \tag{6.47}$$

along with Corollary 1.1, it follows that for **P**-almost all ω,

$$P^\omega(A_t) \ge \left(1 - \exp\left[-\frac{n^2(t)}{2t}\right]\right)^{\exp[\beta_2 t - f(t) + \epsilon]} - o(1)$$

$$= (1 - \exp[-\beta_2 t + f(t)])^{\exp[\beta_2 t - f(t) + \epsilon]} - o(1) \longrightarrow e^{-e^\epsilon},$$

as $t \to \infty$. Consequently, for **P**-almost all ω, $\liminf_{t\to\infty} P^\omega(A_t) > 0$. $\square$

6.7 More general branching and further problems

It should also be investigated, what happens when the dyadic branching law is replaced by a general one (but the random branching rate is as before). In a more sophisticated population model, particles can also die — then the obstacles do not necessarily reduce the population size as they sometimes prevent death.

(i) Supercritical branching:
When the offspring distribution is supercritical, the method of our chapter seems to work, although when the offspring number can also be zero, one has to *condition on survival* for getting the asymptotic behavior.

(ii)(Sub)critical branching:
Critical branching requires an approach very different from the supercritical one, since taking expectations now does not provide a clue: $E^\omega|Z_t(\omega)| = 1$, $\forall t > 0$, $\forall \omega \in \Omega$.

Having the obstacles, the first question is whether it is still true that

$$P^\omega(\text{extinction}) = 1 \; \forall \omega \in \Omega.$$

The answer is yes. To see this, note that since $|Z|$ is still a martingale, it has a nonnegative a.s. limit. This limit must be zero; otherwise $|Z|$ would stabilize at a positive integer. This, however is impossible because following one Brownian particle it is obvious that this particle experiences branching events for arbitrarily large times. (Cf. the proof of Theorem 7.2 in the next chapter.)

Setting $\beta_1 = 0$, the previous argument still goes through. Let τ denote the almost surely finite extinction time for this case. One of the basic questions is the decay rate for $P^\omega(\tau > t)$. Will the tail be significantly heavier[14] than $\mathcal{O}(1/t)$? The critical case, in a discrete setting, will be addressed in the next chapter.

The subcritical case can be treated in a similar fashion. In particular, the total mass is a supermartingale and $P^\omega(\text{extinction}) = 1 \ \forall \omega \in \Omega$.

We conclude with two further problems.

Problem 6.2 (Strong Law). The end of the proof for the lower estimate in Theorem 6.1 is basically a version of the Weak Law of Large Numbers. Using SLLN instead (and making some appropriate changes elsewhere), can one get

$$\liminf_{t \to \infty} (\log t)^{2/d}(r_t - \beta_2) \geq -c(d, \eta) \quad a.s. \ ?$$

Problem 6.3 (Shape). The question investigated in this chapter was the (local and global) growth rate of the population. The next step can be the following: Once one knows the global population size $|Z_t|$, the model can be rescaled (normalized) by $|Z_t|$, giving a population of fixed weight. In other words, one considers the discrete probability measure valued process

$$\tilde{Z}_t(\cdot) := \frac{Z_t(\cdot)}{|Z_t|},$$

Then the question of the *shape* of the population for Z for large times is given by the limiting behavior of the random probability measures $\tilde{Z}_t$, $t \geq 0$. (Of course, not only the particle mass has to be scaled, but also the spatial scales are interesting — see last paragraph.)

Can one for example locate a *unique dominant branch* for almost every environment, so that the total weight of its complement tends to (as $t \to \infty$) zero?

The motivation for this question comes from our proof of the lower estimate for Theorem 6.1. It seems conceivable that for large times the

---

[14]Of course $1/t$ would be the rate without obstacles.

'bulk' of the population will live in a clearing within distance $\ell(t)$ and with radius

$$\hat{\rho}(t) := R_0[\log \ell(t)]^{1/d} - [\log \log \ell(t)]^2, \ t \geq 0,$$

where

$$\lim_{t \to \infty} \ell(t) = \infty \ \text{and} \ \lim_{t \to \infty} \frac{\ell(t)}{t} = 0 \ \text{but} \ \lim_{t \to \infty} \frac{\log t}{\log \ell(t)} = 1.$$

6.8   Superprocesses with mild obstacles

A further goal is to generalize the setting by defining *superprocesses with mild obstacles* analogously to the BBM with mild obstacles.

Recall the definition of the $(L, \beta, \alpha; \mathbb{R}^d)$-superdiffusion. The definition of the superprocess with mild obstacles is straightforward: the parameter α on the (random) set K is smaller than elsewhere.

Similarly, one can consider the case when instead of α, the 'mass creation term' β is random, for example with β defined in the same way (or with a mollified version) as for the discrete branching particle system. Denote now by P^ω the law of this latter superprocess for given environment. We suspect that the superprocess with mild obstacles behaves similarly to the discrete branching process with mild obstacles when $\lambda_c(L + \beta) > 0$ and $P^\omega(\cdot)$ is replaced by $P^\omega(\cdot \mid X \text{ survives})$. The upper estimate can be carried out in a manner similar to the discrete particle system, as the expectation formula is still in force for superprocesses.

As we have already pointed out, there is a large amount of ongoing research on catalytic superprocesses; α is usually taken as a thin (sometimes randomly moving) set, or even another superprocess. In those models, one usually cannot derive sharp quantitative results. In a very simple one-dimensional model, introduced in [Engländer and Fleischmann (2000)], β was spatially varying but deterministic and non-moving — in fact it was the *Dirac delta at zero*. Nevertheless, already in this simple model it was quite challenging to prove the asymptotic behavior of the process (Theorem 2 in [Engländer and Turaev (2002)]). Fleischmann, Mueller and Vogt suggest, as an open problem, the description of the asymptotic behavior of the process in the *three-dimensional* case [Fleischmann, Mueller and Vogt (2007)]; the two-dimensional case is even harder, as getting the asymptotics of the *expectation* is already difficult. Again, the randomization of β may help in the sense that β has some 'nice' properties for almost every environment ω.

6.9   The distribution of the splitting time of the most recent common ancestor

In this section we give the proof of the bound (6.38). In fact we prove a precise formula[15] for the distribution of the splitting time of the most recent common ancestor (denoted by S here), which is of independent interest.

For simplicity we set $\beta_2 = 1$; the general case is similar. Let us fix $t > 0$. Then for $0 < u < t$, one has

$$Q_0^{(t)}([0,u]) = \frac{1 - 2ue^{-u} - e^{-2u} + e^{-t}(2u - 3 + 4e^{-u} - e^{-2u})}{(1 - e^{-t})(1 - e^{-u})^2};   (6.48)$$

and so the density (with respect to Lebesgue measure) for S on $(0,t)$ is

$$f^{(t)}(u) := \frac{dQ_0^t}{dl}(u) = 2\frac{e^{-u}(u - 2 + (u+2)e^{-u}) + e^{-t}(1 - 2ue^{-u} - e^{-2u})}{(1 - e^{-t})(1 - e^{-u})^3}.$$

Proof of (6.48): Consider the Yule population $Y_t := |Z_{t,x}^B|$ and recall that $Q_0^{(t)}$ corresponds to $P(\cdot \mid Y_t \geq 2)$. The first observation concerns the Yule *genealogy*. Let us pick a pair of individuals from the Yule population at time t, assuming that $Y_t = j$, $j \geq 2$. Denote by I the size of the population *just before* the coalescence time of the two ancestral lines (where 'before' refers to backward time): $I := Y_{S+}$. We now show that

$$P(I = i) = \frac{j+1}{j-1} \cdot \frac{2}{(i-1)i} \cdot \frac{i-1}{i+1}.   (6.49)$$

The paper [Etheridge, Pfaffelhuber, and Wakolbinger (2006)] considers, more generally, $\mathcal{Y}$, a generic Yule's process, viewed as an infinite tree, and $\mathcal{Y}_n$, another, smaller random tree which arises by sampling $n \geq 2$ lineages from $\mathcal{Y}$ (see Section 3.5). Let $I = I(t)$ be the number of lines of $\mathcal{Y}$ extant at time t and let K_i be the number of lines extant in $\mathcal{Y}_n$, while $I = i$. Consider now the Yule's process $K = (K_i)$. Viewing the index i as time ('Yule time'), in the paper above it was shown that K is a Markov chain, and the forward/backward transition probabilities were derived.

Since we are only interested in the most recent common ancestor of two particles, we set $n = 2$, and use (4.11) of that paper, yielding in our case that

$$P(K_{i-1} = 1 \mid K_i = 2) = \frac{2}{i(i-1)}.$$

---

[15]The formula and its proof are due to W. Angerer and A. Wakolbinger.

Now define the 'hitting time' F by $F := \min\{l : K_l = 2\}$; this is the Yule time of the sought splitting time. Then

$$P(F \leq i) = P(K_i = 2) = \frac{i-1}{i+1}, \qquad i \geq 2,$$

where the second equality is (2.3) of the above paper. For $i \leq j$ one has

$$
\begin{aligned}
P(I = i) = P(F = i \mid F \leq j) &= P\big(K_{i-1} = 1, K_i = 2 \mid F \leq j\big) \\
&= P\big(K_{i-1} = 1, K_i = 2 \mid K_j = 2\big) \\
&= \frac{P\big(K_{i-1} = 1, K_i = 2\big)}{P\big(K_j = 2\big)}.
\end{aligned}
$$

Using the last three displayed formulae one obtains immediately (6.49).

Let us now *embed* the 'Yule time' into *real time*. Since, by Lemma 1.4, a Yule population stemming from i ancestors has a negative binomial distribution, therefore, using the Markov property at times u and $u + \mathrm{d}u$, one can decompose

$$P(Y_u = i - 1, Y_{u+\mathrm{d}u} = i, Y_t = j) = p_1 \cdot p_2 \cdot p_3, \tag{6.50}$$

where

$$p_1 := e^{-u}(1 - e^{-u})^{i-2},$$

$$p_2 := (i - 1)\,\mathrm{d}u \quad \text{and}$$

$$p_3 := \binom{j-1}{i-1} e^{-(t-u)i}(1 - e^{-(t-u)})^{j-i}.$$

Since the pair we have chosen coalesce independently of the rest of the population, the random variables $\mathcal{S}$ and I are independent. Using that $I := Y_{\mathcal{S}+}$ first, and then the independence remarked in the previous sentence, and finally (6.49) and (6.50),

$$
\begin{aligned}
&P\big(\mathcal{S} \in [u, u + \mathrm{d}u], Y_{\mathcal{S}+\mathrm{d}t} = i, Y_t = j\big) \\
&= P\big(I = i, Y_u = i - 1, Y_{u+\mathrm{d}u} = i, Y_t = j\big) \\
&= P(I = i)P(Y_u = i - 1, Y_{u+\mathrm{d}u} = i, Y_t = j) \\
&= \binom{j-2}{i-2} \frac{2(j+1)}{i(i+1)} e^{-(t-u)i}(1 - e^{-(t-u)})^{j-i} e^{-u}(1 - e^{-u})^{i-2}\,\mathrm{d}u,
\end{aligned}
$$

for $0 < u < t$. Now, summing from $j = i$ to ∞, and from $i = 2$ to ∞, and then dividing by $P(Y_t \geq 2) = 1 - e^{-t}$, one obtains (after doing some algebra) that for $0 < u < t$,

$$
\begin{aligned}
Q_0^{(t)}\big((u, u + \mathrm{d}u)\big) &= \sum_{i=2}^{\infty} e^{-u} \frac{2(2e^{-(t-u)} + i - 1)(1 - e^{-u})^{i-2}}{(1 - e^{-t})i(i+1)}\,\mathrm{d}u \\
&= 2 \cdot \frac{e^{-u}(u - 2 + (u + 2)e^{-u}) + e^{-t}(1 - 2ue^{-u} - e^{-2u})}{(1 - e^{-t})(1 - e^{-u})^3}\,\mathrm{d}u\,.
\end{aligned}
$$

Equivalently, in integrated form, one has (6.48). ∎

6.10   Exponential growth when $d \leq 2$ and $\beta_1 \geq 0$

Assume that $d \leq 2$. Let us now replace the assumption $\beta_1 > 0$ by $\beta_1 \geq 0$. Below we are going to give an a priori exponential estimate on the growth for this case as well. In the proof and in the remarks following it, we will use a number of results from the literature, and so we will hardly be 'self-contained.' But perhaps the reader will forgive this, given that what follows was not needed for the proof of our main result (Theorem 6.1).

Be that as it may, using the next result, it is possible to upgrade it to the quenched LLN for the global mass (Theorem 6.1) for $\beta_1 \geq 0$, just like we did it for $\beta_1 > 0$. We conjecture that Theorem 6.1 holds true for $\beta_1 = 0$ as well, in any dimension.

Theorem 6.3. *Let $0 < \delta < \beta_2$. Then on a set of full* **P***-measure*

$$\lim_{t \to \infty} P^\omega(|Z_t| \geq e^{\delta t}) = 1. \tag{6.51}$$

Proof.   We invoke the definition of the function V from Subsection 6.3.1: $V \in C^\gamma$ $(\gamma \in (0,1])$ with

$$\beta_2 1_{(K^*)^c} \leq V \leq \beta. \tag{6.52}$$

(Recall that $\beta(x) := \beta_1 1_K(x) + \beta_2 1_{K^c}(x)$.) By comparison, it is enough to prove (6.51) for the 'smooth version' of the process, where β is replaced by V. The law of this modified process will be denoted by P^V (and the notation Z is unchanged).

Considering the operator $\frac{1}{2}\Delta + V$ on $\mathbb{R}^d$ we have seen in Subsection 6.3 that its generalized principal eigenvalue is $\lambda_c(\frac{1}{2}\Delta, \mathbb{R}^d) + \beta_2 = \beta_2$ for every ω. Take $R > 0$ large enough so that $\lambda_c = \lambda_c\left(\frac{1}{2}\Delta + V, B(0,R)\right)$, the principal eigenvalue of $\frac{1}{2}\Delta + V$ on $B(0,R)$ satisfies

$$\lambda_c > \delta.$$

Let $\hat{Z}^R$ be the process obtained from Z by introducing killing at $\partial B(0,R)$ (the corresponding law will be denoted by $P_x^{(R)}$). Then

$$\lim_{t \to \infty} P^V(|Z_t| < e^{\delta t}) \leq \lim_{t \to \infty} P^{(R)}(|\hat{Z}_t^R| < e^{\delta t}). \tag{6.53}$$

Let $0 \leq \phi = \phi^R$ be the Dirichlet eigenfunction (with zero boundary data) corresponding to λ_c on $B(0,R)$, and normalize it by $\sup_{x \in B(0,R)} \phi(x) = 1$. Then we can continue inequality (6.53) with

$$\leq \lim_{t \to \infty} P^{(R)}(\langle \hat{Z}_t^R, \phi \rangle < e^{\delta t}),$$

where $\langle \hat{Z}_t^R, \phi \rangle := \sum_i \phi(\hat{Z}_t^{R,i})$ and $\{\hat{Z}_t^{R,i}\}$ is the 'ith particle' in $\hat{Z}_t^R$. Similarly to Subsection 2.4.1, we notice that $M_t = M_t^\phi := e^{-\lambda_c t} \langle \hat{Z}_t^R, \phi \rangle$ is a non-negative martingale, and define

$$N := \lim_{t \to \infty} M_t.$$

Since $\lambda_c(\frac{1}{2}\Delta, B(0,R)) > \delta$, and thus

$$\lim_{t \to \infty} P^{(R)} \left(M_t < e^{(\delta - \lambda_c)t} \cap \{N > 0\} \right) = 0,$$

the estimate is then continued as

$$= \lim_{t \to \infty} P^{(R)} (M_t < e^{(\delta - \lambda_c)t} \mid N = 0) \, P^{(R)}(N = 0) \leq P^{(R)}(N = 0).$$

We have that

$$\lim_{t \to \infty} P^V (|Z_t| < e^{\delta t}) \leq P^{(R)}(N = 0)$$

holds for *all* R large enough. Therefore, in order to prove (6.51), it is sufficient to show that

$$\lim_{R \to \infty} P^{(R)}(N > 0) = 1. \tag{6.54}$$

Consider now the elliptic boundary value problem (which of course depends on K),

$$\begin{aligned} &\frac{1}{2}\Delta u + V(u - u^2) = 0 \text{ in } B(0,R), \\ &\lim_{x \to \partial B(0,R)} u(x) = 0, \\ &u > 0 \text{ in } B(0,R). \end{aligned} \tag{6.55}$$

The existence of a solution follows from the fact that $\lambda_c > 0$ by an analytical argument given in [Pinsky (1996)] pp. 262–263. In fact, existence relies on finding so-called lower and upper solutions.[16]

Uniqueness follows by the *semi-linear elliptic maximum principle* (Proposition 7.1 in [Engländer and Pinsky (1999)]; see also [Pinsky (1996)]); for the same reason, if $w_R(x)$ denotes the unique solution, then $w_R(\cdot)$ is monotone increasing in R. Using standard arguments[17], one can show that $0 < w := \lim_{R \to \infty} w_R$ too solves the first equation in array (6.55).

Applying the well-known *strong maximum principle* to $v := 1 - w$, it follows that w is either one everywhere or less than one everywhere. We now suppose that $0 < w < 1$ and will get a contradiction.

---

[16]The assumption $\lambda_c > 0$ enters the stage when a positive lower solution is constructed.
[17]See the proof of Theorem 1 in [Pinsky (1996)].

Since we assumed that $d \leq 2$, this is simple. We have

$$\frac{1}{2}\Delta w = V(w^2 - w) \lneq 0 \text{ in } \mathbb{R}^d,$$

(Δw is nonnegative and not identically zero) and this contradicts the recurrence of the Brownian motion in one and two dimensions, because of Proposition 1.9. (The symbol $\lneq 0$ means non-negative and not identically vanishing.) This contradiction proves that in fact $w = 1$ and it consequently proves the limit (6.51). □

Remark 6.4 (General $d \geq 1$). The nonexistence of solutions to the problem

$$\begin{aligned}
\frac{1}{2}\Delta u + V(u - u^2) &= 0 \text{ in } \mathbb{R}^d, \\
0 &< u < 1,
\end{aligned} \tag{6.56}$$

in the general $d \geq 1$ case is more subtle than for $d \leq 2$. Assuming that $\beta_1 > 0$, it follows from the fact that β is bounded from below along with Theorem 1.1 and Remark 2.4 in [Engländer and Simon (2006)] (set $g \equiv \beta_1$ in Remark 2.4 in [Engländer and Simon (2006)]).   ◇

Remark 6.5 (Probabilistic solution). The argument below gives a *probabilistic construction* for $w_R(x)$. Namely, we show that $w_R(x) := P_x^{(R)}(N > 0)$ solves (6.55). To see this, let $v = v_R := 1 - w_R$. Let us fix an arbitrary time $t > 0$. Using BMP for Z at time t, it is straightforward to show that

$$P^{(R)}(N = 0 \mid \mathcal{F}_t) = \prod_i P_{\hat{Z}_t^{R,i}}^{(R)}(N = 0). \tag{6.57}$$

Since the left hand side of this equation defines a $P^{(R)}$-martingale in t, so does the right-hand side. That is

$$\widetilde{M}_t := \prod_i v\left(\hat{Z}_t^{R,i}\right)$$

defines a martingale. From this, it follows by Theorem 17 of [Engländer and Kyprianou (2001)] that v solves the equation obtained from the first equation of (6.55) by switching $u - u^2$ to $u^2 - u$. Consequently, $w_R(x) := P_x^{(R)}(N > 0)$ solves the first equation of (6.55) itself. That w_R solves the second equation, follows easily from the continuity of Brownian motion. Finally its positivity (the third equation of (6.55)) follows again from the fact that $\lambda_c > 0$ (see Lemma 6 in [Engländer and Kyprianou (2004)]).   ◇

6.11 Exercises

(1) Give a rigorous proof of the comparison in Remark 6.1.
(2) Derive (6.57).
(3) Let $k \in \{3, 4, ...\}$. How should one modify Theorem 6.1 if each particle has precisely k offspring instead of two?
(4) And how about having a fix supercritical offspring distribution? (Consider first the case when there is no death.)

6.12 Notes

M. Kac [Kac (1974)] considered a Poisson point process with intensity $c(x)$ under the probability $\mathbf{P}$, and a Brownian motion killed by the corresponding hard Poissonian obstacle configuration. When ϵ, the size of the Poissonian obstacles tends to zero, but at the same time, their intensity $c(x)$ is scaled up as $\gamma_d(\epsilon)c(x)$ (with an appropriate γ_d), the limiting distribution of the particle's ('annealed') lifetime is that of a Brownian motion killed by the potential $k_d c(x)$, where $k_d > 0$ only depends on the dimension.

Following Kac's idea, in [Véber (2012)] the author considers a *superprocess* among Poissonian hard obstacles, i.e. mass is annihilated on the boundary of the obstacle configuration Γ_ϵ, and an analogous question is asked. Keeping the 'particle picture' in mind, intuitively, it is plausible that applying a similar scaling, and letting $\epsilon \to 0$, one obtains a limiting model, where instead of obstacles, the superprocess has an additional negative mass creation term $-k_d c(x)$, i.e. the potential $-k_d c(x)$ is added to the corresponding semi-linear operator. The paper [Véber (2012)] formulates and proves this intuition. (Note that one way of defining the negative potential for the superprocess is through the motion rather than the branching mechanism, namely one considers the same branching mechanism but the underlying motion is Brownian motion with killing.)

A *discrete* catalytic model was investigated in [Kesten and Sidoravicius (2003)], where the branching particle system on $\mathbb{Z}^d$ was so that its branching was catalyzed by another autonomous particle system on $\mathbb{Z}^d$. There are two types of particles, the A-particles ('catalyst') and the B-particles ('reactant'). They move, branch and interact in the following way. Let $N_A(x, s)$ and $N_B(x, s)$ denote the number of A- (resp. B-)particles at $x \in \mathbb{Z}^d$ and at time $s \in [0, \infty)$. (All $N_A(x, 0)$ and $N_B(x, 0)$, $x \in \mathbb{Z}^d$ are independent Poisson variables with mean μ_A (μ_B).) Every A-particle (B-particle) performs independently a continuous-time random walk with jump rate D_A (D_B). In addition a B-particle dies at rate δ, and, when present at x at time s, it splits into two in the next ds time with probability $\beta N_A(x, s)ds + o(ds)$. Conditionally on the system of the A-particles, the jumps, deaths and splitting of the B-particles are independent. For large β the existence of a critical δ is shown separating local extinction regime from local survival regime.

A further example of the discrete catalytic setting is given in [Albeverio and Bogachev (2000)]. See also the references for branching random walks in random environments at the notes for the next chapter.

In conclusion, below are some ideas on how our mathematical setting relates to some *models of biology*. First, the following two biological interpretations come immediately to one's mind:

(i) **Migration with infertile areas (Population dynamics):** Population migrates in space and reproduces by binary splitting, but randomly located reproduction-suppressing areas modify the growth.

(ii) **Fecundity selection (Genetics):** Reproduction and mutation takes place. Certain randomly distributed genetic types have low fitness: even though they can be obtained by mutation, they themselves do not reproduce easily, unless mutation transforms them to different genetic types. In genetics this phenomenon is called '*fecundity selection*'. Of course, in this setting 'space' means the space of genetic types rather than physical space.

One question of interest is of course the (local and global) growth rate of the population. Once one knows the global population size, the model can be rescaled (normalized) by the global population size, giving a population of unit mass (somewhat similarly to the fixed size assumption in the Moran model or many other models from theoretical biology) and then the question becomes the *shape* of the population.

In the population dynamics setting this latter question concerns whether or not there is a preferred spatial location for the process to populate. In the genetic setting the question is about the existence of a certain kind of genetic type that is preferred in the long run that lowers the risk of low of fecundity caused by mutating into less fit genetics types.

Of course, the *genealogical structure* is a very intriguing problem to explore too. For example it seems quite possible that for large times the 'bulk' of the population consists of descendants of a single 'pioneer' particle that decided to travel far enough (resp. to mutate many times) in order to be in a less hostile environment (resp. in high fitness genetic type area), where she and her descendants can reproduce freely.

For example, a related phenomenon in marine systems [Cosner (2005)] is when hypoxic patches form in estuaries because of stratification of the water. The patches affect different organisms in different ways but are detrimental to some of them; they appear and disappear in an 'effectively stochastic' way. This is an actual system that has some features that correspond to the type of assumptions built into our model.

It appears [Fagan (2005)] that a very relevant existing ecological context in which to place our model is the so-called '*source-sink theory*'. The basic idea is that some patches of habitat are good for a species (and growth rate is positive) whereas other patches are poor (and growth rate smaller, or is zero or negative). Individuals can migrate between patches randomly or according to more detailed biological rules of behavior.

Another kind of scenario where models such as ours would make sense is in systems that are subject to periodic local disturbances [Cosner (2005)]. Those would include forests where trees sometimes fall creating gaps (which have various effects on different species but may harm some) or areas of grass or brush which are subject to occasional fires. Again, the effects may be mixed, but the burned areas can be expected to less suitable habitats for at least some organisms.

Finally, for a modern introduction to population models from the PDE point of view, see the excellent monograph [Cantrell and Cosner (2003)].

Chapter 7

Critical branching random walk in a random environment

In the previous chapter we have studied a spatial branching model, where the underlying motion was a d-dimensional ($d \geq 1$) Brownian motion, the particles performed dyadic branching, and the branching rate was affected by a random collection of reproduction suppressing sets; the obstacle configuration was given by the union of balls with fixed radius, where the centers of the balls formed a Poisson point process.

Consider now the model where the offspring distribution is *critical*. One can easily prove (cf. Theorem 7.2 below) that, despite the presence of the obstacles, the system still dies out with probability one.

In this chapter we are going to investigate this model in a *discretized* setting. More precisely, we consider a modified version of the model, by replacing the Poisson point process with IID probabilities on the lattice $\mathbb{Z}^d$, as described in the next section.[1] Continuous time will also be replaced by discrete time $n = 1, 2, \ldots$.

The problem posed in the previous chapter now takes the following form.

Problem 7.1. *What is the rate of decay for the survival probability of the particle system as $n \to \infty$? Is it still of order C/n as in the obstacle-free (non-spatial) case?*

7.1 Model

Consider a model when, given an environment, the initial ancestor, located at the origin, first moves according to a nearest neighbor simple random

---

[1] Recall from the previous chapter that the discrete setting has the advantage over its continuous analogue (Poisson trap configuration on $\mathbb{R}^d$) that the difference between the sets K and K^c is no longer relevant (self-duality).

walk, and immediately afterwards, the following happens to her:

(1) If there is no obstacle at the new location, the particle either vanishes or splits into two offspring particles, with equal probabilities.
(2) If there is an obstacle at the new location, nothing happens to the particle.

The new generation then follows the same rule in the next unit time interval and produces the third generation, etc.

Let $p \in [0,1]$. In the sequel $\mathbb{P}_p$ will denote the law of the obstacles (environment), and P^ω will denote the law of the branching random walk, given the environment ω. So, if $\mathbf{P}_p$ denotes the 'mixed' law in the environment with obstacle probability p, then one has

$$\mathbf{P}_p(\cdot) = \mathbb{E}_p P^\omega(\cdot).$$

Just like before, P^ω and $\mathbf{P}_p$ will be called *quenched* and *annealed* probabilities, respectively.

Warning: Almost all of the results (with the exception of the first two theorems) we are going to present here are based on computer simulations. Hence, the style of this final chapter will be significantly different from that of the previous, more 'theoretical' chapters.

7.2   Monotonicity and extinction

Even though most of what we are going to present are based on simulations, there are two simple statements which are fairly easy to verify. Let S_n denote the event of *survival* for $n \geq 0$. That is, $S_n = \{Z_n \geq 1\}$, where Z_n is the population size at time n.

Theorem 7.1 (Monotonicity). *Let $0 \leq p < \widehat{p} \leq 1$ and fix $n \geq 0$. Then*

$$\mathbf{P}_p(S_n) \leq \mathbf{P}_{\widehat{p}}(S_n).$$

Proof.   First notice that it suffices to prove the following statement:

> *Assume that we are given an environment with some 'red' obstacles and some additional 'blue' obstacles. Then the probability of S_n with the additional obstacles is larger than or equal to the probability without them.*

Indeed, one can argue by coupling as follows. Let $q := 1 - p$, $\delta := \widehat{p} - p$. First let us consider the obstacles that are received with IID probabilities and parameter p; these will be the 'red' obstacles. Now with probability δ/q, at each site independently, add a blue obstacle. Then the probability for any given site, that there is at least one obstacle there, is $p + \delta/q - p\delta/q = p + \delta = \widehat{p}$. Delete now those blue obstacles where there was a red obstacle too. This way, the red obstacles plus the additional blue obstacles together correspond to parameter $\widehat{p}$.

In light of the argument in the previous paragraph, we are going to prove[2] the statement in italics now. To this end, consider the generating functions of 'no branching' and critical branching: $\varphi_1(z) = z$ and $\varphi_2(z) = \frac{1}{2}(1 + z^2)$, respectively, and note that $\varphi_1 \leq \varphi_2$ on $\mathbb{R}$. Fix an environment and define

$$u(n, x, N) := P_{n,x}(S_N^c),$$

that is, the probability that the population emanating from a single particle, which is at time $n \geq 0$ is located in $x \in \mathbb{Z}^d$, becomes extinct at time $N \geq n$. If the particle moves to the random location ξ_{n+1}, then one has

$$u(n, x, N) = E \sum_{i=0}^{2} p_i(\xi_{n+1}) \left[P_{n+1,\xi_{n+1}}(S_N^c) \right]^i$$

$$= E \sum_{i=0}^{2} p_i(\xi_{n+1}) \left[u(n+1, \xi_{n+1}, N) \right]^i = E\varphi^{\xi_{n+1}}[u(n+1, \xi_{n+1}, N)],$$

where $p_i(\xi_{n+1})$ is the probability[3] of producing i offspring ($0 \leq i \leq 2$) at the location ξ_{n+1}, and $\varphi^{\xi_{n+1}}$ is either φ_1 or φ_2.

Consider now two environments: one with red obstacles only, and another one, where there are some additional blue obstacles as well, and let us denote the corresponding functions by u_1 and u_2. We have

$$u_1(n, x, N) = E\varphi_1^{\xi_{n+1}}[u_1(n+1, \xi_{n+1}, N)]$$

and

$$u_2(n, x, N) = E\varphi_2^{\xi_{n+1}}[u_2(n+1, \xi_{n+1}, N)].$$

Clearly, $u_1(N, x, N) = u_2(N, x, N) = 0$. Hence, using that $\varphi_1 \leq \varphi_2$ along with backward induction, $u_2 \geq u_1$ for all $n = 0, 1, ..., N - 1$. In particular,

$$u_1(0, x, N) \leq u_2(0, x, N),$$

finishing the proof. $\qquad\qquad\qquad\qquad\qquad\qquad\qquad\qquad\qquad\qquad\qquad$ $\square$

---

[2]The argument was provided by S. E. Kuznetsov.

[3]So either $p_1 = 1$ or $p_0 = p_2 = 1/2$, according to whether there is no obstacle at this location or there is one.

We now give a precise statement and a rigorous proof concerning eventual extinction.

Theorem 7.2 (*Extinction*). *Let $0 \leq p < 1$ and let A denote the event that the population survives forever. Then, for $\mathbb{P}_p$-almost every environment, $P^\omega(A) = 0$.*

Proof. Let again Z_n denote the total population size at time n for $n \geq 1$. Then Z is a martingale with respect to the canonical filtration $\{\mathcal{F}_n; n \geq 1\}$. To see this, note that just like in the $p = 0$ case, one has $E(Z_{n+1} - Z_n \mid \mathcal{F}_n) = 0$, as the particles that do not branch (due to the presence of obstacles) do not contribute to the increment. Being a nonnegative martingale, Z converges a.s. to a limit Z_∞, and of course Z_∞ is nonnegative integer valued. We now show that for $\mathbb{P}_p$-almost every environment, $P^\omega(Z_\infty = 0) = 1$. Introduce the events

- $A_k := \{Z_\infty = k\}$ for $k \geq 1$,
- B: branching occurs at infinitely many times $0 < \sigma_1 < \sigma_2 < \ldots$

Clearly, $A = \cup_{k \geq 1} A_k = \{Z_\infty \geq 1\}$. We first show that

$$\text{for } \mathbb{P}_p\text{-a.e. environment, } P^\omega(B^c A) = 0. \tag{7.1}$$

Obviously, it is enough to show that $\mathbf{P}_p(B^c A) = 0$.

Now, $B^c A \subset C$, where C denotes the event that there exists a first time N such that for $n \geq N$, there is no branching and particles survive and stay in the region of obstacles. On C, one can pick randomly a particle starting at N, and follow her path; this path visits infinitely many points P^ω-a.s., whatever ω is.[4] Since this path is independent of ω, and since $p < 1$, the $\mathbb{P}_p$-probability that it contains an obstacle at each of its sites is zero. Hence $\mathbf{P}_p(C) = 0$, and (7.1) follows.

On the other hand, for each $k \geq 1$, there is a $p_k < 1$, such that the probability that the population size remains unchanged (it remains k) at σ_m is not more than p_k for every $m \geq 1$, uniformly in ω. Thus,

$$P^\omega(B A_k) = P^\omega(B \cap \{Z_{\sigma_m} = k \text{ for all large enough } m\}) = 0,$$

whatever ω is. Using this along with (7.1), we have that for $\mathbb{P}_p$-almost every ω, $P^\omega(A_k) = P^\omega(B^c A_k) + P^\omega(B A_k) = 0$, $k \geq 1$, and so $P^\omega(A) = 0$. □

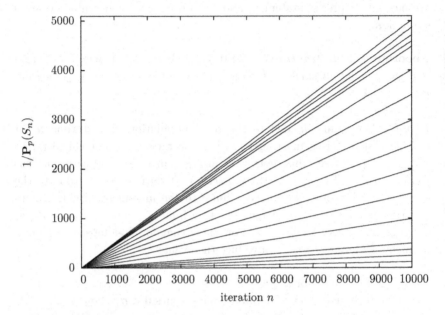

Fig. 7.1 Results for an annealed two-dimensional simulation. The graph shows the reciprocal of the survival probability as a function of the number of iterations. Each line represents a different obstacle probability. One such line is the result of 10^8 runs of the simulation with a newly generated obstacle landscape.

7.3 Simulation results

All simulations were programmed and executed by N. Sieben,[5] to whom the author of this book owes many thanks!

The annealed simulation ran on $\mathbb{Z}^d$ with $d \in \{1, 2, 3\}$. The one-dimensional case turned out to be the most challenging, and so we start the description of our results with the two-dimensional case. The three-dimensional case produced essentially the same output as the two-dimensional case.

---

[4]Because for every ω, it is true P^ω-a.s., that every particle that does not branch, has a path that visits infinitely many points.

[5]Department of Mathematics and Statistics, Northern Arizona University, Flagstaff.

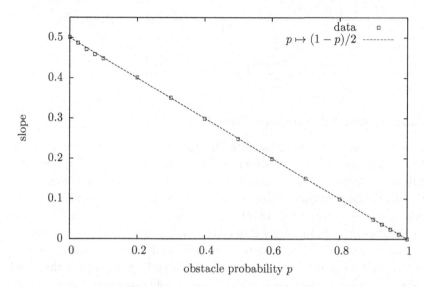

Fig. 7.2   Results for an annealed two-dimensional simulation. The graph shows the apparent slopes in Figure 7.1 (i.e. the limits of the functions of Figure 7.6) as a function of the obstacle probability together with the graph of $p \mapsto (1 - p)/2$.

7.3.1   *Annealed simulation on $\mathbb{Z}^2$*

We executed 10^8 runs allowing a maximum of $n_{\max} = 10000$ iterations with $p \in \{0, 0.025, 0.05, 0.075, 0.1, 0.2, \ldots, 0.9, 0.925, 0.95, 0.975\}$. For $p = 1$ we used the obvious fact that the survival probability is one. Preliminary results made it clear that the simulation was more sensitive for small and large values of p; this is why we picked more of these values instead of a uniformly placed set of values. The reciprocal of the calculated survival probabilities are shown in Figure 7.1. The figure suggests that $n \mapsto 1/\mathbf{P}_p(S_n)$ is asymptotically linear. We calculated the slopes for these curves from the values at $4n_{\max}/5$ and $n_{\max}$. These slope values are presented in Figure 7.2.

To verify the correctness of our simulation, we computed the exact theoretical survival probabilities after the first two iterations. It is easy to see (and is left to the reader to check) that $\mathbf{P}_p(S_1) = 1/2 + p/2$ and

$$\mathbf{P}_p(S_2) = 3/8 + 11p/32 + 3p^2/16 + 3p^3/32. \tag{7.2}$$

The next table compares some of the exact and simulated values.

p	0	.5	.975	0	.5	.975
n	1	1	1	2	2	2
exact	.5	.75	.9875	.375	.605469	.975292
simulated	0.50005	.74998	.98749	.37501	.605465	.97528

7.3.2   Annealed simulation on $\mathbb{Z}^1$

A one-dimensional simulation with 10^8 runs and $n_{\max} = 10,000$ produces less satisfactory results, as shown in Figure 7.3. The reasons behind this will be explained in Subsection 7.4.2 below, with a discussion concerning the fluctuations of the empirical curves in the figures. Essentially, in the annealed case, small values of $\mathbf{P}_p(S_n)$ result in large errors[6] and therefore we modified the original algorithm by introducing a *stopping rule:* when the estimated value of $\mathbf{P}_p(S_n)$ reaches a certain small threshold value, we stop and do not simulate more iterations. Fortunately, when larger threshold values are needed, they are actually large: we obtained slower convergence for large values of p, and, clearly, for those values, the probability $\mathbf{P}_p(S_n)$ is large. The threshold value was set $1/4000$, based on trial and error. This way, we stopped the iteration at $n_{\text{stop}}(p)$; the slopes were then calculated from the values at $4n_{\text{stop}}(p)/5$ and $n_{\text{stop}}(p)$. See Figure 7.4.

Having adjusted the algorithm, using the above stopping rule, the curve indeed straightened out and the picture became very similar to the two-dimensional one in Figure 7.2.

To verify the correctness of our simulation we computed the exact theoretical survival probabilities after the first two iterations. It is easy to see (and is again left to the reader to check) that $\mathbf{P}_p(S_1) = 1/2 + p/2$ and

$$\mathbf{P}_p(S_2) = 3/8 + 5p/16 + p^2/4 + p^3/16. \tag{7.3}$$

The next table compares some of the exact and simulated values.

p	0	.5	.975	0	.5	.975
n	1	1	1	2	2	2
exact	.5	.75	.9875	.375	.601563	.975272
simulated	0.50002	.7499992	.98749	.37502	.601569	.975269

---

[6]See Subsection 7.4.2 for more explanation.

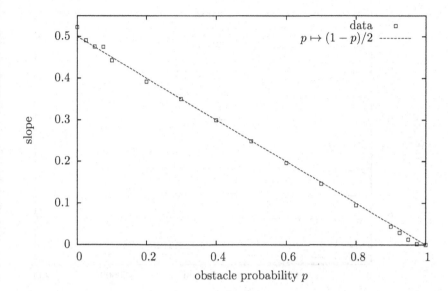

Fig. 7.3   Results for an annealed one-dimensional simulation. Every parameter for this simulation is chosen to be the same as that of Figure 7.2 except the dimension.

7.3.3   *Quenched simulation*

From the annealed simulation it has been clear that convergence is much faster in two dimensions than in one dimension. Therefore, in the quenched case we chose to present our results for $d = 2$. In fact, qualitatively similar results have been obtained for $d = 1$ as well.

In Figure 7.5 we see three 'bundles' corresponding to three values of p. Those bundles are very thin, so essentially the same thing happens for every realization; the slopes of the lines are roughly $3/8$, $1/4$ and $1/8$ from top to bottom, corresponding to $p = 0.25$, $p = 0.5$ and $p = 0.75$, respectively. That is, for each one of these values of p, the slope is the same as in the annealed case.

Although Figure 7.5 is about the $d = 2$ case, we have a similar simulation result for $d = 1$; in fact *we conjecture that this qualitative behavior* (that is, the coincidence of the first order asymptotics of the quenched and annealed survival probability) *will hold for all $d \geq 1$*.

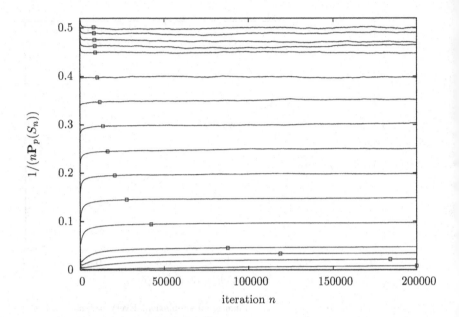

Fig. 7.4 Annealed one-dimensional simulation with $959,965,800$ runs. For small values of p (lines at the top), a small iteration number would actually give better results, because otherwise the survival probability $\mathbf{P}_p(S_n)$ becomes too small, even with a huge number of runs. On the other hand, for large values of p (lines at the bottom) one needs large iteration numbers because the convergence is apparently slow. The squares represent the iteration thresholds after which $\widehat{\rho}_n < \frac{1}{4000}$.

7.4 Interpretation of the simulation results

7.4.1 *Main finding*

Recall Kolmogorov's result (Theorem 1.14) for critical unit time branching, and as a particular case, let us now consider a non-spatial toy model as follows. Suppose that branching occurs with probability $q \in (0,1)$, and then it is critical binary, that is, consider the generating function

$$\varphi(z) = (1-q)z + \frac{1}{2}q(1+z^2).$$

It then follows that, as $n \to \infty$,

$$P(\text{survival up to } n) \sim \frac{2}{qn}. \tag{7.4}$$

Returning to our spatial model, the simulations suggest (Figures 7.1 and 7.5) the *self-averaging* property of the model: as explained in the previous

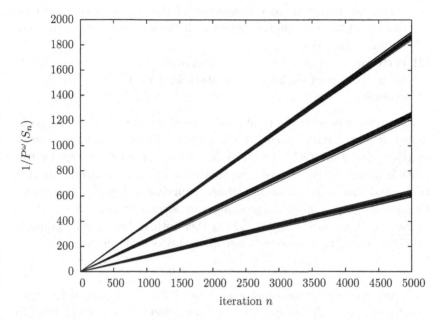

Fig. 7.5   Results for a quenched two-dimensional simulation with 10^8 runs. Each line represents a different obstacle landscape. One such line is the result of 10^8 runs of the simulation. The lines are in three groups corresponding to three different obstacle probability. Each group has 50 lines. The obstacle probabilities from top to bottom are 0.25, 0.5 and 0.75. The total number of simulations required for this graph is $3 \cdot 50 \cdot 10^9 = 15 \cdot 10^{10}$; the total running time was about 29 hours.

section, the asymptotics for the annealed and the quenched case are the same. In fact, this asymptotics is *the same as the one in (7.4)*, where $p = 1 - q$ is the probability that a site has an obstacle. In other words, despite our model being spatial, in an asymptotic sense, the parameter q simply plays the role of the branching probability of the above non-spatial toy model. To put it yet another way, q only introduces a 'time-change,' that is, time is 'slowed down.'

To get an intuitive picture behind this asymptotics, we will use the jargon of large deviation theory. Namely, *there is nothing that either the environment or the BRW could do to increase the chance of survival,* at least as far as the leading order term is concerned (unlike in the case when a single Brownian motion is placed into Poissonian medium). Hence,

(1) given any environment (quenched case), the particles move freely and

experience branching at q proportion of the time elapsed (quenched case), and the asymptotics agrees with the one obtained in the non-spatial setting as in (7.4).

(2) Furthermore, creating a 'good environment' (annealed case) and staying in the part of the lattice with obstacles for very long would be 'too expensive.'

Note that whenever the total population size reduces to one, the probability of that particle staying in the region of obstacles is known[7] to be of lower order than $\mathcal{O}(1/n)$ as $n \to \infty$. So the optimal strategy for this particle to survive is obviously not to attempt to stay completely in that region and thus avoid branching. Rather, survival will mostly be possible because of the potentially large family tree stemming from that particle. In fact, the formula $\mathbf{P}_p(S_n) \sim \frac{2}{qn}$, together with the martingale property of $|Z_n|$, implies linear expected growth, conditioned on survival:

$$\mathbf{E}_p(|Z_n| \mid S_n) \sim \frac{q}{2} \cdot n \text{ as } n \to \infty.$$

Notice that the straight lines on Figures 7.2 and 7.3 start at the value $1/2$, that is, as $p \downarrow 0$, one gets the well-known non-spatial asymptotics $2/n$ as $n \to \infty$, which is a particular case of Theorem 1.14. We conclude that there is apparently *no discontinuity* at $p = 0$ (no obstacles) for the quantity $\lim_{n \to \infty} nP$ (survival up to n).

7.4.2 *Interpretation of the fluctuations in the diagrams*

What can be the source of the apparent fluctuations in the diagrams?

Since we estimated the *reciprocal* of the survival probabilities and not the probabilities themselves, both in the annealed and the quenched case (Figures 7.1 and 7.5), we cannot expect good approximation results when those probabilities are small. Indeed, in the annealed case, if $\rho_n := \mathbf{P}_p(S_n)$ (with p being fixed) and $\widehat{\rho}_n$ denotes the relative frequency obtained from simulations, then LLN only asserts, that if the number of runs is large, then the difference $|\rho_n - \widehat{\rho}_n|$ is small. However, looking at the difference of the reciprocals

$$\left| \frac{1}{\rho_n} - \frac{1}{\widehat{\rho}_n} \right| = \frac{|\rho_n - \widehat{\rho}_n|}{\rho_n \widehat{\rho}_n},$$

---

[7]This is the 'hard obstacle problem for random walk.' Hard and soft obstacles, and quenched and annealed survival probabilities have been studied for random walks as well, similarly to the case of Brownian motion. IID distributed obstacles at lattice points play the role of PPP.

it is clear that a small ρ_n value magnifies the error; in fact the effect is squared as $\hat{\rho}_n$ is close to ρ_n, exactly because of the LLN. This effect is the reason of the 'zigzagging' of the line on Figure 7.3 for small values of p. In fact, small values of p result in small ρ_n values in light of Theorem 7.1 and the continuity property mentioned at the end of the previous subsection. Clearly, there is a competition between ρ_n being small (as a result of p being small and n being large) on the one hand, and the large number of runs on the other. The first makes the denominator small in $\frac{|\rho_n - \hat{\rho}_n|}{\rho_n \hat{\rho}_n}$, while the second makes the numerator small, as dictated by LLN.

Looking at Figure 7.3, one notices another peculiarity in the one-dimensional setting. For large values of p, the empirical curve is slightly under the straight line. The explanation for the relatively poor fit is simply that the iteration number is not large enough for the asymptotics to 'kick in.'

These arguments are bolstered by the experimental findings that increasing the number of runs helps for small values of p, whereas increasing the number of iterations helps for large ones. For example, in Figure 7.4 we increased the maximal iteration number $n_{\max}$ to $200,000$ and plotted $n \mapsto (n\mathbf{P}_p(S_n))^{-1}$. One can see that for small values of p, it is beneficial to stop the iterations earlier, but for large values, large iteration numbers give better results.

We do not have an explanation, however, for the deviation *downward* from the straight line (for large values of p) in Figure 7.3. Finding at least a heuristic explanation for this phenomenon would be desirable.

Interestingly, for higher dimensions there is apparently a perfect fit for large values of p, indicating that for higher dimensions the convergence in the asymptotics is much more rapid than for $d = 1$. Figure 7.6 checks the assumption (for $d = 2$, annealed) that the reciprocal of the survival probability is $\frac{qn}{2} + o(n)$ as $n \to \infty$. We divide the reciprocal of the survival probability by n, and the graphs convincingly show the existence of a limit, which depends on the parameter p.

7.5 Beyond the first order asymptotics

We will now attempt to draw conclusions about more delicate phenomena beyond the first order asymptotics, and the conclusions will necessarily be less reliable than the ones in the previous sections.

We start with the planar case.

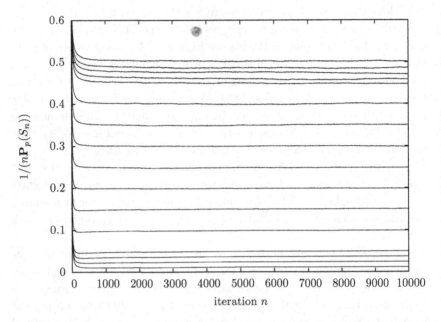

Fig. 7.6 Results for an annealed two-dimensional simulation with 10^8 runs. The graph presents the reciprocal of the survival probability divided by the number of iterations, as a function of the number of iterations. The data used to create the graph has been the same as that of Figure 7.1.

(a) <u>Two dimensions:</u>

Consider again Figure 7.6. Zooming in gives Figure 7.7. Looking at Figures 7.4, 7.6 and 7.7, for small values of p (top lines) the convergence seems to be from above, and for large values of p, it seems to hold from below.

(b) <u>One dimension:</u>

For $d = 1$, figures somewhat similar to the two-dimensional ones were obtained; we summarize them below without actually providing them.

Simulation seems to suggest that for 'not too small' values of p, the convergence is also from below; this is in line with the fact that, as we have already discussed, in Figure 7.3 the one-dimensional empirical curve is *below* the straight line for large values of p. For 'very small' p's, the direction of the convergence is not clear from the pictures. Although the convergence is apparently quicker, the effects are 'blurred,' due to the magnification of error explained earlier.

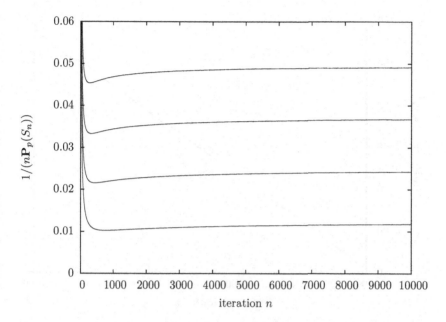

Fig. 7.7 Zooming in at the bottom part of Figure 7.6 (i.e. large values of p): the convergence is apparently from below.

The following conjecture concerning the second-order asymptotics is based on Figure 7.8. It says that the difference $n\mathbf{P}_p(S_n) - \frac{2}{q}$ is on the order $1/\sqrt{n}$ as $n \to \infty$. The reader is invited to think about a proof, or at least a heuristic explanation.

Conjecture 7.1 (Second order asymptotics). *For $d = 1$, the annealed survival probability obeys the following second order asymptotics:*

$$\mathbf{P}_p(S_n) = \frac{2}{nq} + f(n),$$

where $\lim_{n\to\infty} f(n) \cdot n^{3/2} = C > 0$, *and C may depend on p.*

7.5.1   Comparison between one and two dimensions

The annealed convergence to the limit $2/q$ (as $n \to \infty$) seems to be quite different for $d = 1$ and $d = 2$. Figure 7.9 shows this difference, and in particular, it illustrates that in one dimension, the convergence is slower, and it is apparently from below for $p = 0.5$.

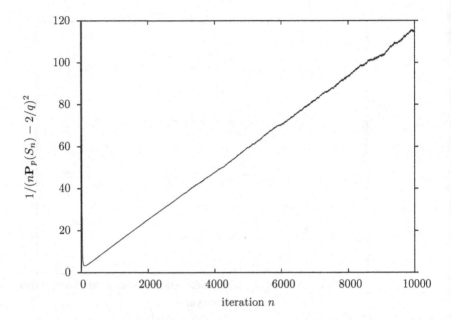

Fig. 7.8 Annealed one-dimensional simulation with 7,259,965,800 runs and $p = 0.5$.

7.6 Implementation

This last section is for the reader familiar/interested in computer simulations.

The code for the simulations was written in the programming language C++, using the MPIqueue parallel library [Neuberger, Sieben and Swift (2014)]. The code was run on 96 cores, using a computing cluster containing Quad-Core AMD Opteron(tm) 2350 CPU's. An implementation [Wagner (2014)] of the *Mersenne Twister* [Matsumoto and Nishimura (1998)] was used to generate random numbers.[8] The total running time for the simulations was several months.

7.6.1 *Annealed simulation*

Algorithm 1 shows the C++ function that runs a single annealed simulation. One essentially implements a 'depth-first search.' Below is a detailed

---

[8]Of course, as usual, these 'random' numbers are only pseudo-random. The Mersenne Twister is, by far, the most widely used pseudo-random number generator.

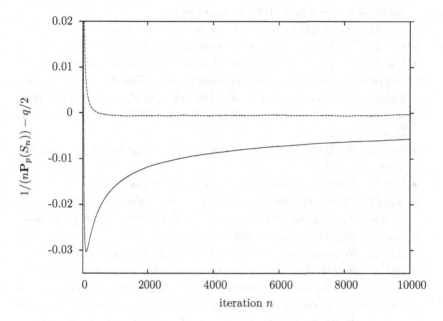

Fig. 7.9   Annealed simulation with $p = 0.5$. The solid curve shows the one-dimensional result, the dashed curve shows the two-dimensional result.

description of the code.

- line 1: We define a data type to store particles.
- line 2: The location of the particle is stored in the *cell* field, that is a vector with the appropriate dimensions.
- line 3: The *iter* field stores the number of iterations survived by the particle.
- line 5: We define a data type to store all the particles alive.
- line 7: The simulation function takes three input variables and one output variable.
- line 8: The dimension of the space is the first input.
- line 9: The probability of an obstacle at any given location is the second input.
- line 10: The maximum number of allowed iterations is the third input.
- line 11: The output of the function is the maximum number of iterations any particle survived.
- line 13: We erase all the obstacles from the board. Every run of the

simulation uses a new obstacle placement.

- line 14: The initial value of the output must be zero.
- line 15: We define a variable to store all our alive particles.
- line 16: We reserve some space to store the particles. Making the reserved space too small results in unnecessary reallocation of the variable which degrades performance. On the other hand, reserving too much space can be a problem too since different CPU's compete with each other for RAM.
- lines 18–19: The initial particle starts at the origin before the iterations start.
- line 20: At the beginning we only have the initial particle.
- line 21: We run the simulation while we have alive particles and none of them stayed alive for the maximum allowed number of iterations.
- lines 22–23: We generate a random direction.
- line 24: We move the last of our alive particles in this random direction.
- line 25: We call the obstacle function to check if there is an obstacle at the new location of the particle. The obstacle function checks in the global variable *board* if any particle already visited this location and as a result we know already whether there is an obstacle there. If no particle visited this location before, then the function uses the obstacle probability to decide whether to place an obstacle there or leave the location empty. This information is then stored for future visitors.
- line 26: If there is an obstacle at the new location, then the particle has survived one more iteration, so we increment the *iter* variable.
- line 27: It is possible that this is the longest surviving particle so far, so we update the output variable.
- line 29: If there is no obstacle at the new location, then the particle splits or dies.
- line 30: We generate a random number to decide what happens.
- lines 31–32: If the particle splits, then it survives, so we update information about the number of iterations.
- line 33: The particle splits, so we place a copy of it into our collection of particles as the last particle.
- lines 35–36: If the particle dies, then we remove it from our collection of particles.

The rest of the code takes care of the parallelization, data collection and the calculation of survival probabilities. The program splits the available nodes into a 'boss node' and several 'worker nodes.' The boss assigns simulation jobs to the workers. The workers call the simulation function several times. The boss node collects the results of these jobs and calculates the survival probabilities using all the available simulation runs. More precisely, $\mathbf{P_p}(S_n)$ is estimated as:

$$\frac{\#\{\text{Simulation runs with longest survival value} \geq n\}}{\#\{\text{All simulation runs}\}}.$$

7.6.2   *Quenched simulation*

The code for quenched simulation is essentially the same with only minor modifications. In this version, line 13 of the simulation function is missing, since we do not want to replace the board at every simulation.

The other change in the simulation function is at line 25. In the annealed case, every worker node has a local version of the board and the obstacle function can create the board on the fly. In the quenched case, the worker nodes need to use the same board, so the obstacle function cannot generate the board locally. The new version of the obstacle function still stores information about the already visited locations. On the other hand, if a location is not visited yet, then the worker node asks the boss node whether this new location has an obstacle. The boss node first checks whether the location was visited by any other particle at any other worker node. If the location was visited, then the boss already has a record of this location. Otherwise, the boss node uses the obstacle probability to decide whether the location should have an obstacle. Essentially, the boss node has the ultimate information about the board, but the worker nodes keep partial versions of the board and only consult the boss node when it is necessary.

Remark 7.1. In the quenched case, note that if $\widehat{\rho}_n$ denotes the relative frequency of survivals (up to n) after r runs for a fixed environment ω, that is,

$$\widehat{\rho}_n = \widehat{\rho}_n^\omega := \frac{|\text{survivals}|}{r},$$

then using our method of simulation, the random variables $\widehat{\rho}_n$ and $\widehat{\rho}_m$ are not independent for $n \neq m$, because the data are coming from the same r runs.

Similarly, in the annealed case, for a fixed environment and a fixed run, the random variables $\mathbf{1}_{S_n}$ and $\mathbf{1}_{S_m}$ are not independent for $n \neq m$.

7.7   Exercises

(1) Reformulate the statement of Theorem 7.2 for the continuous model (Poissonian obstacles and critical BBM). Is the assertion still true?
(2) Prove (7.2).
(3) Prove (7.3).
(4) Try to give, at least at a heuristic level, an explanation for the deviation *downward* from the straight line (for large values of p) in Figure 7.3. (Note: We do not have one.)

7.8   Notes

This chapter follows very closely the paper [Engländer and Sieben (2011)]. Since the results are based on simulations (except the two, intuitively evident statements), it would obviously be desirable to find rigorous proofs. As far as the first order asymptotics, and the 'self-averaging property' are concerned, recently Y. Peres has outlined for me a method for a proof.

The concept of self-averaging properties of a *disordered system* was introduced by the physicist I. M. Lifshitz. Roughly speaking, a property is self-averaging if it can be described by averaging over a sufficiently large sample.

The shape and local growth for multidimensional branching random walks in random environments (BRWRE) were analyzed in [Comets and Popov (2007)]. Local/global survival and growth of a BRWRE has been studied in [Bartsch, Gantert, and Kochler (2009)]. Earlier, phase transitions for local and global growth rates for BRWRE have been investigated in [Greven and den Hollander (1992)].

Appendix A

Path continuity for Brownian motion

In this appendix we explain what exactly path continuity means for Brownian motion. In fact the path continuity for more general diffusion processes, and even for superdiffusions, should be interpreted in a similar manner. (For superdiffusions, though, continuity is meant in the weak or vague topology of measures.)

Let us recall from the first chapter the basic problem: as we will prove shortly, the set of continuous paths, as a subset of $\mathbb{R}^{[0,\infty)}$, is not measurable.

The reader might first think naively that replacing $\mathbb{R}^{[0,\infty)}$ by Ω at the very beginning would serve as a simple remedy, however after a second thought one realizes that we cannot use Kolmogorov's Consistency Theorem for that space, simply because it does not hold for that space. In fact, in general, just because each finite dimensional measure is σ-additive, it does not necessarily follow that we have σ-additivity on the family of *all* cylindrical sets. For instance, for $k \geq 1$, let

$$\nu_{t_1,\dots,t_k}(A_1 \times \dots \times A_k) = 1, \text{if and only if } 0 \in A_j, \text{for all } 1 \leq j \leq k,$$

when $0 < t_1 < \dots < t_k$, and

$$\nu_{t_1,\dots,t_k}(A_1 \times \dots \times A_k) = 1, \text{if and only if } 0 \in A_j, \text{for all } 2 \leq j \leq k, \text{and } 1 \in A_1,$$

when $0 = t_1 < \dots < t_k$.

These equations describe all the finite dimensional distributions in a consistent way; they attempt to describe a deterministic process, which is everywhere zero, except at $t = 0$, when it is one.

Since $\nu_{1/n}(\{X. \in \Omega \mid X_{1/n} = 0\}) = 1$ for $n \geq 1$ and $\nu_0(\{X. \in \Omega \mid X_0 = 1\}) = 1$, thus for any $N \geq 1$, the set

$$E_N := \bigcap_{n=1}^{N} \{X. \in \Omega \mid X_{1/n} = 0\} \cap \{X. \in \Omega \mid X_0 = 1\}$$

has measure one. On the other hand,

$$E_N \downarrow E := \bigcap_{n=1}^{\infty} \{X_. \in \Omega \mid X_{1/n} = 0\} \cap \{X_. \in \Omega \mid X_0 = 1\} = \emptyset,$$

where the last equality follows from continuity of the paths. This contradicts σ-additivity.[1]

After this brief detour on why we cannot just work on Ω directly, let us see now why Ω is not measurable in $\mathbb{R}^{[0,\infty)}$.

Proposition A.1. $\Omega \notin \mathcal{B}'$.

Proof. Let $(\Omega_0, \mathcal{F}_0, P_0)$ be an arbitrary probability space and consider X defined by $X(\omega, t) = 0$ for $\omega \in \Omega_0, t \geq 0$. That is, $X : \Omega_0 \to \widehat{\Omega}$ is the deterministically zero random element (stochastic process). Let μ be the law of this process on $\widehat{\Omega}$ (i.e. the Dirac-measure on the constant zero path). Now, 'destroy' the path continuity by changing the values at time N, where N is an independent, non-negative, absolutely continuous random variable, defined on, say, $(\Omega^*, \mathcal{F}, Q)$. To be more rigorous, equip the product space $\widetilde{\Omega} := \Omega_0 \times \Omega^*$ with the product probability measure $\mathbf{P} := P_0 \times Q$, and let

$$X'_t(\omega, \omega^*) := \mathbf{1}_{\{N(\omega^*) = t\}}.$$

We now have an $\widehat{\Omega}$-valued random element X' on the probability space $\widetilde{\Omega}$, and the distribution of X' is exactly μ, that is,

$$\mathbf{P}(X'_{t_1}(\omega, \omega^*) \in B_1, ..., X'_{t_k}(\omega, \omega^*) \in B_k) = P(X_{t_1}(\omega) \in B_1, ..., X_{t_k}(\omega) \in B_k)$$
$$= \mu_{t_1,...,t_k}(B_1 \times ... \times B_k),$$

for any $k \geq 1$, $0 \leq t_1 < t_2 < ... < t_k$, and $B_1, ..., B_k \in \mathcal{B}(\mathbb{R})$, because N is absolutely continuous and therefore the cylindrical sets 'do not feel the change:'

$$Q(N(\omega^*) \in \{t_1, t_2, ..., t_k\}) = 0.$$

However, X' has discontinuous paths $\mathbf{P}$-almost surely. Thus, $\Omega \in \mathcal{B}'$ would lead to the following contradiction:

$$1 = P_0(X^{-1}(\Omega)) = \mu(\Omega) = \mathbf{P}(X'^{-1}(\Omega)) = 0. \qquad \square$$

Now, even if Ω isn't a measurable subset of $\widehat{\Omega}$, it definitely has an outer measure:

$$\nu^*(\Omega) := \inf \left\{ \sum_{i=1}^{\infty} \nu(A_i) \mid A_1, A_2, ... \in \mathcal{B}', \Omega \subset \bigcup_{i=1}^{\infty} A_i \right\}.$$

---

[1]Consider the sets $\Omega \setminus E_1, E_1 \setminus E_2, ...$, which are disjoint zero sets.

The trouble with the non-measurability of Ω disappears if we can work somehow with ν^* instead of ν. (Of course ν^* is not a measure, that is, it lacks σ-additivity.) Indeed, we will see below that, instead of our original question (Q.1), the correct question one should ask is:

Q.2: What is the *outer* measure of Ω?

To answer this latter question, we will invoke the Kolmogorov-Čentsov Continuity Theorem. But before doing so, let us note a few definitions and facts.

Definition A.1. The stochastic processes X and $\widetilde{X}$, defined on the common probability space $(\Omega, \mathcal{F}, P)$, are called *versions* (or modifications) of each other, if, for all $t \geq 0$, $X_t = \widetilde{X}_t$, P-almost surely. In such a case, even the two vectors $(X_{t_1}, X_{t_2}, ..., X_{t_k})$ and $(\widetilde{X}_{t_1}, \widetilde{X}_{t_2}, ..., \widetilde{X}_{t_k})$ agree P-a.s. for any choice of $0 \leq t_1 < t_2 < ... < t_k$ and $k \geq 1$. Note that this is stronger than just having that the finite dimensional distributions ('fidi's' or 'fdd's' for short) of the two processes agree.

Definition A.2 (Trace σ-algebra). *Let $\mathcal{A}$ be a σ-algebra of subsets of a set A and $E \subset A$. Then $\mathcal{A}_E$ will denote the trace σ-algebra, that is* $\mathcal{A}_E := \{A \cap E \mid A \in \mathcal{A}\}$.

(The reader can easily check that $\mathcal{A}_E$ is indeed a σ-algebra.)

Note that Ω is meant to be equipped with the topology of uniform convergence on bounded t-intervals. It is well known that this topology yields a nice metrizable space (in fact, a complete separable one), where one compatible metric ρ may be defined by

$$\rho(X., X'.) := \sum_{n=1}^{\infty} 2^{-n} \frac{\sup_{0 \leq t \leq n} |X_t - X'_t|}{1 + \sup_{0 \leq t \leq n} |X_t - X'_t|}.$$

The σ-algebra one considers is of course the Borel σ-algebra, which we will denote by $\mathcal{B}$. It is also well known that $\mathcal{B}$ is generated by the cylindrical sets of the form

$$A := \{X. \in \Omega \mid X_{t_1} \in B_{t_1}, X_{t_2} \in B_{t_2}, ..., X_{t_k} \in B_{t_k}\},$$

where B_{t_m}, $m = 1, 2, ..., k$; $k \geq 1$ are Borels of the real line. Recall that $\mathcal{B}'$ is also generated the same way: all one has to do is to replace Ω by $\widehat{\Omega}$ in the definition of cylindrical sets. Using this, it is an easy exercise (left to the reader) to show that

$$\mathcal{B} = \mathcal{B}'_\Omega. \tag{A.1}$$

After this preparation, we state the continuity theorem, which settles all our questions.

Theorem A.1 (Kolmogorov-Čentsov Continuity Theorem). *Assume that X, a stochastic process on $(\widehat{\Omega}, \widehat{B}, \mathbb{P})$ with corresponding expectation $\mathbb{E}$, satisfies for all times $T > 0$, that there exist positive constants $\alpha = \alpha_T, \beta = \beta_T, K = K_T$ such that*

$$\mathbb{E}\left[|X_t - X_s|^{\alpha}\right] \leq K|t - s|^{1+\beta}$$

for all $0 \leq s, t \leq T$. Then

(i) Ω has outer measure one,

(ii) X has a version, say $\widetilde{X}$, with continuous paths.

(For a proof of (i), see Chapter 2 in [Stroock and Varadhan (2006)], in particular Corollary 2.1.5; for a proof of (ii) and the existence of a locally Hölder-continuous version, see [Karatzas and Shreve (1991)], Theorem 2.2.8.) It is easy to check that the condition of this theorem is satisfied with $\alpha = 4, \beta = 1, K = 3$ in our case.

Although (ii) immediately shows that Brownian motion has a version with continuous paths, we now finish the train of thoughts concerning non-measurability and outer measure, using (i).

The following simple argument shows that one can actually 'transfer' our measure from $(\widehat{\Omega}, B')$ to (Ω, B). What we mean by this is that for $A \in B = B'_{\Omega}$ we define $\mu(A) := \nu(A')$ for $A = A' \cap \Omega$, $A' \in \widehat{\Omega}$. This is indeed the natural way to do it, provided that it makes sense, that is, that the definition is independent of the choice of A'. This, however, isn't really an issue, exactly because of (i). Indeed, if A' is replaced by A'', then $A' \triangle A'' \in B'$ is a subset of $\widehat{\Omega} \setminus \Omega$ and therefore[2] $\nu(A' \triangle A'') = 0$, which means that $\nu(A') = \nu(A'')$.

Whatever argument one chooses, the point is that we can now forget our original probability space $(\widehat{\Omega}, B', \nu)$, and work with the corresponding probability measure μ on (Ω, B), called *Wiener measure*.

Remark A.1 (Doob's method; separability). This discussion would not be complete without mentioning Doob's ingenious solution to the problem of the non-measurability of Ω in $\widehat{\Omega}$. The notion of *separable processes* was introduced by him. Suppose that we do not require a stochastic process to have continuous paths, only that it has paths which are 'not too wild' as follows. Fix a countable set $S \subset [0, \infty)$. Let $E^S \in \widehat{\Omega}$ denote the set of

---

[2]Otherwise $A_1 = A_2 = \ldots = \widehat{\Omega} \setminus (A' \triangle A'')$ would constitute a measurable cover with total measure less than 1, contradicting that $\nu^*(\Omega) = 1$.

paths satisfying that

$$\forall t \geq 0 : \liminf_{S \ni s \to t} X_s \leq X_t \leq \limsup_{S \ni s \to t} X_s.$$

Call a stochastic process X separable if there exists a countable set $S \subset [0, \infty)$ such that the paths of the process belong to E^S almost surely.

Doob proved that *any* stochastic process has a version which is separable. Just like in the case of a continuous modification, this also means that the law of the process can be transferred from $(\widehat{\Omega}, \widehat{\mathcal{B}})$ to $(E^S, \widehat{\mathcal{B}}_{E^S})$. For a separable process, however, it can be shown, that our non-measurability problem disappears, because the set Ω is measurable! (Meaning that $\Omega \in \widehat{\mathcal{B}}_{E^S}$.) In fact, some other important sets, for example the set of all bounded functions and the set of all increasing functions, become measurable as well.

Thus, one alternative way of defining Brownian motion is to first take a separable version of it and then to *prove* that its paths are almost surely continuous. ◇

Appendix B

Semilinear maximum principles

When dealing with branching diffusions and superdiffusions, one frequently uses the following parabolic semilinear maximum principle, proved in [Pinsky (1996)]:

Proposition B.1 (Parabolic semilinear maximum principle). *Let* L *satisfy Assumption 1.2 on* $D \subset \mathbb{R}^d$, *let* β *and* α *be in* $C^\eta(D)$, *and let* $D' \subset\subset D$. *Let* $0 \leq v_1, v_2 \in C^{2,1}(D' \times (0, \infty)) \cap C(\overline{D'} \times (0, \infty))$ *satisfy*

$$Lv_1 + \beta v_1 - \alpha v_1^2 - \dot{v}_1 \leq Lv_2 + \beta v_2 - \alpha v_2^2 - \dot{v}_2$$

in $D' \times (0, \infty)$, $v_1(x, 0) \geq v_2(x, 0)$ *for* $x \in D'$, *and* $v_1(x, t) \geq v_2(x, t)$ *for* $x \in \partial D'$ *and* $t > 0$. *Then* $v_1 \geq v_2$ *in* $D' \times [0, \infty)$.

Even though in [Pinsky (1996)] the setting was more restrictive, the proof goes through for our case without difficulty. (See Proposition 7.2 in [Engländer and Pinsky (1999)].)

For the less frequently used, but still handy, *elliptic* semilinear maximum principle, see again [Pinsky (1996); Engländer and Pinsky (1999)].

Bibliography

Albeverio, S. and Bogachev, L. V. (2000) *Branching random walk in a catalytic medium. I. Basic equations.* Positivity, **4**, 41–100.

Asmussen, S. and Hering, H. (1976a) *Strong limit theorems for general supercritical branching processes with applications to branching diffusions.* Z. Wahrscheinlichkeitstheorie und Verw. Gebiete **36**(3), 195–212.

Asmussen, S. and Hering, H. (1976b) *Strong limit theorems for supercritical immigration-branching processes.* Math. Scand. **39**(2), 327–342.

Asmussen, S. and Hering, H. (1983) *Branching processes.* Progress in Probability and Statistics **3**, Birkhäuser.

Athreya, K. B. (2000) *Change of measures for Markov chains and the $L \log L$ theorem for branching processes.* Bernoulli **6**(2) 323–338.

Athreya, K. B. and Ney, P. E. (1972) *Branching processes.* Springer-Verlag (reprinted by Dover, 2004).

Balázs, M., Rácz, M. Z. and Tóth, B. (2014) *Modeling flocks and prices: Jumping particles with an attractive interaction.* Ann. Inst. H. Poincaré Probab. Statist. **50**(2), 425–454.

Bartsch, C., Gantert, N. and Kochler, M. (2009) *Survival and growth of a branching random walk in random environment.* Markov Process. Related Fields **15**(4), 525–548.

Berestycki, H. and Rossi, L. (2015) *Generalizations and properties of the principal eigenvalue of elliptic operators in unbounded domains.* To appear in Comm. Pure Appl. Math. Preprint available at http://arxiv.org/pdf/1008.4871v5.pdf

Berestycki, J., Kyprianou, A. E. and Murillo-Salas, A. (2011) *The prolific backbone for supercritical superdiffusions.* Stoch. Proc. Appl. **121**(6), 1315–1331.

Biggins, J. D. (1992) *Uniform convergence of martingales in the branching random walk.* Ann. Probab. **20**(1), 137–151.

Biggins, J. D. and Kyprianou A. E. (2004) *Measure change in multitype branching.* Adv. in Appl. Probab. **36**(2), 544–581.

Billingsley, P. (2102) *Probability and measure.* Anniversary edition. Wiley Series in Probability and Statistics. John Wiley and Sons.

Bramson, M. D. (1978) *Minimal displacement of branching random walk.* Z. Wahrsch. Verw. Gebiete **45**(2), 89–108.

Bramson, M. D. (1978) *Maximal displacement of branching Brownian motion.* Comm. Pure Appl. Math. **31**(5), 531–581.

Breiman, L. (1992) *Probability.* Corrected reprint of the 1968 original. Classics in Applied Mathematics, 7. Society for Industrial and Applied Mathematics (SIAM).

Cantrell, R. S. and Cosner C. (2003) *Spatial ecology via reaction-diffusion equations.* Wiley Series in Mathematical and Computational Biology. John Wiley and Sons.

Champneys, A., Harris, S. C., Toland, J., Warren, J. and Williams, D. (1995) *Algebra, analysis and probability for a coupled system of reaction-diffusion equations.* Phil. Trans. R. Soc. Lond. **350**, 69–112.

Chaumont, L. and Yor, M. (2012) *Exercises in probability: A guided tour from measure theory to random processes, via conditioning.* (Cambridge Series in Statistical and Probabilistic Mathematics **35**), Cambridge Univ. Press, 2nd edition.

Chauvin, B. and Rouault, A. (1988) *KPP equation and supercritical branching Brownian motion in the subcritical speed area. Application to spatial trees.* Probab. Theory Related Fields **80**, 299–314.

Chen, L. H. Y. (1978) *A short note on the conditional Borel-Cantelli lemma.* Ann. Probab. **6**(4), 699–700.

Chen, Z-Q., Ren, Y. and Wang, H. (2008) *An almost sure scaling limit theorem for Dawson-Watanabe superprocesses.* J. Funct Anal. **254**, 1900 2010.

Chen, Z-Q. and Shiozawa, Y. (2007) *Limit theorems for branching Markov processes.* J. Funct Anal. **250**, 374–399.

Comets, F. and Popov, S. (2007) *Shape and local growth for multidimensional branching random walks in random environment.* ALEA Lat. Am. J. Probab. Math. Stat. **3**, 273–299.

Cosner, C. (2005) *Personal communication.*

Dawson, D. A. (1993) *Measure-valued Markov processes.* Ecole d'Eté Probabilités de Saint Flour XXI., LNM **1541**, 1–260.

Dawson D. A. and Fleischmann, K. (2002) *Catalytic and mutually catalytic super-Brownian motions,* in Proceedings of the Ascona '99 Seminar on Stochastic

Analysis, Random Fields and Applications (R. C. Dalang, M. Mozzi and F. Russo, eds.), 89–110, Birkhäuser.

den Hollander, F. and Weiss, G. (1994) *Aspects of trapping in transport processes*, in: Contemporary Problems in Statistical Physics, SIAM.

Donsker, M. and Varadhan, S. R. S. (1975) *Asymptotics for the Wiener sausage.* Comm. Pure Appl. Math. **28**, 525–565.

Dynkin, E. B. (1991) *Branching particle systems and superprocesses.* Ann. Probab. **19**(3), 1157–1194.

Dynkin, E. B. (1994) *An introduction to branching measure-valued processes.* CRM Monograph Series **6**. American Mathematical Society.

Durrett, R. (1995) *Probability: Theory and examples.* Duxbury press, 2nd edition.

Eckhoff, M., Kyprianou, A. and Winkel, M. (2014) *Spines, skeletons and the Strong Law of Large Numbers for superdiffusions*, to appear in Ann. Probab.

Engländer, J. (2000) *On the volume of the supercritical super-Brownian sausage conditioned on survival.* Stoch. Proc. Appl. **88**, 225–243.

Engländer, J. (2007) *Branching diffusions, superdiffusions and random media.* Probab. Surv. **4**, 303–364.

Engländer, J. (2008) *Quenched law of large numbers for branching Brownian motion in a random medium.* Ann. Inst. H. Poincaré Probab. Statist. **44**(3), 490–518.

Engländer, J. (2009) *Law of large numbers for superdiffusions: The non-ergodic case.* Ann. Inst. H. Poincaré Probab. Statist. **45**(1), 1–6.

Engländer, J. (2010) *The center of mass for spatial branching processes and an application for self-interaction.* Electron. J. Probab. **15**, paper no 63, 1938–1970.

Engländer, J. and den Hollander, F. (2003) *Survival asymptotics for branching Brownian motion in a Poissonian trap field.* Markov Process. Related Fields **9**(3), 363–389.

Engländer, J. and Fleischmann, K. (2000) *Extinction properties of super-Brownian motions with additional spatially dependent mass production.* Stochastic Process. Appl. **88**(1), 37–58.

Engländer, J. and Kyprianou, A. E. (2001) *Markov branching diffusions: Martingales, Girsanov-type theorems and applications to the long term behavior*, Preprint 1206, Department of Mathematics, Utrecht University, 39 pp. Available electronically at http://www.math.uu.nl/publications

Engländer, J. and Kyprianou, A. E. (2004) *Local extinction versus local exponential growth for spatial branching processes.* Ann. Probab. **32**(1A), 78–99.

Engländer, J., Harris, S. C. and Kyprianou, A. E. and (2010) *Strong law of large numbers for branching diffusions.* Ann. Inst. H. Poincaré Probab. Statist. **46**(1), 279–298.

Engländer, J. and Pinsky, R. (1999) *On the construction and support properties of measure-valued diffusions on $D \subseteq R^d$ with spatially dependent branching.* Ann. Probab. **27**(2), 684–730.

Engländer, J., Ren, Y. and Song, R. (2013) *Weak extinction versus global exponential growth for superdiffusions corresponding to the operator $Lu + \beta u - ku^2$.* to appear in Ann. Inst. H. Poincaré Probab. Statist.

Engländer, J. and Sieben, N. (2011) *Critical branching random walk in an IID random environment.* Monte Carlo Methods and Applications **17**, 169–193.

Engländer, J. and Simon, P. L. (2006) *Nonexistence of solutions to KPP-type equations of dimension greater than or equal to one.* Electron. J. Differential Equations No. 9, 6 pp. (electronic).

Engländer, J. and Turaev, D. (2002) *A scaling limit theorem for a class of superdiffusions.* Ann. Probab. **30**(2), 683–722.

Engländer, J. and Winter, A. (2006) *Law of large numbers for a class of superdiffusions.* Ann. Inst. H. Poincaré Probab. Statist. **42**(2), 171–185.

Etheridge, A. (2000) *An introduction to superprocesses.* AMS lecture notes.

Etheridge, A., Pfaffelhuber, P. and Wakolbinger, A. (2006) *An approximate sampling formula under genetic hitchhiking.* Ann. Appl. Probab. **16**(2), 685–729.

Ethier, S. N. and Kurtz, T. G. (1986) *Markov processes. Characterization and convergence.* Wiley Series in Probability and Mathematical Statistics. John Wiley and Sons.

Evans, S. N. (1993) *Two representations of a conditioned superprocess.* Proc. Royal. Soc. Edin. Sect. A. **123**(5), 959–971.

Evans, S. N. and Steinsaltz, D. (2006) *Damage segregation at fissioning may increase growth rates: A superprocess model.* Theor. Popul. Biol. **71**, 473–490.

Fagan, B. *Personal communication.*

Feng, J. and Kurtz, T. G. (2006) *Large deviations for stochastic processes.* Mathematical Surveys and Monographs, **131**. American Mathematical Society.

Fleischmann, K., Mueller, C. and Vogt, P. (2007) *The large scale behavior of super-Brownian motion in three dimensions with a single point source.* Commun. Stoch. Anal. **1**(1), 19–28.

Freidlin, M. (1985) *Functional integration and partial differential equations.* Annals of Mathematics Studies **109**, Princeton University Press.

Friedman, A. (2008) *Partial Differential Equations of Parabolic Type*, Dover.

Gill, H. (2013) *Super Ornstein-Uhlenbeck process with attraction to its center of mass*. Ann. Probab. **41**(2), 445–1114.

Git, Y., Harris, J. W. and Harris, S. C. (2007) *Exponential growth rates in a typed branching diffusion*. Ann. Appl. Probab. **17**(2), 609–653.

Gradshteyn, I. S. and Ryzhik, I. M. (1980) *Table of integrals, series, and products*. Academic Press.

Greven, A. and den Hollander, F. (1992) *Branching random walk in random environment: phase transitions for local and global growth rates*. Probab. Theory Related Fields **91**(2), 195–249.

Hardy, R. and Harris, S. C. (2006) *A conceptual approach to a path result for branching Brownian motion*. Stoc. Proc. Appl. **116**(12), 1992–2013.

Hardy, R. and Harris, S. C. (2009) *A spine approach to branching diffusions with applications to $\mathcal{L}^p$-convergence of martingales*. Séminaire de Probabilités, XLII, 281–330.

Harris, S. C. (2000) *Convergence of a "Gibbs–Boltzmann" random measure for a typed branching diffusion*. Séminaire de Probabilités XXXIV. Lecture Notes in Math. **1729**, 239–256. Springer.

Harris, J. W. and Harris S. C. (2009) *Branching Brownian motion with an inhomogeneous breeding potential*. Ann. Inst. H. Poincaré Probab. Statist. **45**(3), 793–801.

Harris T. E. (2002) *The theory of branching processes*. Dover Phoenix Editions, Corrected reprint of the 1963 original, Dover.

Harris, J. W., Harris S. C. and Kyprianou, A. E. (2006) *Further probabilistic analysis of the Fisher-Kolmogorov-Petrovskii-Piscounov equation: One sided travelling-waves*. Ann. Inst. H. Poincaré Probab. Statist. **42**(1), 125–145.

Harris, S. C., Hesse, M. and Kyprianou, A. E. (2013) *Branching Brownian motion in a strip: Survival near criticality*, preprint.

Harris, S. C. and Roberts, M. (2012) *The unscaled paths of branching Brownian motion*. Ann. Inst. H. Poincaré Probab. Statist. **48**(2), 579–608.

Harris, S. C. and Roberts, M. (2013a) *A strong law of large numbers for branching processes: almost sure spine events*, preprint.

Harris, S. C. and Roberts, M. (2013b) *The many-to-few lemma and multiple spines*. Preprint available at http://arXiv:1106.4761v2.

Jacod, J. and Shiryaev, A. N. (2003) *Limit theorems for stochastic processes*. Grundlehren der Mathematischen Wissenschaften **288**. Springer-Verlag, 2nd edition.

Kac, M. (1974) *Probabilistic methods in some problems of scattering theory.* Rocky Mountain J. Math. **4**(3), 511–538.

Kallenberg, O. (1977) *Stability of critical cluster fields.* Math. Nachr. **77**, 7–43.

Karatzas, I. and Shreve, S. E. (1991) *Brownian motion and stochastic calculus.* Graduate Texts in Mathematics, Springer, 2nd edition.

Karlin, S. and Taylor, M. (1975) *A First course in stochastic processes.* Academic Press.

Kelley, J. L. (1975) *General topology.* Reprint of the 1955 edition [Van Nostrand, Toronto, Ont.]. Graduate Texts in Mathematics **27**, Springer-Verlag.

Kesten, H. and Sidoravicius, V. (2003) *Branching random walk with catalysts.* Electron. J. Probab. **8**(5) (electronic).

Kingman, J. F. C. (1993) *Poisson processes.* Oxford Studies in Probability **3**. Oxford Science Publications. Oxford Univ. Press.

Klenke, A. (2000) *A review on spatial catalytic branching.* Stochastic models (Ottawa, ON, 1998), 245–263, CMS Conf. Proc. **26**, American Mathematical Society.

Kyprianou, A. E. (2004) *Travelling wave solutions to the K-P-P equation: Alternatives to Simon Harris' probabilistic analysis.* Ann. Inst. H. Poincaré Probab. Statist.**40**(1), 53–72.

Kyprianou, A. E. (2005) *Asymptotic radial speed of the support of supercritical branching and super-Brownian motion in $\mathbb{R}^d$.* Markov Process. Related Fields. **11**(1), 145–156.

Kyprianou, A. E. (2014) *Fluctuations of Lévy processes with applications; Introductory lectures,* Universitext, Springer, 2nd edition.

Kyprianou, A. E., Liu, R.-L., Murillo-Salas, A. and Ren, Y.-X. (2012) *Supercritical super-Brownian motion with a general branching mechanism and travelling waves.* Ann. Inst. H. Poincaré Probab. Statist. **48**(3), 661–687.

Lee, T. Y. and Torcaso, F. (1998) *Wave propagation in a lattice KPP equation in random media.* Ann. Probab. **26**(3), 1179–1197.

Le Gall, J.-F. and Véber, A. (2012) *Escape probabilities for branching Brownian motion among soft obstacles.* Theor. Probab. **25**, 505–535.

Liggett, T. M. (2010) *Continuous time Markov processes.* Graduate Studies in Mathematics **113**, American Mathematical Society.

Lyons, R., Pemantle R. and Peres, Y. (1995) *Conceptual proofs of L log L criteria for mean behavior of branching processes.* Ann. Probab. **23**(3), 1125–1138.

Marcus, M., Mizel, V. J. and Pinchover, Y. (1998) *On the best constant for Hardy's inequality in $\mathbb{R}^n$.* Trans. Amer. Math. Soc. **350**(8), 3237–3255.

Matsumoto, M. and Nishimura, T. (1998) *Mersenne twister: a 623-dimensionally equidistributed uniform pseudo-random number generator.* ACM Trans. Model. Comput. Simul. **8**(1), 3–30.

McKean, H. P. (1975) *Application of Brownian motion to the equation of Kolmogorov-Petrovskii-Piskunov.* Comm. Pure Appl. Math. **28**, 323–331.

McKean, H. P. (1976) *A correction to "Application of Brownian motion to the equation of Kolmogorov-Petrovskii-Piskunov."* Comm. Pure Appl. Math. **29**, 553–554.

Merkl, F. and Wüthrich, M. V. (2002) *Infinite volume asymptotics of the ground state energy in a scaled Poissonian potential.* Ann. Inst. H. Poincaré Probab. Statist. **38**(3), 253–284.

Mörters, P. and Peres, Y. (2010) *Brownian motion. With an appendix by Oded Schramm and Wendelin Werner.* Cambridge Series in Statistical and Probabilistic Mathematics. Cambridge Univ. Press.

Neuberger, J. M., Sieben, N. and Swift, J. W. (2014) *MPI queue: A simple library implementing parallel job queues in C++.* unpublished manuscript.

Øksendal, B. (2010) *Stochastic differential equations: An introduction with applications* (Universitext), corrected 6th printing of the 6th edition, Springer.

Öz, M., Çağlar, M., Engländer, J. (2014) *Conditional Speed of Branching Brownian Motion, Skeleton Decomposition and Application to Random Obstacles.* Preprint.

Pinchover, Y. (1992) *Large time behavior of the heat kernel and the behavior of the Green function near criticality for nonsymmetric elliptic operators.* J. Functional Analysis **104**, 54–70.

Pinchover, Y. (2013) *Some aspects of large time behavior of the heat kernel: An overview with perspectives*, Mathematical Physics, Spectral Theory and Stochastic Analysis, Oper. Theory Adv. Appl. **232**, Birkhäuser, 299–339.

Pinsky, R. G. (1995) *Positive harmonic functions and diffusion.* Cambridge Univ. Press.

Pinsky, R. G. (1996) *Transience, recurrence and local extinction properties of the support for supercritical finite measure-valued diffusions.* Ann. Probab. **24**(1), 237–267.

Révész, P. (1994) *Random walks of infinitely many particles.* World Scientific.

Revuz, D and Yor, M. (1999) *Continuous martingales and Brownian motion.* Grundlehren der Mathematischen Wissenschaften **293**. Springer-Verlag, 3rd edition.

Roberts, M. I. (2013) *A simple path to asymptotics for the frontier of a branching Brownian motion.* Ann. Probab. **41**(5), 3518–3541.

Salisbury, T. S. and Verzani, J. (1999) *On the conditioned exit measures of super Brownian motion.* Probab. Theory Related Fields **115**(2), 237–285.

Schied, A. (1999) *Existence and regularity for a class of infinite-measure* (ξ, ψ, K)-*superprocesses.* J. Theoret. Probab. **12**(4), 1011–1035.

Sethuraman, S. (2003) *Conditional survival distributions of Brownian trajectories in a one dimensional Poissonian environment.* Stochastic Process. Appl. **103**(2), 169–209.

Shnerb, N. M., Louzoun, Y., Bettelheim, E. and Solomon, S. (2000) *The importance of being discrete: Life always wins on the surface.* Proc. Nat. Acad. Sciences **97**, 10322–10324.

Shnerb, N. M., Louzoun, Y., Bettelheim, E. and Solomon, S. (2001) *Adaptation of autocatalytic fluctuations to diffusive noise.* Phys. Rev. E **63**, 21103–21108.

Stam, A. J. (1966), *On a conjecture by Harris.* Z. Wahrsch. Verw. Gebiete **5**, 202–206.

Stroock, D. W. (2011), *Probability theory. An analytic view.* Cambridge Univ. Press, 2nd edition.

Stroock, D. W. and Varadhan, S. R. S. (2006) *Multidimensional diffusion processes.* Reprint of the 1997 edition. Classics in Mathematics. Springer-Verlag.

Sznitman, A. (1998) *Brownian motion, obstacles and random media.* Springer-Verlag.

Tribe, R. (1992) *The behavior of superprocesses near extinction.* Ann. Probab. **20**(1), 286–311.

van den Berg M., Bolthausen E. and den Hollander F. (2005) *Brownian survival among Poissonian traps with random shapes at critical intensity.* Probab. Theory Related Fields **132**(2), 163–202.

Veber, A. (2009) *Quenched convergence of a sequence of superprocesses in* $\mathbb{R}^d$ *among Poissonian obstacles.* Stoch. Process. Appl., **119**, 2598–2624.

Wagner, R. (2014) *An online document on 'Mersenne Twister,'* at http://www-personal.umich.edu/~wagnerr/MersenneTwister.html.

Watanabe, S. (1967) *Limit theorem for a class of branching processes.* Markov Processes and Potential Theory (Proc. Sympos. Math. Res. Center, Madison, Wis.), 205–232, Wiley.

Watanabe, S. (1968) *A limit theorem of branching processes and continuous state branching processes.* J. Math. Kyoto Univ. **8**, 141–167.

Wentzell, A. D. (1981) *A course in the theory of stochastic processes.* McGraw-Hill.

Xin, J. (2000) *Front propagation in heterogeneous media.* SIAM Rev. **42**(2), 161–230.

Printed in the United States
By Bookmasters